RECLAIMING THE WEST:

The Coal Industry and Surface-Mined Lands

RECLAIMING THE WEST:

The Coal Industry and Surface-Mined Lands

by
Daniel Philip Wiener

with
John DiStefano
Ron Lanoue

preface by
the Hon. Morris K. Udall

editor
Joseph Mohbat

an INFORM book

INFORM, Inc.
25 Broad Street
New York City 10004
212/425-3550

ISBN D-918780-16-0

Library of Congress #80-81777

Cover design by Eric Schwartzman

Production Manager: Catherine Seigerman

Contents

Preface

The story of the six-year battle to enact a national strip mine reclamation law is the story of perhaps the single greatest environmental struggle of the 1970s. I am proud to have been a part of it.

But that struggle is over now, and a new challenge faces a nation intent as ever on economic growth and more intent than ever on preservation of its unique natural heritage. That challenge is being joined in the courts, where the Surface Mining Control and Reclamation Act and its regulatory structure are under attack, in the state capitals, where programs to implement the federal law are being written, and most importantly, in the coal fields.

Ultimately, the success or failure in meeting this awesome challenge will be judged by what happens on the land and not what happens in the committee rooms, board rooms or court rooms. And what happens on the land will be determined essentially by the coal operators themselves. For unless the coal industry adopts the attitude that for the right to remove economic value from the land they must return economic and environmental value to the land, and learn the pride and satisfaction that comes from such an accomplishment, then our children will not have the benefit of the inheritance we intended to leave them.

This fine new volume from INFORM is about the challenge that lies ahead. This non-partisan group has sent its researchers out into the western coal fields--where the great majority of surface-mineable coal lies and from which the United States already derives 28 percent of its coal production. Through on-site visits, interviews with operators, scientists, government officials and citizens, and outside evaluations at 15 different western strip mines, INFORM has prepared an extremely valuable assessment of corporate reclamation performance. The organization's analysis covers the full range of reclamation issues from earth-moving to reseeding and revegetation and from agricultural productivity to wildlife enhancement.

INFORM's case studies demonstrate clearly that the federal reclamation act is not a yoke hung around industry's neck. It does not present companies with burdens they cannot carry. Instead, it is what it was intended to be--a manageable law, the lowest common denominator of good industrial practice.

Furthermore, these case studies indicate that reclamation is not primarily limited by cost. Rather, it is a matter of corporate will and the availability of scientific expertise. Some companies, INFORM has found, are simply more anxious to do a good job of reclamation than others. Their attitude is manifested not only in the resources they devote to post-mining activities, but just as importantly in the selection of mining sites and in the hiring of aggressive, competent reclamation staff.

Western mining companies are not faced with a simple task. INFORM's case studies indicate that for the most part, successful reclamation is still something of an unknown, an advancing theoretical skill. Furthermore, it is apparent that the scientific expertise necessary to guide even conscientious companies is in short supply. For example, long-term revegetation has not been demonstrated yet and water availability problems may prove severe. In some parts of the Southwest, reclamation may not be at all possible.

Nevertheless, in recent hearings before the House Subcommittee on Energy and the Environment, I was encouraged to hear every industry representative that was asked the question, agree with the proposition, "If you can't reclaim it, you shouldn't mine it."

American industry faces a very tough challenge in meeting this goal. I am grateful to INFORM for the effort it has made to move forward and clarify the debate over strip mine reclamation in the American West.

> -- The Hon. Morris K. Udall
> Chairman
> House Committee on Interior and Insular Affairs
> U.S. House of Representatives
> April 24, 1980

Acknowledgments

This study, compiled over more than two years, is the re-sult of much dedicated effort on the part of many. Foremost, the direction, encouragement, and stamina of INFORM's former senior editor, Frank Stella, were instrumental in shaping this project from the beginning. His influence is felt from the first page through the last.

INFORM's executive director, Joanna Underwood, provided invaluable insight and support for this study, reinforcing mor-ale and refreshing the author's outlook. James Cannon, both the director of Citizens for a Better Environment and an INFORM Fellow, carried out his many roles as adviser and mentor to the very end. In addition, the project's Advisory Board members' responsiveness and patience, with their attentiveness to detail, helped guide me in the early stages and later, as the work gelled. Thus, a warm thanks to: Genevieve Atwood, staff officer, National Academy of Sciences study on *Rehabilitation Potential of Western Coal Lands* and member of the Utah House of Representatives; Roger Beers, coal specialist for the Natural Resources Defense Council; Russell Boulding, member of the Mining Task Force of the National Coal Policy Project and former coal consultant to the Environmental Defense Fund; George Dials, vice president for technical services and research at the Mining and Reclamation Council of America; Edward Dobson, coal spec-ialist and Montana representative for Friends of the Earth; How-ard Hardesty, partner, Rose, Schmidt, Dixon, Halsey, White & Hardesty; Donald Hawes, energy consultant for the Equitable Life Assurance Society; David Masttann, coal specialist for the Environmental Defense Fund; Steve Shapiro, president, United Mine Workers Local 6025, Virginia Bishop, and Stewart Udall, former secretary of the Interior and Phoenix, Arizona lawyer.

For the months of research and review of many of the chapters in this book, sincere thanks to the officials of the Montana Department of State Lands, North Dakota Public Service Commission, Colorado Mined Land Reclamation Board, and the New Mexico Surfacemining Commission. Also, the Navajo Tribe, the Office of Surface Mining (both in Washington, D.C., and Denver), and the U.S. Environmental Protection Agency, Geological Survey, and Soil Conservation Service.

It would take a book itself to list all the farmers, ranchers, public-interest representatives and concerned citizens who took me into their homes, fed me and broadened my visions with clear, concise documentation of their concerns. As well, the representatives of the 12 coal companies that chose to work closely with INFORM to see all sides of any given issue, and dispelled the myth that the concerns for increased coal development are not recognized by all involved. Also, many thanks to INFORM's Board of Directors, which stood by and contributed encouragement and guidance throughout.

Of course, there is always one shining light that leads a study this complex from its initial form of scribbled sheets and shorthand footnotes, to a fine, well-designed publication. In this regard, the author is deeply indebted to Catherine Seigerman, the finest production coordinator that ever graced the hallways of INFORM. As the work piled up, so did the hours, and Cathy was never without a cheerful smile. Several student interns also provided valuable assistance; they include Ruth Kennedy, Sarah Palmer, Rhodi Margesson, Susan Jakoplic, Lisa King, Bettina Lewis and Sally Frye. Certain members of INFORM staff deserve special recognition, especially our editor-typist extraordinaire Mary Ferguson. Pat Holmes helped type us through the last-minute crunches; Viviane Arzoumanian and Virginia Jones kept the books straight, the finances clear, and the author within limits. Also, thanks to Vincent Trivelli for has advice and invaluable interrogation; and Perrin Stryker for giving the study its final reading before publication. Greg Cohen, Dick Griffin and Jean Halloran also deserve a cheer. As for the rest of the INFORM staff, I couldn't have done it without all your support.

For the financial support that made this project possible, INFORM and the author are deeply grateful to the Needmor Fund, the German Marshall Fund, the Equitable Life Assurance Society, and one anonymous donor.

Introduction

• A western rancher awakens to find that the previous evening's rainstorm has washed away soils at a nearby surface coal mine, depositing up to five feet of sediment on his field and almost burying a spring used to water his livestock.

• A congressional report on state mining and reclamation programs finds enforcement agencies understaffed, underfunded, and underpaying their inspectors, whose offices are usually so far from the mines that effective inspection and enforcement are impossible.

• Under new ownership, more than 700 acres of land left ungraded and unreclaimed are leveled and seeded by the new mine operator.

• A program to turn pasture and cropland into a golf course begins with the planting of numerous trees along a road. All the trees die.

• A herdsman watches as 13 sheep, which have drunk from waters impounded at a nearby surface mine, drop dead. The next morning he finds that nine more have succumbed overnight.

• Lands once growing with grasses, forbs, shrubs and trees are reclaimed. But the new dominant plant is Russian thistle—commonly known as tumbleweed—which is useless for grazing.

• At one mine, innovative design and construction of a sedimentation pond has led to good control of surface water quality.

In these and countless other ways, the U.S. and its coal industry have upset and tried to restore the delicate balance between developing the nation's coal potential and maintaining the health and stability of its other natural resources. The national energy plan for the 1980s holds dear the desire to wean the United States from dependence upon foreign oil, in large part by

increasing use of its own most abundant fossil fuel energy re-
source--coal. However, each link in the chain of steps involved
in putting this fuel to greater use--from coal prospecting and
mining, to coal transportation, to storage and combustion--in-
volves a multitude of environmental effects. It is these effects
that must be dealt with by land reclamation programs, coal cleaning processes, stack scrubbing techniques and programs to dis-
pose of combustion waste.

In this book, INFORM examines the mining and reclamation
practices at each of 15 mine sites operated by 13 coal producers
in Montana, North Dakota, Wyoming, New Mexico, Colorado and
Arizona. They are judged according to well-grounded agricultural principles which are embodied in INFORM's Guidelines described in a later chapter of this report. INFORM's study focuses on the company practices that create the specific links in
the coal production and use chain. We start with the land's destruction, and conclude with the prospects for re-creation of the
premining topography, plant and animal communities, and ecosystem balance.

Prospecting for the coal can disturb the land with roads
built for drilling equipment, destruction of vegetation, and harm
to air and water quality through erosion and sedimentation. Mining further destroys the land and vegetation, with the potential
for damage to both surface water and groundwater resources.
The effect of massive energy development upon the social and
political structure in boom-towns is still being quantified. And
what of the condition of the land, water and air while mining
continues, as well as when it is ended? Will plant communities
return, and will the land be as productive and self-sustaining as
it was before mining? Will wildlife return? Will groundwater resources have been permanently polluted or destroyed?

Once coal is removed from the earth, it must be transported
to the coal-burning facility. Cleaning may occur along the way.
How much water and other resources will cleaning require? How
much land will be used for the construction of coal-slurry pipe-
lines or railroad tracks? What will become of the waste produced
by coal cleaning plants?

The actual end use--the burning of the coal--raises more
questions. Even in this process it must be cleaned further before combustion gases are released into the atmosphere. Will
particulates continue to dirty our skies? Will sulfur rains and
foul air become greater and greater hazards? Will there be increasing concentrations of carbon dioxide and nitrous oxides in
the atmosphere? Do current technologies promise to clean the
gases before release into the atmosphere, only to find that the
dirtiest components are left behind for disposal or storage in a
manner that in itself will pollute?

We can assess the cumulative effects of expanded U.S. reliance on coal--and compare them to the energy benefits to be gained--only with a thorough understanding of each of the links in the chain.

To understand one of the critical links, INFORM undertook this study of surface-mine land reclamation in the coal industry. Until now, no thorough comparative analysis of the environmental protection practices of some of the country's largest coal producers has ever been made. In the West--where coal leasing and production have been growing fast since the early 1960s--the relationship between coal producer and the environment is essential to an understanding of the potential benefits and pitfalls of expanded coal use.

Coal provided more than 70 percent of the country's demand for energy in 1923. By 1972 it had hit its low point, providing less than 18 percent.[1] The cleaner fossil fuels, oil and gas, provided about 25 percent of the country's energy in 1923. But they were the sources of 70 *percent* of our energy in 1972.[2] In 1973 came the Middle East oil embargo, to change the course of a nation's energy history. In the last decade, coal use for the production of electricity has risen more than 49 percent. The amount that comes from United States surface mines has risen by 58 percent. Today, more than 60 percent of all the coal mined in the United States comes from surface operations. The federal government estimates that between 1980 and 1985 surface mining for coal will disturb more than 40,700 acres in the six western states studied by INFORM.[3] But in the long term, more than 91 billion tons of mineable coal underlies more than *3.8 million acres* of western land. That is fully 65 percent of all the surface-mineable coal in the entire country.[4]

Surface mining and land reclamation in the West are new phenomena. Though some mining occurred in the early half of the 20th century, only in the last 10 years has coal production from surface mines skyrocketed in this region. In 1971, surface coal mining produced 31,382,000 tons in North Dakota, Montana, Wyoming, Colorado, Arizona and New Mexico. By 1977, almost four times as much coal--116,430,000 tons--had been surface-mined in those states.[5] The geographic shift in production is also startling. In 1972, of the 15 largest mines operating in the United States (which produced 67.5 million tons of coal), only 3 were western operations.[6] Six years later, 13 of the 15 largest mines were in the West, and all were surface mines. That year, the 15 largest producers dug 89.7 million tons of coal.[7]

The coal industry in the West is a new industry--in terms of the kinds of companies involved, and the scale and sophistication of their individual approaches to the problems of surface mining and land reclamation. In the past, western coal was mined by

the giants of industry, including railroads, steel manufacturers
and coal companies, as well as family-owned and operated businesses. In the West today, coal is being produced predominantly
by energy conglomerates like EXXON, Gulf, Mobil and CONOCO
--side by side with such utilities as Pacific Power & Light, Montana-Dakota Utilities, Idaho Power, and Montana Power. Other
corporate giants managing or owning mining interests include
General Electric, Peter Kiewit, AMAX, Boeing, and the Equitable
Life Assurance Society. The western coal industry also faces a
significantly new environment within which to operate--both climatic and regulatory.

The arid and semi-arid West presents a totally new set of
environmental challenges to mining and reclamation. In direct
contrast to the eastern mining regions, rainfall is low and its
distribution hard to predict; soils are poorly developed, and
groundwater resources are critical to an area that is short of
water. To the benefit of the mine operator, the terrain is less
steep, diminishing the problems of erosion and of returning
mined land to natural contours. Surface mining can undo the
centuries-old product of nature in hours--or minutes.

Regulation plays a key role in coal development in the West.
In the six states where mines studied by INFORM are located,
state regulatory programs range from strict (Montana) to non-
existent (Arizona). The Environmental Protection Agency
and state water quality boards regulate water discharges from
surface mines, and a variety of federal agencies, including the
Bureau of Land Management and U.S. Geological Survey, are involved in certain aspects of coal mining and reclamation. With
the signing of the federal Surface Mining Control and Reclamation
Act in August 1977 came the mandate for nationwide scrutiny of
the everyday practices of the coal industry.

But there is only so much land. There is only so much water. According to the Congress of the United States, surface
mining operations can be condemned on a great many counts:

> [They] adversely affect commerce and the public
> welfare by destroying or diminishing the utility
> of land for commercial, industrial, residential,
> recreational, agricultural, and forestry purposes,
> by causing erosion and landslides, by contribut-
> ing to floods, by polluting the water, by destroy-
> ing fish and wildlife habitats, by impairing natu-
> ral beauty, by damaging the property of citizens,
> by creating hazards dangerous to life and prop-
> erty, by degrading the quality of life in local
> communities and by counteracting governmental
> programs and efforts to conserve soil, water, and
> other natural resources.[8]

Vast areas of the agricultural West and Midwest are underlain by coal. Where should it be mined? What land uses will suffer in the short term? What will be the effect upon the long-range quality of life? Is one part of the country to be "sacrificed" to the effects--both good and bad--of expanding coal development, while other areas are reserved solely for wheat production, or cattle grazing? Will some lands be opened for development that were never accessible before? Can land reclamation as practiced today assure us that the environmental questions have all been answered, and that their solutions are being implemented?

With so many questions left to be answered, we may be moving too fast. As we shall see, western land reclamation is still in the stage of research and development, with no clear answers. Knowing a great deal more about the numerous links in the chain of coal utilization will help toward planning energy independence for the United States. But INFORM's research surely shows that the U.S. cannot afford *not* to slow down and chart its course more rationally.

Findings

If the 13 western surface coal mining companies surveyed in this book have anything in common in their reclamation programs, it is this:

- All say they plan to reclaim fully--return to its former uses and productivity--the land they are disturbing.

- None has yet proved that successful reclamation can be achieved.

Even of the most dedicated reclaimers in this report, the best that can be said at this point is that it is still too early to tell whether they can ever fully undo the damage that their mining has wrought to western land and water resources.

Some mining companies have had many years in which to try and reclaim their mined lands. Some are trying hard. Others seem to have placed reclamation at the bottom of a priority list that begins with production and profit. Many face conditions of climate and terrain that militate heavily against any kind of meaningful reclamation.

Of the 15 mines studied, 8 were producing coal before 1970 and have theoretically had adequate time to achieve some measure of reclamation success. Yet three of these mines are in the arid southwest--Black Mesa-Kayenta; Navajo; and McKinley. Even if all the basic reclamation practices referred to in this study had been fully implemented during these years, site conditions might have precluded full reclamation. At the other five mines, more than ten years in operation, effective reclamation programs could have yielded results providing documentation of what can be achieved. However, at none of these five sites--Seneca II (Colorado); Colstrip (Montana); Gascoyne, Glenharold and Indian Head (North Dakota)--have more than 2/3rds of INFORM's recommended basic reclamation practices been used--in most cases far less than that. Many steps have been initiated only since 1977 under pressure of the federal Surface Mining Control and Reclamation Act. Reclamation achievements vary greatly at these 5 sites as they do at the seven newer mines INFORM studied.

INFORM's Sample

INFORM's sample of 15 surface mines in the West covers mines in six states--North Dakota, Montana, Wyoming, Colorado, New Mexico and Arizona. Climate, soil conditions, ground and surface water systems, and vegetative communities vary widely, from the desert and near-desert areas of the Southwest to the fertile farmlands of North Dakota. The mines sampled range in size from NACCO's Indian Head Mine--which produced less than 1 million tons of coal in 1978--to the largest coal mine in the country, AMAX's Belle Ayr Mine, which produced more than 18 million tons in the same year. The 15 mines dug more than 84 million tons of coal in 1978--about two-thirds of all coal produced in those six western states. Twelve of the 15 mines are among the top 20 producing mines in the nation. They represent some of the oldest as well as some of the newest mines in the West.

Since the mines began operation, they have disturbed more than 35,000 acres of land--more than 55 square miles--and all are still active. (See Chart A.) Most of the land was used for grazing cattle and sheep, with some, mostly in North Dakota, devoted to grain production.

INFORM chose its 13 sample operators because they offer a broad and representative cross-section of the western coal industry. In this sample are 9 of the 15 largest coal producers in the United States, four utilities, three oil companies, three mining and minerals concerns, one manufacturing company, one large construction firm, and only one firm whose business is strictly coal mining. (See Chart B.)

Of the 13 companies that operate the 15 mines studied, 12 demonstrated considerable openness and interest in sharing information on their program efforts with INFORM. Only one company--Pittsburg & Midway--refused to cooperate in any way. It is noteworthy that P&M's parent--Gulf Oil--has touted the success of P&M's reclamation efforts in advertisements in national newspapers and magazines. It is the only company in the INFORM sample found to do so, yet the only company that provided no information to this study to document its claims. P&M would not allow INFORM to visit its mine or speak with its reclamation (or other) personnel.

TABLE A

MINE HISTORY AND 1978 COAL PRODUCTION

INFORM's 15 Sample Mines

Region/Mine	Operator	First Year of Production	Acres Disturbed*	1978 Coal Production (tons)
The Southwest				
Black Mesa-Kayenta	Peabody	1966	4,707	9,287,588
McKinley	Pittsburg & Midway	1959	over 1,500	2,993,000
Navajo	Utah International	1963	5,573	6,200,000
Colorado				
Energy Fuels	Energy Fuels	1972	2,516	3,770,000
Seneca II	Peabody	1969	978	1,372,251
Southern Wyoming				
Jim Bridger	NERCO	1973	1,850	5,200,000
Rosebud	Peter Kiewit	1973	3,493	2,868,048
Seminoe II	Arch Mineral	1973	1,247	2,800,000
The Northern Plains				
Absaloka	Westmoreland	1974	516	4,500,000
Belle Ayr	AMAX	1973	1,273	18,065,664
Colstrip	Western Energy	1968	3,158	10,576,000
Decker	Peter Kiewit	1972	2,540	9,073,592
North Dakota				
Gascoyne	Knife River	1950	1,105	2,871,839
Glenharold	CONSOL	1966	2,998	3,686,094
Indian Head	NACCO	1957	1,392	914,207

*The dates of these figures vary from February 1978 to April 1980 (see individual profiles).

TABLE B
OWNERSHIP OF THE MINES
INFORM's 15 Sample Mines

Sector/Parent Company	Mines	1978 Net Income ($ million)
Utilities		
Idaho Power	Jim Bridger	33.5
Montana-Dakota Utilities	Gascoyne	14.8
Montana Power	Colstrip	31.5
Pacific Power & Light	Decker/Jim Bridger	105.8
Mining and Minerals		
AMAX	Belle Ayr	160.0
Energy Fuels	Energy Fuels	NA
Westmoreland Resources	Absaloka	1.4 (coal only)
Oil		
Ashland Oil	Seminoe II	245.0
CONOCO	Glenharold	451.0
Gulf Oil	McKinley	513.0
Coal		
NACCO	Indian Head	9.4
Other		
General Electric*	Navajo	1,230.0
Hunt Family†	Seminoe II	NA
Peabody Holding Company△	Black Mesa-Kayenta/ Seneca II	NA
Peter Kiewit▽	Decker/Rosebud	NA

NA = Not available

*Manufacturing

†Private family

△Consortium of six companies: Newmont Mining Corporation; Williams Companies; Bechtel Corporation; Boeing Company; Fluor Corporation; and Equitable Life Assurance Society

▽Construction--privately held

Judging the Results

The central question of this report is whether the land so disturbed can ever be fully returned to its original uses.

Successful reclamation is now commonly judged against three criteria:

•The land must resist erosion.

•It must be covered by a native, diverse and self-regenerating plant community.

•It must be returned to productivity equal to or greater than that which the land was capable of sustaining before mining began.

None of the 15 mines in INFORM's sample has met all three of these criteria.

State and federal regulators are not satisfied that these criteria have been met. At the 12 mines where reclamation bonds are held by state or federal agencies to guarantee compliance with these criteria, none of that bond has yet been returned to the mining companies.

As examples of the difficulty of judging the success of reclamation, two mines in the INFORM sample (both in Montana) have sought to validate their productivity claims by conducting grazing trials on reclaimed lands. But at both Peter Kiewit's Decker Mine and Western Energy's Colstrip Mine, the experiments were poorly controlled, and the data failed to reflect accurately the true productivity of the reclaimed land as compared to that of undisturbed lands. This is because the spayed or emasculated cattle used in the grazing trials were incapable of reproducing, so that consumed food was directly transferred to weight gains instead of fueling biological processes necessary for reproduction. Thus the animals gained more weight than they would have, had they been reproductively intact. In "real-world" practice, most Montana ranchers graze cattle first for their ability to calve, secondly for meat production. If their achievements are to be fairly assessed, mine operators must test their reclamation work against land uses that will actually occur once the miners have departed. Short of that, the industry's claims are questionable at best, deceptive at worst.

The quality of reclamation achieved by each of the 13 companies at their mines is the result of a combination of two factors: good practices and favorable site conditions. However, INFORM evaluated the land reclamation programs at each of the 15 mine sites against a set of 33 criteria which are combined under five broad headings. Thirty guidelines comprise the basic steps that should make up a good reclamation program.

The success of these steps will be greatly affected by the

specific site characteristics--including the climate, soil conditions, plant communities and water resources. Indeed, INFORM found that the climates and soils at six of the 15 mines studied (three in the Southwest and three in southern Wyoming) *may virtually preclude successful reclamation.* Nevertheless, mining continues in almost every conceivable area--combining some of the worst soil, driest conditons, and most delicate ecosystems in the nation.

Judged according to the guidelines set up by INFORM:

- Only one of the 15 mines emerges as far superior in its reclamation practices.

- Only one mine is far inferior to all the others in reclaiming the land.

- Six mines faced crucial revegetation problems.

- Six mines faced crucial water problems.

- Only one mine had minor geographic and climatic problems to contend with.

The 15 Mines: A Mixed Record

Each of the companies studied was found to have several distinguishing characteristics in terms of the quality of its reclamation program, the site-specific conditions the program addresses and the results to date.

- The company with the most impressive performance overall was *Energy Fuels,* whose three-mine complex, which has disturbed over 2,500 acres since mining began in 1972, is in the mountainous region of northwestern Colorado. Attention to many details in its reclamation program, coupled with a full staff of experts located at the mine site (along with the company president), allows decisions to be made with the fullest and clearest grasp of their consequences. The company has seeded over 55 percent of the lands disturbed by mining and is using a more diverse seed mixture than any other company in the study. It has one of the two best records in the direct haulback of topsoil materials essential for successful revegetation (of the 14 companies using dragline techniques for mining). It has carefully controlled erosion and aided water retention on seeded lands by use of furrows along the contour. Energy Fuels has also shown particular innovation in tree and shrub transplanting. The company meets over two-thirds of INFORM's criteria for a sound reclamation program.

- The worst showing in the INFORM sample was that of Knife River, at its *Gascoyne Mine* in North Dakota. This mine, opened in 1950, is the oldest in this study. Though blessed with more topsoil, flatter slopes, and greater precipitation than most of the other mines, reclamation efforts have been weak. Only 4 percent of the land disturbed by mining and related activities has been seeded here--less than any other mine studied. Knife River is one of four companies that does not conduct its own groundwater monitoring program, and its failure to do so is more critical because of the size and importance of the aquifer system its mining affects. Company officials have drawn up thoughtful seeding and amendment plans. Extraordinarily, however, these plans have never been implemented. Coal production has overridden all other concerns. An inappropriate experiment, in which trees were transplanted with the hope of building a golf course on reclaimed lands, met with total failure. This symbolizes Knife River's superficial approach to reclamation.

•An average site in INFORM's study is Peabody's *Seneca II Mine* in northwestern Colorado (near Energy Fuels). At this mine the climate does not impair the potential of revegetation like it does in southern Wyoming, Arizona and New Mexico, nor are groundwater problems a serious consideration for local communities. Reclamation is improving from rather poor beginnings, but the entire program still suffers from distant management and little on-site expertise. In one of the easier western areas to reclaim, vegetation to date is dominated by introduced legumes, and grading and topsoil salvage have been poor. Some practices are improving, such as contour furrowing, better topsoil use, an above-average seed mixture, and shallower grading which will all benefit the company's program. Far from the best overall mine, approximately 50 percent of disturbed acreage has been seeded.

Water Troubles

The other 12 company sites face a variety of reclamation challenges and have responded to these with differing levels of effort, imagination and success. At six sites the central issue is water. The threat of damage--if not long-term destruction--of important regional water supplies at these sites is severe. In five cases damage is well documented already. Three mines are in Montana: Westmoreland's Absaloka Mine; Western Energy's Colstrip Mine; and Peter Kiewit's Decker Mine.

•Westmoreland's *Absaloka Mine* faces one of the most serious problems of any company in INFORM's sample-- one that threatens its very future. It may be the site of a showdown over provisions in the federal law protecting valuable agricultural lands known as alluvial valley floors. These fertile lands are only now beginning to be identified in the West. A large portion of the Absaloka site, could be so classified, in which case it may be closed to mining altogether. With only about 500 acres already disturbed since this new mine opened in 1974, and with company plans to mine almost 3,000 more acres in the next 20 years, such classification would severely inhibit planned expansion. Aside from this potential problem, Westmoreland has set up a fairly good (if relatively new) overall reclamation program. It is attempting to re-establish trees, incorporate a diverse seed mixture (3rd best in the study) and grade slopes to avoid erosion and drainage problems.

•At *Colstrip,* mining disturbs an extremely valuable ground-water system (the coal seam being mined supplies some of the best water in the area--water relied on by ranchers and farmers.) At nearby *Decker,* the same is true. To date there is no evidence that aquifer reconstruction is possible, or that groundwaters will eventually find a cleaner path through which to flow. While both Colstrip and Decker face groundwater problems, there are many differences in these companies' performances to date. At Colstrip, WECo is using the second most diverse seed mixture in the study, and has seeded an encouraging 49 percent of all lands disturbed. At Decker, only 9 percent of the land disturbed by mining has been seeded, and the company's seed mixture is below average. Yet, Decker has shown innovation in the design and construction of sedimentation ponds to control surface water and prevent sedimentation. Neither company conducts contour furrowing as a basic part of its reclamation program. This would also be helpful in revegetation and surface water control.

Two other mines with damaged water supplies are AMAX's Belle Ayr Mine in Wyoming and NACCO's Indian Head Mine in North Dakota:

•AMAX's *Belle Ayr Mine* outside Gillette, Wyoming, only a few hours to the south of the Montana mines by automobile, is known for many reasons. It is located next to one of the most famous (or infamous) boom-towns in the nation. The mine produces more coal (over 18 million tons in 1978) than any other in the United States. Though the problems with water are not as severe as seen at other mines, groundwater levels have dropped in the general mine area with no clear evidence that it will return. Water control practices have led to erosion in drainage diversions, and portions of a creek which AMAX is diverting may be considered an alluvial valley floor. With its exclusive use of the truck and shovel technology for coal removal (unique among the mines studied by

INFORM), the advantages of selective handling of soils and other earthen materials can best be found. More topsoil-- approximately 75 percent--is directly hauled back than at any other sample mine. But the otherwise good reclamation program is marred by poorly controlled experimentation, and a seed mixture which is of only average diversity (for the INFORM sample). Further, more than one-third of the land already seeded will be disturbed again in the mine's life as boxcut spoils are moved in order to gain the coal which they cover.

•At NACCO's small (less than one million tons of coal produced in 1978) *Indian Head Mine,* research by North Dakota officials has found that groundwater quality has been severely degraded as waters have flowed from disturbed soils into groundwater systems. Degradation is more concretely documented here than at any other site in the INFORM sample. State officials claim that salts and minerals are polluting valuable water supplies, and that some salts are also moving into root zones in the supper soil layers, a fact the company denies. Salt migration could result in a serious decline in plant growth, rendering the land much less useful. While NACCO has seeded 70 percent of all disturbed lands at its site (the best record of any site studied), the company's current reclamation program suffers from a poor seed mixture (better only than that at Knife River's Gascoyne Mine) and a minimal tree and shrub transplanting program.

•At another North Dakota site, CONSOL's *Glenharold Mine,* major erosion and sedimentation problems have occurred but are being corrected. While groundwater impacts are unclear at present, there is good reason to believe that problems found at the nearby Indian Head Mine may eventually occur here as well. Careful monitoring is needed. CONSOL's mine is slated for quite rapid expansion. Hence, it is especially important that its practices, weak in several areas, be improved. Currently, CONSOL has seeded just 21 percent of all disturbed lands at the site, (ranking 10th). The seed mixture used is poor. It is more diverse than only two others in the study, and contour furrowing--which would help plant establishment as well as controlling erosion and sedimentation-- is not conducted. Especially valuable wildlife habitat--wooded draws--are being destroyed at Glenharold. But CONSOL has begun some attempts to recreate these important resources.

Revegetation Troubles

At six other INFORM sites the central land reclamation challenge facing coal mine operators is whether they will be able to re-establish self sustaining vegetative growth in their arid climates. Three of these sites are in New Mexico and Arizona: Utah International's Navajo Mine; Pittsburg and Midway's McKinley Mine; and Peabody's Black Mesa-Kayenta complex:

- •Utah International's *Navajo Mine* has disturbed more land since it began operating in 1963 than any other mine in INFORM's sample; over 5,500 acres. Revegetation is being attempted using a massive irrigation program. But the seed mixture being used is below average, and with 45 percent of the land disturbed at the site seeded, more than 3,000 acres remain to be reclaimed. With an extremely arid climate--the lowest annual average precipitation of any mine in the sample--the long term results of reclamation cannot be predicted. Whether plants grown using irrigation will survive and reproduce once maintenance ends is a critical question.

- •Pittsburg & Midway, as observed above, did not allow INFORM to visit its *McKinley Mine,* but data from outside sources indicate that--like the other southwestern mines INFORM studied--McKinley's reclamation program is up against severe climate and site conditions. P&M has also relied on a reclamation program that is only fair at best. Using many introduced plant species, and expansive irrigation and fertilizer to make them grow, makes it uncertain whether the plant communities established will be able to survive once artificial maintenance is ended.

- •At *Black Mesa-Kayenta* in northern Arizona, Peabody has made a strong recent effort to grade and seed--by plane and on the ground--mined lands neglected from a reclamation point of view for many years. By 1980, 45 percent of the disturbed lands had been seeded, the 6th best record in the study. However, the seed mixture used is average, and the risk of seed loss as wind blows seeds distributed from the air is considerable. Reclamation at this arid Peabody site, if at all possible, will be long in coming. A separate and important issue here involves the company's relationship with the Native American population. Controversies over land use, surface water quality and relocation of families need to be better addressed.

The other three mines with revegetation troubles are in Southern Wyoming: Peter Kiewit's Rosebud Mine, Arch Mineral's Seminoe II Mine; and NERCO's Jim Bridger Mine:

•Outside Hanna, Wyoming, the *Rosebud Mine* and *Seminoe II
Mine*--operated by Peter Kiewit and Arch Mineral, respective-
ly--are battling the odds in an area replete with conditions
that inhibit successful revegetation. Poor soils and dry climates combine to create an uphill battle for these mines.
(Peter Kiewit's reclamation program at Rosebud--though designed by the same individual responsible for the Decker
Mine--is not as good. This may partially reflect the fact
that the company's offices are less than 20 miles from Decker--
while over 200 miles from Rosebud.) Reclamation lags far
behind disturbance at both Seminoe II and Rosebud, owing
to steeply pitching coal seams that do not allow grading and
reclamation to take place concurrently with disturbance and
production. At both sites also, vegetative establishment on
reclaimed lands has been only fair to poor. Although the
two mines are using contour furrowing to protect soils against
erosion and to aid in water retention and plant growth, the
seed mixtures are less diverse than the average for the IN-
FORM sample (Rosebud's ranks 13th of 15). It is not at all
certain that reclamation will be possible here.

•At NERCO's *Jim Bridger Mine,* which lies at the edge of the
Red Desert, in an effort to make use of scarce water and to
encourage vegetative growth contour furrowing is used and
the reclamation staff has been innovative in adapting machinery to better seed and mulch on the site. But the seed mixture used lacks important diversity and little land (only 13
percent) has been seeded. NERCO is relying on a massive
irrigation program to introduce vegetative communities to
lands disturbed by what will be one of the largest surface
mines in the country. Plans call for extending the mine more
than 15 miles in length in the next few years. It is the size
of the operation, coupled with dry site conditions, little soil,
and a mine layout that is not conducive to reclamation that
will present the most serious obstacle to returning the Jim
Bridger lands to productive grazing use.

Regulation: More Help Than Harm

Government regulation has had a positive effect upon many
of the operations in the West, where state regulatory programs
have ranged from fairly strict in Montana to nonexistent in Arizona. Montana's detailed set of regulations--established before
the federal law was passed--in fact served as a partial model for
that law. Its inspection and enforcement is comprehensive. The
three Montana mines in INFORM's survey--Colstrip, Absaloka and
Decker--have been inspected virtually bimonthly. Two of these
mines are the only sites where measurement of reclamation results
was even attempted through use of controlled grazing (though

testing was somewhat poorly controlled, as mentioned earlier).

Arizona, by contrast, even 3 years after passage of the federal law, still has no state standards. Nor is there any state level agency inspecting or enforcing federal guidelines. It is true that special legislation is being developed which will define requirements for Native American lands (on which the one Arizona Mine is located--Black Mesa-Kayenta). Nonetheless, the lack of any interim state involvement is unfortunate. Improvements at Peabody's operation have resulted primarily from federal pressures by the OSM.

Regulations of the Federal Office of Surface Mining (established by SMCRA--the 1977 Surface Mining Control and Reclamation Act) have served to establish a "lowest common denominator" for regulating and enforcing prompt and suitable reclamation in all states, thus reducing the large differences among them. SMCRA has helped to prod the states into creating and enforcing better reclamation programs. Improved performance is all but guaranteed as a result. Virtually every company studied wants more awareness of site-specific conditions by the Office of Surface Mining that might justify special innovative efforts or deviations from particular requirements. Contrary to industry forecasts when the law was passed, coal production has not been stifled, and by equalizing the reclamation requirements imposed upon companies, regardless of where they are operating, regulation should help to make the industry more competitive.

Reclamation Cost: Still a Question

One important factor still eludes us: the cost of land reclamation. It appears to elude the industry as well. While every company studied could define costs of specific aspects of reclamation work, not one could provide a clear overall picture of total reclamation expenditures. Coal companies' cost reporting on reclamation is sketchy, and available information is presented in such a variety of forms that any comparison between companies is virtually impossible. Furthermore, owing to differing geographical factors, costs can vary considerably from area to area. For example, NACCO told INFORM that reclamation at its Indian Head Mine in North Dakota ranged in price from $200-$12,000 per acre, depending on such variables as regulation and site-specific characteristics. Tipping the scales, Peabody reports that reclamation at its Black Mesa-Kayenta complex ranges from $12,000-$18,000 per acre. From available data, INFORM determined that earthmoving costs make up more than 70 percent of whatever a company spends on reclamation, yet it is unclear which earthmoving costs are truly attributable only to reclamation, and which are necessary expenses for coal production.

The 5 Phases of Reclamation Programs

INFORM established 30 guidelines for land reclamation practices and three criteria for the resulting vegetation. These are
grouped into five distinct categories, including:

Earthmoving

This phase of a reclamation program involving 8 guidelines
relates to returning the land to its approximate original contour,
restoring water drainage patterns, salvaging and re-using soils
conducive to plant growth, and handling especially toxic or sodic
materials encountered during mining.

Adequate attention to earthmoving issues is the critical first
step in any effort to return land to a productive state. In the
West, probably the two most important earthmoving considerations
are the return of land contours to a gently sloping, stable topography, and the immediate re-use of salvaged topsoil; both can
have a profound influence upon later attempts at revegetation.

At every one of the 15 sites in INFORM's study, return to
approximate original contour of lands is being achieved. Although
this was certainly not so in the past, it is standard practice today.
Even in the steeper areas of the West--such as Routt County, Colorado, site of the Seneca II and Energy Fuels Mines--land is being
returned to original contour. On the older sections of mines owned
by Peter Kiewit, Peabody, and Utah International, slopes left too
steep by earlier grading practices are now being reduced.

All sites are using proper grading techniques, but the speed
with which grading is done varies widely. The smallest mine in
the INFORM sample--NACCO's Indian Head in North Dakota--ranks
first, having graded approximately 85 percent of all its disturbed
lands. Colstrip and Navajo come next, having graded 66 percent
each. By contrast, Peter Kiewit has graded proportionately less
land at its two mines--only 9 percent at Decker, and 18 percent
at Rosebud--than any other company. Whether the reason for
good or poor records relates to the size of the mining operation
or the layout of the mines, the fact remains that the more lands
that have not been graded, the more there is that cannot be seeded
and revegetated.

Possibly the most important first step toward successful revegetation is the direct re-use (without stockpiling) of topsoil
materials. Plant establishment and diversity are enhanced by
this practice; moreover, one-step handling of material is more
economical than multiple handling, storage and stabilization of
stockpiles. But, of INFORM's 15 sites, only Amax's Belle Ayr in
Wyoming--a truck and shovel operation (see Techniques of Surface Mining and Reclamation)--is designed to maximize this practice. At only two other mines--Energy Fuels and Absaloka (and
perhaps also at McKinley)--is more than 35 percent of topsoil be-

TABLE C: PRINCIPAL FINDINGS
GRADING OF LAND AND TOPSOIL REPLACEMENT
INFORM's 15 Sample Mines

Region/Mine	Total Acres Disturbed*	Percentage of Disturbed Land Graded to Date*	Grading to Approximate Original Contour	Direct Haulback of Topsoil
The Southwest				
Black Mesa–Kayenta	4,707	57%	yes	some
McKinley	over 1,500	NA	yes	over 35%
Navajo	5,573	66%	yes	some
Colorado				
Energy Fuels	2,516	NA (over 55%)	yes	over 35%
Seneca II	978	64%	yes	10%–15%
Southern Wyoming				
Jim Bridger	1,850	NA (over 13%)	yes	none
Rosebud	3,493	18%	yes	very little
Seminoe II	1,247	37%	yes	5%
The Northern Plains				
Absaloka	516	49%	yes	50%
Belle Ayr	1,273	NA (over 43%)	yes	75%
Colstrip	3,158	66%	yes	some
Decker	2,540	9%	yes	some
North Dakota				
Gascoyne	1,105	23%	yes	none
Glenharold	2,998	53%	yes	none
Indian Head	1,392	85%	yes	none

NA = Not available
*The dates of these figures vary from February 1978 to April 1980 (see individual profiles).

ing hauled back directly. The other 11 mines do little or none.
A North Dakota law actually prevents the three mines in that
state (Glenharold, Indian Head and Gascoyne) from direct haulback by prohibiting transfer of topsoil from one surface-owner's
land to another's; a change in this law would encourage this important practice.

A critical factor in maximizing chances for successful reclamation is the *initial* designing of mines so that practices that will
enhance reclamation are incorporated into the day-to-day operations on the site. A continuous flow in the transfer of topsoil
from land to be mined to already graded lands increases mining
and reclamation efficiency. Selective materials handling is easier
to accomplish with trucks and shovels--as at Belle Ayr--or with
bucket-wheel excavators (used extensively in West Germany,
where all topsoil is directly re-used) than with the draglines used
at 14 INFORM sites. If more mines could be designed to use Belle
Ayr's truck and shovel techniques, the coal industry could move
topsoil more efficiently. These techniques not only reduce the
loss of topsoil (a scarce and valuable resource in the West) but
aid in differential movement and replacement of toxic, sodic or
otherwise undesirable earthen matter encountered in mining. A
matter for continuing debate among regulators, industry and western communities is whether mining should even be permitted where
steeply pitching coal seams, such as occur in Wyoming's Hanna
Basin (Seminoe II and Rosebud Mines) prevent direct haulback
and reduce chances for reclamation success.

Water Quality

Seven of INFORM's guidelines relate to protecting the water
supply, which is the West's most precious natural resource, and
one of the least abundant. Over the entire course of operations
at a surface mine, control of surface waters flowing on and from
the mine site, and monitoring of both surface and groundwaters,
are crucial. Control of surface waters minimizes erosion, sedimentation, and damage to the land and water quality both on and off
the mine site.

Alluvial valley floors (AVFs), which have higher levels of
plant productivity owing to their superior soils, and shallow, subsurface aquifers that provide irrigation to root systems, require
special protection. Under the SMCRA, AVFs may not be mined at
all, unless their area is small; the alluvial system must be protected throughout mining, so that hydrological or agricultural
degradation does not occur. Coal seams act as the major underground transportation networks (aquifers) for much of the high-
quality groundwater in the West. In mining, these pathways--
which also filter the water--are destroyed.

With the mining of the land, layers of rock and dirt compressed over thousands of years are being broken apart and the strata destroyed. Minerals and dirt are exposed, loosely compacted, and thus prone to increased infiltration of water from the surface. Groundwaters, formerly kept pure in their tightly protected underground formations, are then easily contaminated by the increased intrusion of transportable pollutants. A surface mine may cover only a few thousand acres of land, but the groundwater system running beneath it may underlie the entire state. While one surface mine may have limited discernible effects upon water in that system on a state-wide basis, local users may be severely affected.

Federal law now requires that groundwater monitoring be conducted wherever mining operations may harm aquifer systems. Such harm could be occurring at virtually all of INFORM's sites. Yet INFORM found only nine of the 13 companies monitoring groundwater impacts to any degree. Of the four firms *not* doing their own monitoring, Knife River and NACCO rely on the U.S. Geological Survey for information: Utah International has abandoned its program on Indian lands, claiming that it is waiting for specific requirements to be spelled out by tribal regulators, and Arch Mineral has no monitoring program at all.

Available information documents that extensive mining-related degradation has occurred to aquifer systems at at least four sites. The most serious situation exists at NACCO's Indian Head Mine. The North Dakota Geological Survey's tests indicate groundwater quality has been severely degraded by salts and other minerals loosened by mining in the area. One rancher in the area lost water in a well serving his ranch, and although water returned to the well later, another well had to be dug (by NACCO) for his use. Hydrogeologist Gerold Groenwold has deemed the NACCO mine one of the worst in the state for its contamination of groundwater resources. In Montana, where valuable coal aquifers are also being damaged, WECo's Colstrip Mine presents the most alarming case. On some parts of the Colstrip site, water for both domestic use and cattle watering has been degraded, and new wells have had to be dug. On other lands near the mine, water tables have been rising, causing fields to flood. WECo told INFORM that the aquifer is only of secondary value, but the Bureau of Land Management has said the coal seam being mined is one of the most valuable water sources for livestock in the area. Problems are similar at Decker, and at AMAX's Belle Ayr Mine in Wyoming, the company reports that groundwater has been lost near the mine, that it may be at least five years before it returns (if at all), and that relevant data are still very limited. With available information, it is entirely unclear what the current and near-term detrimental effects--not to mention those that would accompany greatly expanded mining operations--may be on vital ranching-agricultural

TABLE D: PRINCIPAL FINDINGS
IMPACT ON WATER RESOURCES
INFORM's 15 Sample Mines

Region/Mine	First Year of Production	Company Water Resource Monitoring	Identified or Potential AVFs	Documented Groundwater Change
The Southwest				
Black Mesa–Kayenta	1966	yes	potential	none
McKinley	1959	yes	no	unclear
Navajo	1963	no	no	unclear
Colorado				
Energy Fuels	1972	yes	potential	none
Seneca II	1969	yes	no	none
Southern Wyoming				
Jim Bridger	1973	yes	potential	unclear
Rosebud	1973	yes	potential	unclear
Seminoe II	1973	no	no	unclear
The Northern Plains				
Absaloka	1974	yes	identified	unclear
Belle Ayr	1973	yes	potential	quantity; quality unclear
Colstrip	1968	yes	potential	quantity and quality
Decker	1972	yes	potential	quantity; quality unclear
North Dakota				
Gascoyne	1950	no	no	unclear
Glenharold	1966	yes	no	unclear
Indian Head	1957	no	no	quantity unclear; quality

uses of land, locale by locale. This issue transcends local boundaries. Montana now ranks second in the nation in barley production, third in wheat production, seventh in beef-cow production
and eighth in sheep and lamb production. North Dakota ranks
first in barley production and in the production of two types of
wheat. Wyoming is a leading wool and sugar producer. With a
severe risk that the lifeblood of the western region--the water
used by both people and livestock--may be impaired, much more
extensive study of mining's effects on groundwater is needed.

Seeding and Amendments

The seeding and amendment program, evaluated on the basis
of 13 INFORM guidelines, is the backbone of revegetation efforts
at mine sites. Many factors influence the quality of these programs: the type and amount of seeds (plants) used; how they
are planted; maintenance of the land before and after planting;
using "amendments" such as fertilizer, irrigation and mulches;
the manipulation of the soil surface to enhance water retention
and help decrease erosion and sedimentation, and the replacement
of more developed vegetative species, such as shrubs or trees.
Successful revegetation, it should be emphasized, involves re-
creation of a diverse plant community that can survive without
constant maintenance by man. With revegetation, the mine operator attempts to accelerate greatly the processes of nature, creating an appropriate, sustainable plant community of grasses, forbs,
shrubs and trees.

All of the many facets of seeding and amendment programs
must be carefully tailored to the needs of individual sites. The
combined result of well-focused planning of each aspect determines the success of revegetation. Experimentation also plays a
key role.

Though all companies in this study have conducted some
form of data collection, it is difficult to identify those that have
the best experimental programs. Unfortunately, some have ex-
pended effort and resources on rather meaningless work. At
Peabody's Black Mesa-Kayenta complex, spoil piles were paved
with asphalt in an attempted water-harvesting project to grow
fruit trees. At Knife River's Gascoyne Mine, tree transplants
were conducted in vain to create a golf course on mined lands.
Most research conducted by coal companies, the U.S. Department
of Agriculture and state universities is more to the point. But
at each site, the value of the experiments conducted and the
means by which data are put to use in the reclamation program
require further evaluation.

An overall assessment of the individual components of a company's seeding practice begins with the amount of land that has
been seeded. This at least indicates an attempt at post-mining

TABLE E: PRINCIPAL FINDINGS
COMPANY SEEDING PROGRAMS
INFORM's 15 Sample Mines

Region/Mine	Total Acres Disturbed	Percentage of Disturbed Land Seeded	Number of Plants in Seed Mixture	Pounds of Seed Mixture Applied Per Acre
The Southwest				
Black Mesa–Kayenta	4,707	45%	12--average	26.5
McKinley	over 1,500	70%*	12--average	26.5
Navajo	5,573	45%	10--below	17.5
Colorado				
Energy Fuels	2,516	55%	27--above average	12.9
Seneca II	978	50%	13--above average	18.0
Southern Wyoming				
Jim Bridger	1,850	13%	8--below average	12.5
Rosebud	3,493	20%	7--below average	15.0–16.5
Seminoe II	1,247	24%	11--below average	18.5
The Northern Plains				
Absaloka	516	25%	18--above average	19.5
Belle Ayr	1,273	43%†	11--below average	14.6
Colstrip	3,158	49%	21--above average	20.0
Decker	2,540	9%	11--below average	28.5
North Dakota				
Gascoyne	1,105	4%	5--below average	17.0
Glenharold	2,998	21%	7--below average	21.0–25.0
Indian Head	1,392	70%	6--below average	17.0

*Much of this is being re-mined.
†16% will be re-mined.

revegetation. In this regard, the small Indian Head Mine has
the best record in the sample, having seeded approximately 70
percent of all lands disturbed. Energy Fuels combines a good
record of seeding (approximately 55 percent) with the most diverse seed mixture in the INFORM sample. By contrast two companies stand out for the minimal amounts of land that have seeded:
Knife River (whose Gascoyne Mine also has one of the worst grading records) has seeded only 4 percent of site lands disturbed by
mining and related activities. Peter Kiewit has seeded a mere 9
percent of the more than 2,500 acres of land disturbed at its
Decker Mine. (At Decker the design of the mine, layout and
transportation facilities, and expanding production all have contributed to a dearth of graded (and thus seedable) land). Peabody, at its vast 4,700 acre Black Mesa/Kayenta complex in Arizona, undertook a massive seeding program in 1979 in an all-out
effort to compensate for years of mining when grading or seeding
practices lagged far behind disturbance. Seeds were strewn from
airplanes as well as being distributed by ground machinery. However, seeding over many seasons, using techniques that spread
out but do not bury seeds, risk yielding meager results.

While the extent of seeding offers a measure of the speed
with which companies have moved to initiate revegetation, the
mix of seeds chosen tells more about the quality and long-term
viability of the effort. The most extensive seed mixtures in the
INFORM sample are those utilized by WECo at Colstrip and by
Energy Fuels, the most limited by NACCO and Knife River. More
research and experimentation is needed not only to achieve increased variety in grasses but also in the planting and establishment of forbs, shrubs and trees on much of the rangeland now
being mined. Broad diversity of the plant community serves several purposes, including the supply of more and varied sources
of forage for wildlife and grazing animals, better protection of
soils and reduced erosion, better protection against major variations in climate on a year-to-year basis, and habitat for wildlife.

Performance by the coal industry in establishing forbs, shrubs
and trees has varied. Some of the best tree establishment can be
found at the Energy Fuels site in Colorado, where groves of transplanted trees are growing and multiplying. The company's selection of trees, its common-sense approach to the timing of their
transplanting, and the use of natural rainfall as an irrigant enabled it to achieve excellent results. Several other companies
are making marked recent improvements--CONSOL, Peter Kiewit,
Western Energy and Westmoreland, which has instituted a complex program for tree establishment. Montana regulators have
been the most concerned of all state reclamation agencies in pushing for increased forb use; this emphasis is reflected in the work
being done at Colstrip. Failure--due to poor knowledge and plan-

TABLE F: PRINCIPAL FINDINGS
COMPANY AMENDMENT PROGRAMS
INFORM's 15 Sample Mines

Region/Mine	Contour Furrowing	Mulch	Fertilizer	Irrigation
The Southwest				
Black Mesa–Kayenta	yes	straw	yes	no
McKinley	yes	straw	yes	yes
Navajo	no	straw	no	yes
Colorado				
Energy Fuels	yes	standing grain	no	no
Seneca II	yes	hay	no	no
Southern Wyoming				
Jim Bridger	yes	standing grain	no	yes
Rosebud	yes	standing grain/hay	yes	no
Seminoe II	yes	standing grain/hay	yes	no
The Northern Plains				
Absaloka	no	standing grain/straw	yes	no
Belle Ayr	no	hay/hydromulch/wood chips	yes	no
Colstrip	no	standing grain/hay/straw	yes	no
Decker	no	wood fiber	yes	no
North Dakota				
Gascoyne	no	standing grain	no	no
Glenharold	no	standing grain/straw	yes	yes*
Indian Head	no	standing grain	yes	no

*Used on approximately one-half the seeded areas for establishment of warm-season grasses only.

ning--has occurred at Gascoyne and the Black Mesa-Kayenta complex (as described earlier).

Since most of the coal industry in the West rides the middle ground, as far as performance in this area goes more effort will be required if re-establishment of diverse plant communities is to be achieved.

Can the new plant community thrive on its own? Particular concern arises at those mines where irrigation is used extensively to establish and grow plants on graded lands. Excluding CONSOL, where irrigation is being used to encourage the initial establishment of particular types of plants but will not be needed for their continued growth, three mines are using substantial amounts of water in efforts to get something--other than weeds--to grow on lands being reclaimed. Two are in the Southwest--P&M's McKinley, and Utah International's Navajo mines; the third, NERCO's Jim Bridger Mine, is in Wyoming at the edge of the Red Desert. Average annual precipitation is only 8.7 inches at Jim Bridger, 6.1 inches at Navajo (the lowest in the INFORM sample), and 14 inches at McKinley (where much of the water falls as snow, and rains come in brief, intense summer storms). At all three, supplemental irrigation is part of the basic reclamation plan for the site. But using an irrigation system creates an artificial climate, upon which the resulting vegetation relies. Although the process may prove successful in establishing vegetation, it is unclear whether this vegetation will be able to be weaned from the water and continue to thrive in such fragile land environments. In some areas at Navajo, irrigation has been stopped, but there is no evidence here, either, that the plant growth achieved so far will survive at productive grazing levels in the long run.

Given all the possible factors to be considered in a seeding and amendment program, Energy Fuels ranks the highest for several reasons--its use of a diverse seed mixture (even though it does use a number of introduced grasses), its good use of contour furrowing and mulching, and its continual testing to determine changes needed in the overall program. The tree transplants at Energy Fuels are the best found by INFORM. By far the worst performance is at Knife River's Gascoyne Mine--the only mine without any active program. Although the company's conservationist, David Jordan, has proposed a good plan for reclamation, nothing has been done.

Wildlife

Based on INFORM's two guidelines on wildlife, the main issue concerning wildlife at surface mining operations is not the protection of a specific endangered species, or a handful of particular animals. It involves the ecological balance between plant and animal communities--as well as between different animal species within a given area--that creates the total environmental system.

Surface mining--removing vegetation, tearing up the land and
disturbing the air and water--causes major disruptions to wildlife habitat. Both shelter and food sources are disturbed--often
destroyed--and animal populations must find other lands on which
to survive. This changes the nature of animal population densities and places additional demands on surrounding lands and animals.
    It is difficult to ascertain the impact of mining on animal life
in western coal areas. Wildlife takes a back seat in virtually all
reclamation considerations, as grazing is the primary land use in
most areas studied by INFORM. Industry efforts to attract and
provide a useful setting for the wildlife population have been limited to piling large rocks together as shelter and providing forage
in the basic plant mixture used on the site. More attention needs
to be given tree and shrub growth and to developing ways to re-
create or save unique wildlife habitats. Western Energy, NERCO
and Westmoreland have made this effort. Success with tree transplants by Energy Fuels is also beneficial. Water impoundments at
Indian Head (NACCO) and Black Mesa-Kayenta (Peabody) are at-
tracting waterfowl. But more work must be done to measure these
effects if wildlife populations--crucial to creation of a balanced
and sustainable ecosystem--are to survive.

Vegetation

    Three criteria were used by INFORM in evaluating the success
or failure of revegetation programs. (These are not "guidelines"
since they do not involve specific recommended practices for at-
taining reclamation programs).
    The end result of reclamation at surface mines (excluding
water impacts, discussed earlier) is to be found on the ground.
Barring major flaws--such as steep grading, poor seeding, etc.--
erosion can be controlled even if only with weeds such as Russian
thistle (tumbleweed). Beyond this, the visual appearance of lands
called "reclaimed" may be deceiving. Sites that appear to be lush
and green may reflect heavy irrigation and fertilization, and the
lack of any grazing activity on the vegetation. Reclaimed sites
may also exhibit a large volume of vegetation (plant biomass) and
hence seem to be reclaimed, but if the biomass includes twigs,
weeds and unpalatable species, it may be actually low in productivity for the proposed post-mining land uses.
    Real reclamation success is measured by the extent to which
vegetative communities stabilize the land--preventing erosion and
sedimentation--and are made up of a diverse, native selection of
plants. Success is further measured by the level of the land's
productivity, which should be equal to or greater than that which
the land could sustain before mining began. Of the 15 sites in
INFORM's sample, not one yet meets all three of these tests. Two

companies have made strong efforts to verify the productivity of
their reclaimed lands by grazing cattle on them. However, the
controls placed on these experiments have led to misleadingly
positive results. As described at the beginning of this chapter,
both WECo at its Colstrip Mine and Peter Kiewit at the Decker
Mine have conducted experiments that favored weight gains by
using animals that cannot reproduce. At Decker, shelters were
also provided so that animals could stay cool in the hottest hours
of the day--losing less weight--and water was within easy reach
on the small test areas--unlike the distances that cattle must
travel to water on most rangelands in the area. More realistic
testing is required.

Geographical Constraints

In addition to evaluating a company's reclamation program on
the basis of its guidelines, INFORM recognizes the importance of
geography in determining the success or failure of any program.
One company's successful reclamation program might well fail
if transposed to another site with different soil, precipitation and
vegetative characteristics. The same level of effort does not guarantee the same results from site to site. In the past, coal mining
areas have been selected on the basis of coal quality--Btu (energy
content) per pound, sulfur and ash content, etc.; ease of gaining
access to the land, and proximity to markets and transportation.
Reclamation potential has, unfortunately, never been a major
factor in a company's decision on where--or even whether--to dig
for coal. Many areas in the West may be--if mined--all but doomed
to ruin by their basic environmental conditions. The arid Southwest, southern Wyoming, and parts of Montana may never enjoy
pre-mining levels of plant diversity, water-resource integrity and
land without continuous need for post-mining maintenance. Even
in North Dakota and Colorado, where rainfall is higher or soils
are better, no proof exists that total land reclamation can be
achieved.
In planning for coal production, reclamation potential must
be a major factor in determining not just how, but where to put
the surface mine. To choose an area for mining without considering how well land and water resources will function as a unit once
mining is complete is to invite environmental degradation and
destruction.
Many environmental leaders in the West feel the region best
suited for massive coal development--from many points of view,
including reclaimability--is the Gillette, Wyoming area, where at
least 24 mines are either operating or planned for the near future.
Thick coal seams and thin overburden allow the maximum amount
of coal to be produced while disturbing the minimum amount of
land. Although land reclamation is uncertain, at least the lands
surrounding Gillette are not the mainstay of the agricultural base
in the West, nor are groundwater resources the most highly val-
ued in the area.

AVERAGE ANNUAL PRECIPITATION
INFORM's 15 Sample Mines

Region/Mine	County/State	Annual Precipitation (Inches)
The Southwest		
Black Mesa-Kayenta	Navajo, AZ	15.0 -17.0
McKinley	McKinley, NM	14.0
Navajo	San Juan, NM	6.1
Colorado		
Energy Fuels	Routt, CO	16.0
Seneca II	Routt, CO	16.0 -19.0
Southern Wyoming		
Jim Bridger	Sweetwater, WY	8.7
Rosebud	Carbon, WY	8.75- 12.0
Seminoe II	Carbon, WY	10.5 -12.0
The Northern Plains		
Absaloka	Big Horn, MT	16.0
Belle Ayr	Campbell, WY	16.0
Colstrip	Rosebud, MT	15.8
Decker	Big Horn, MT	11.8
North Dakota		
Gascoyne	Bowman, ND	15.0 -16.0
Glenharold	Mercer/Oliver, ND	16.9
Indian Head	Mercer, ND	16.9

Perhaps most significantly, the Gillette area lends itself to the greater use of truck-and-shovel mining technology. This may be the first step toward incorporating reclamation into coal production right from the initial stages of the planning and design of a surface mine. The use of truck-and-shovel mining operations as opposed to dragline mining increases the miner's ability to recreate geological strata in their approximate original positions in the ground. More research into ways to reconstruct aquifer pathways is needed. But the flexibility of handling materials with trucks and shovels, combined with careful planning for aquifer reconstruction, might make full land reclamation possible.

Techniques of Surface Mining and Reclamation

The purpose of surface coal mining is to remove a layer of coal buried deep beneath the earth's surface. This may be done in various fashions depending upon the type of mine intended. The information presented here describes a typical mine--a general surface mine, exhibiting the most common or significant aspects of surface mines in the West. However, because mines vary widely with individual site conditions, the hypothetical description presented here should not be compared to any specific mine.

Surface mining is technologically grand: huge machines excavate the coal and eventually, it is hoped, help reclaim the land to its approximate original condition or a better one. The process involves large-scale and long-term earthmoving and revegetation. Before mining begins, a variety of preliminary activities must be undertaken; these include exploring for coal, obtaining mining rights to the land, and acquiring permits to mine the land and discharge water from it. These preliminary activities, as well as the earthmoving and reclamation that follow, are described in the following pages.

Premining Procedures

A prospective mine operator commonly uses mineralogical maps of coal reserves to locate an area of potential coal formations. He then obtains an exploratory permit from the federal or state agency with jurisdiction over surface mining in the region. The operator usually drills core samples to find where the coal lies.[1] These samples are used to analyze the quality and depth of the coal *seam* and the earth that covers it. Such information ultimately determines the physical size of the mine and the type of equipment needed to mine the coal economically.

NOTE: Definitions of italicized words in this chapter may be found in Glossary.

Once a prospective mine operator locates a commercially viable coal seam, he must obtain the *rights* (see Glossary)--either by purchase or lease--to the surface and minerals on the land to be mined. This can be a relatively simple or complex problem, depending on several factors. For example, different parties may control the surface and mineral rights (or even one set of rights alone). Such is the case at the Absaloka mine in Big Horn County, Montana, where the Crow Indian Nation owns the coal but the company, Westmoreland Resources, owns the surface. These areas of purchase or lease, called *leaseholds*, generally range in size at large western mines from 5,000-30,000 *acres*.

If the prospective mine operator is successful in obtaining the surface and mineral rights, he must next seek a *permit* to mine part or all of the leasehold. The area permitted for mining may be from 5,000-10,000 acres. The mining operator must apply for a permit and present supporting data to whatever federal, state, or Indian Nation agencies regulate surface mining in the area. The permit application is a voluminous document describing both actual mining procedures intended, such as the use of bulldozers, and the impact of these operations, such as depletion of the water table. It describes the location of the coal and the nature and extent of the deposits and the environment of the area, including climate, soils, vegetation, land use, and wildlife. It also describes proposed mining and reclamation procedures, including the mine plan, engineering materials, earthmoving techniques, and safety measures. The mine plan itself is an encyclopedic report. It includes environmental, climatological, mineralogical and archeological data on the land for which the operator is seeking a permit.

The National Environmental Policy Act of 1969 requires that the U.S. Geological Survey and the appropriate state regulatory agency evaluate the potential impact of disturbances such as those created by coal mining. In the case of surface coal mines, the agency that issues the mining permit, usually the federal Bureau of Land Management or a state surface mining regulatory agency, produces an Environmental Impact Statement based on information collected from coal companies and such non-industry sources as ranchers and university researchers. This document describes the pre-mining environment and the changes that mining is likely to produce.[2] The permitting agency develops a draft, which is reviewed by all interested parties. Their comments are incorporated into the Final Environmental Impact Statement, which serves as the basis for determining whether mining should take place.

Surface Mine Regulation

The laws regulating surface mining have changed considerably since their inception in the early 20th century. Before 1977, three federal agencies regulated surface mining on federal lands. The Bureau of Land Management leased all federal lands. The Bureau of Indian Affairs, the agency chartered to protect the interests of the Indian Nations as negotiated with the federal government, helped tribes lease their lands. The U.S. Geological Survey (USGS), in the Department of the Interior, inspected both federal and Indian lands leased for coal mining. USGS reported quarterly on the acreage of land disturbed and reclaimed, the seed mix used to reclaim land, and the equipment inventory. Because the USGS's main interest is in revenue, other issues such as placement of toxic materials, topsoil, water quality, revegetation procedures, and land use were not usually included in these inspections. Before 1977, state agencies leasing and inspecting mines differed significantly from state to state. State programs varied: some had stringent requirements and frequent inspections; others, inadequate programs or none. The surface mine regulatory agency of Montana, for example, required a mine operator to post a $2,500 reclamation bond per acre, while Arizona had no reclamation laws as of June 1979.[3]

Surface Mining Control and Reclamation Act

The passage of the federal *Surface Mining Control and Reclamation Act (SMCRA)* in August 1977 significantly changed the regulatory requirements for the surface mining industry. The law established federal legal controls where few, if any, had existed and provided a uniform program to control surface mining in states where regulations may not have existed. SMCRA created the *Office of Surface Mining (OSM)* to develop and administer federal regulations for surface coal mining.

In May 1978 interim regulations took effect for all surface coal mining in the United States. At the same time, agency officials began inspecting mines[4] unannounced, at least quarterly, to monitor compliance with all permit conditions. After visiting a mine, OSM inspectors write an inspection report and issue notices of violations if any have been detected.[5] These officials check earthmoving techniques, preservation of the water systems, and land productivity. All the mines in the INFORM sample have been inspected by OSM. However, OSM procedures are still taking shape. The agency has not yet begun routine inspections of surface coal mines. In March 1979 OSM established final regulations, which were to become effective in 1980 for all lands surface mined in the United States.

Jurisdiction over surface mining varies according to owner-

ship of the land. OSM regulates federal and state lands unless
the federal agency and the state have reached a cooperative
agreement on jurisdiction or OSM has approved a state regulatory program. Under OSM-approved state programs, the state
has the right to regulate surface mining, but its regulations must
be equal to or more stringent than OSM's.[6] Under coop agreements, the state simply agrees to enforce OSM's regulations
within its boundaries. At present, of the six western states IN-
FORM studied, only Montana and Wyoming have reached cooperative agreements with OSM.[7] Cooperative agreements and approved state programs alike enable states to establish their own
inspection and enforcement programs. Both cooperative agreements and approved state programs are subject to periodic review and intermittent inspections of surface mines in the state
by OSM. Lands in states where neither cooperative agreements
nor state programs have been approved remain totally under
federal jurisdiction.

The question of jurisdiction over Native American lands remains unanswered. Some tribes are trying to establish their own
enforcement agencies, and OSM may allow them to regulate mines
on their lands. But as of today, jurisdiction varies. For example, the current regulatory authority at the Navajo Mine in New
Mexico is the Navajo Environmental Protection Commission.

Water Quality Regulation

Jurisdiction over water quality at surface mines is divided
between OSM and the U.S. *Environmental Protection Agency
(EPA)*. OSM requires mine operators to build *sediment ponds*,
but operators must apply to EPA for discharge permits. (See
Controlling Surface Water, below.) The EPA requires that a
mine operator obtain a *National Pollutant Discharge Elimination
System (NPDES)* permit and submit quarterly *Discharge Monitoring Reports*. An NPDES permit is a lengthy document, which
specifies permissible levels of pollutants in water discharged
from a mine site and the required method and frequency of monitoring these levels. Discharge Monitoring Reports are filed with
the EPA by the mine operator to record the range of water quality and, most important, the levels of total suspended solids
(sediment) in the water.

EPA still grants NPDES permits and collects Discharge Monitoring Reports although, under the current system, NPDES requirements have been incorporated into the procedure for receiving a mining permit under OSM regulations. Therefore, when
a mine fulfills OSM permit procedures, the EPA requirements are
concurrently satisfied. EPA and OSM also cooperate in inspecting mines for water quality; OSM has the primary responsibility,
and EPA has the authority to spot-check compliance.[8]

A final detail about the permitting process: the mine operator must post with the permitting agency a *performance bond*, a
sum of money to ensure that he will adhere to the procedures
described in his permit application. The amount of bond is determined by the reclamation costs that would be incurred if the
company should fail to reclaim.[9] This amount is proportional to
the area disturbed; the average total bond posted is $5 million.
The permitting agency, usually OSM or the appropriate state
agency, determines whether bond may be released. No bonds
have yet been released by OSM, although some state agencies
such as those of Colorado and North Dakota have released mines
from bond.[10]

EARTHMOVING AND COAL MINING

Before digging the first mining pit, a mine operator removes
vegetation and topsoil from the several hundred acres of land at
the area where the pits are to be. The operator stores the *topsoil* in temporary or life-of-mine *stockpiles*. Huge earthmoving
machines then dig the first pit. The earth below the topsoil is
removed and placed on the preceding piece of land until the coal
deposit is uncovered.

At this stage, an operator constructs diversion ditches and
sediment ponds to collect rain and other surface water (see Controlling Surface Water) that will traverse the mine site. Diversion ditches channel water into sediment ponds, unless the ponds
are placed so that natural drainage will carry water directly into
them.

Only after sediment ponds are constructed may the operator
begin the actual mining of the coal.[11] The simplest pattern of
surface mining consists of a series of rows in which earth is removed, in much the same back-and-forth pattern as a lawn is
mowed. Once the first one or two pits have been mined, as each
successive row or cut is made, the preceding cut is filled with
newly removed earth. Thus, the surface mine is constantly refilled as the mine advances.

Because earth, compacted over centuries, expands during
mining, refilling the pits usually leaves ridges above the former
surface level. When this process is complete, there remains a
large pile of earth consisting of overburden from the first pit,
followed by several smaller ridges of *overburden* (on the site of
the refilled mine pits), followed by an empty last pit. At this
stage the production process is complete. The coal is either
crushed and sorted at the pits or taken directly to railway cars
or trucks for removal. (See Figure A.) (For a variation, see
Belle Ayr profile.)

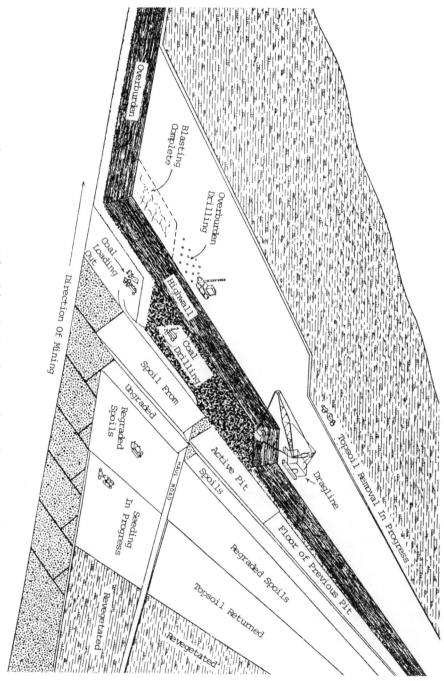

Figure A: An Idealized Surface Mine in Operation (Westmoreland Resources, 1979)

Topsoil Removal

The first step in earthmoving is to remove surface vegetation and topsoil, a relatively light task compared with the total amount of materials that must later be moved. Machines called scrapers carry out this process much as a plane shaves a block of wood. Scrapers usually consist of a vehicle similar to a dump-truck that has a blade protruding from its chassis; the blade scrapes small plants and soil into the truck bed. These trucks can hold between 20 and 50 cubic yards of soil. The topsoil is removed in two separate layers (*horizons*) if that much topsoil exists. Loosely speaking, topsoil is the term applied to the first 6-12 inches of soil. It is often rich in decaying organic material and microbes, which supply nutrients and aerate the soil, making it suitable for plant growth. However, the richness of soil varies in the West. The definition of "topsoil" depends on the quality of the earth in the area. In some southwestern regions, the quality of topsoil is roughly that of beach sand.

Topsoil can be either removed and replaced immediately on graded and contoured spoilpiles (dirt removed from the mine pit), or stockpiled (see Grading and Contouring) for later use. The "immediate replacement" technique is commonly called *direct-haulback*. This procedure requires that grading has reached the the point where land is available upon which the topsoil may be replaced. The topsoil removed in the period when mining begins is placed in a stockpile that remains for the life of the mine or is used as needed. To prevent wind and water *erosion*, a mine operator generally stabilizes these stockpiles with vegetation. In practice, unless otherwise mandated by law, operators generally use the stockpiling technique. When topsoil is stockpiled, it is simply removed and piled in a convenient place for later use. This creates a backlog and necessitates a costly and time-consuming hauling process.[12] Whatever technique is used, OSM requires that topsoil be removed in two horizons (where the two exist) and that the piles be stabilized to prevent erosion.

At this stage, haul roads and sediment ponds are also being built. Haul roads are the width of a three-lane highway, designed for use by dump trucks and other large vehicles on the mine site.

Draglines and Overburden Removal

Removing topsoil exposes the overburden, the next and largest section of earth to be removed. The U.S. Bureau of Mines defines overburden as "material of any nature, consolidated or unconsolidated, that overlies materials, ores or coal, especially those mined from the surface in open cuts."[13] Typically, the overburden is between 100 and 150 feet thick. It is frequently so compacted that blasting or drilling is necessary to loosen it before it can be removed.

Enormous earth-moving machines, called *draglines*, remove the overburden. These machines have roughly the shape of a crane and are about 100 feet long and 60 feet wide, with a boom 300 feet high. Draglines can mine more than 75 cubic yards of earth in a single bucketload. The dimensions of the dragline usually dictate the design of the mine. For example, the size of the boom limits the width of a mine pit, and its speed, usually between 0.1 and 0.15 miles per hour, makes it uneconomical to create a mine where the dragline is often in transit rather than moving dirt.

In the dragline's first cut into the overburden, called the *boxcut*, the dragline removes the overburden and places it behind the newly opened pit in a *spoil ridge*. From this point the mine expands in a series of extended pit rows which reach depths of 150-200 feet and may be 40-50 feet wide and from one to many miles long. Located at one side of the pit is the working or active face, called the *highwall*. The spoil pile created by the boxcut or initial cut is usually located atop the exterior of the pit opposite the highwall. The mine advances in a series of parallel cuts, each later to be refilled by overburden from the next active pit.

Figure B: Overburden Removal and Pit Refilling
(Mathematica Incorporated)

After draglines have exposed the coal seam, it is drilled and
blasted, and the coal is removed by earthmoving equipment. A
truck some 20 feet high, equipped with a large drill, executes
this operation. The drill bores holes six inches in diameter in
the coal seam. Explosives are placed in these holes and detonat-
ed. Large trucks, which would span three to four highway
lanes, commonly carry between 50 and 150 tons of coal from the
seam and out of the pit to nearby crushing facilities. From there
the coal is shipped to various points across the country, unless
it is being used locally.[14]

The spoil piles resulting from these successive cuts and re-
placements of overburden drastically change the surface of the
land. After several pits have been refilled, expansion of the
overburden creates a series of ridges each up to 40 feet high by
a mile long. (See Figure B.)

Controlling Surface Water

Surface water is not difficult to control if sediment ponds
and diversion structures are properly built and maintained.
Surface mining unearths layers of soil that would normally remain
compacted and unexposed. As surface water passes over areas
of freshly mined land, it accumulates loosened dirt, creating ab-
normally high levels of sediment and increasing the potential for
large amounts of *Total Suspended Solids (TSS)* to be carried into
streams and rivers.[15] A *sediment pond* provides a place for surface waters to sit while their sediment collects·and settles before
the purified water is discharged. Most mines have three or four
sediment ponds, although some have as many as 20. Ponds vary
in size; a surface area of 1,000 square feet and a depth of 5 feet
are average. These ponds usually discharge water into streams
and rivers off the mine site. A mine water purification system
also may include man-made drainages to channel water runoff
from active areas of the mine, which may cover more than 100
acres, into sediment ponds. Here the water is monitored, and
when it is sufficiently purified, it is released at a discharge
point offsite into streams, rivers or fields.

Besides using sediment ponds to control water quality, mine
operators often re-route water courses. Water courses fall into
three categories: permanent, intermittent, and ephemeral. Permanent water courses (rivers and streams) flow all year and are
composed of groundwater and surface water. Flowing less frequently are intermittent streams, fed by surface water and by
some groundwater through springs. Ephemeral water courses
flow least frequently, primarily after precipitation, and are composed primarily of surface water.[16] Mine operators divert these
waterways by creating channeling systems to prevent excessive
erosion due to unnatural drainage patterns. If not properly

maintained, these diversions--because of the land disturbance
that mining creates--may destroy vegetation and displace topsoil,
picking up sediment and depleting the soil. After mining and
grading are completed in an area, diverted waterways are reconstructed to preserve the original drainage patterns.

Controlling Groundwater

It is no easy matter to control the effects of mining upon
groundwater. OSM prohibits any permanent disturbance of
groundwater, but little regard has been given to the protection
of this resource until recently, and only now are extensive data
being developed on the changes in groundwater flow, recharge
and quality after mining.

"Groundwater" includes all water penetrating the surface
and filling the openings in rocks and soil to the saturation
point.[17] Rain and snow and other forms of precipitation are not
considered groundwater because they do not remain below
ground long enough to saturate permeable rocks and soil.

Surface mining, quite clearly, can degrade the quality of
groundwater. It may disturb the rock through which groundwater travels beneath the surface. Agriculturally rich *alluvial val-*
ley floors are especially sensitive. An alluvial valley floor is
composed of rocks and subsoil through which water flows in the
form of an intermittent underground stream beneath the surface
that irrigates vegetation.[18] The agricultural productivity of
these alluvial valley floors can be many times higher than that of
the surrounding land. The soils of these floors are easily penetrated by plant roots. They also provide a permeable medium
for water, allowing it to percolate up through the soil and reach
the roots of plants above.[19] Because of their productivity, OSM
regulations restrict mining alluvial valley floors.

Simply removing coal can upset the groundwater balance. As
many coal seams are *aquifers*--water-bearing rock or strata--re-
moving this coal reduces the quantity of groundwater.

In addition, removing the earth above a coal seam, whose
volume is much greater than that of the seam itself, can seriously
interfere with the groundwater balance. Surface mining a 5- to
50-foot coal seam involves cutting through 50-200 feet of overburden. This activity may cut across aquifers. Or it can disrupt water recharge areas, regions of semi-permeable and permeable rock (for example, a sandstone deposit) where surface
water, usually as rain, infiltrates to replenish underlying aquifers. Mining through an aquifer disturbs both the groundwater
physically removed with the coal seam and additional groundwater that gradually seeps out of the remaining portion of the aquifer into the newly dug mine pit and surrounding areas. This
seepage depletes the groundwater level over a region extending

at least 200 feet in all directions from the pit walls.[20] Besides, mining a recharge area can seriously affect the water table. By re-routing underground water from its natural course, some areas are left with an abundance of water and others with a scarcity.[21]

RECLAMATION TECHNIQUES

Grading and Contouring

Grading and *contouring* are simple earthmoving operations using huge and powerful bulldozers to smooth large amounts of soil and rock. After several spoilpiles have been created, they must be contoured and shaped to the surrounding topography. This is the first step in returning the area to a topography consistent with that which existed before mining--although ideally, reclamation planning has begun in the earliest stages of preparing to open a mine. Bulldozers flatten and smooth the ridges created by the spoilpiles. While it is not necessary to replace every knoll, the goal is to ensure that the particulars of an individual topographical feature, such as the slope of a valley, and of the entire site, blend with the surrounding topography (see Figure C).

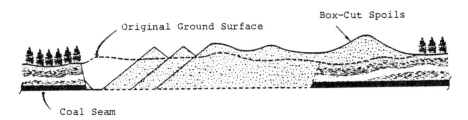

Figure C: Cross-Section of Partially Contoured Spoil-Piles
(Mathematica Incorporated)

Replacing Topsoil

Although it is a simple task to drop topsoil over any area of land with the same machinery that removed it, topsoil replacement is made more difficult by stockpiling. Topsoil in stockpiles becomes compacted and harder to remove from the pile.

Once some contouring has been completed, land is available for scrapers to replace the topsoil to a uniform depth. As the scraper moves over the area where the topsoil is to be replaced, topsoil is mechanically ejected from the interior of the vehicle onto the graded surface. To preserve the segregation and se-

quence in which the topsoil has been removed--a procedure required by law--the topsoil is usually replaced in two horizons.
If most of the topsoil has been spread on the mined land immediately after the land has been graded, only the topsoil from the
first cut remains to be replaced at the end of mining.   However,
if topsoil has been stockpiled extensively, replacing it requires
additional time and expense.

REVEGETATION

Seedbed Preparation

When the topsoil has been replaced, the land is ready for
seeding. However, if the replaced soil becomes compacted before
seeding--a period of up to six months--the topsoil must first be
broken up by a process called *discing*. A discer is a machine
approximately 8 feet wide and 2 feet high. As the discer is
pulled by a tractor, its eight 6- to 8-inch discs turn on a horizontal axis, loosening and breaking clumps of soil to a depth of
approximately one foot.

Seed Mix

Once the topsoil is prepared, seeding normally begins. *Revegetating* mined land to create pre-mining plant diversity or to
accommodate a new use is the most difficult part of reclamation.
No two mine sites are identical. On-site research with plant
communities is necessary to select and apply a seed mixture to
ensure that the recreated plant community can support long-term
land use, whether it be a wildlife habitat, *rangeland* or *pasture*.
A successful seeding program relies on a knowledgeable choice of
plant species, quantity of seed, seeding method, and appropriate
time for sowing the seeds.

Seeding Techniques

There are three basic seeding techniques: *drill seeding*,
broadcast seeding, and *hydroseeding*. Drill-seeding is generally
the most successful, although special circumstances may require
broadcast seeding or hydroseeding. The term "drill seeding" is
misleading, as no drill is used or drill-holes produced in the
process. The drill-seeding device, about the size of a small
tractor, digs small furrows in the ground, channels seed into
them, and rakes surrounding topsoil over the seed to cover it.
The whole process takes place in an unbroken sequence, minimizing seed loss to wind and dehydration. Broadcast seeding is
a process much like fertilizing a lawn. A tractor or similar vehicle tows a seed-filled bin. As the bin is pulled, its wheels turn
a set of rotary blades within the bin to dispense the seed. The

technique does not implant seed in the soil. In hydro-seeding, a water-based mixture of seed and fertilizer or mulch is sprayed onto the land. A small truck carries a container full of this mixture, which is dispensed through a thick hose.

Mulching

Mulch, which generally consists of straw, hay, or wood chips, fosters germination of seeds and growth by stabilizing and preventing evaporation of water from the soil. It can be spread in its natural state or sprayed as part of a water-based mixture. In the first instance, a device shreds the balled mulch and sprays it the way a snowblower distributes snow, or it is grown on revegetated plots and mowed to serve as a mulch. A mulcher is a device, approximately 6 feet wide and 5 feet high, capable of processing a standard hay bale.

Several precautions must be taken to insure effective mulching. The mulch must be anchored to the ground or it may blow away. A *crimper*, a tractor-size vehicle, pulls a drum with hoof-shaped protrusions that press the mulch into the ground and anchor the seed.

Irrigation and Fertilizer

Both *irrigation* and fertilizer can help to grow a ground cover, which will help to prevent erosion more quickly after seeding than if natural precipitation and nutrients are relied on alone. However, once the permanent species are growing, irrigation or fertilizer may foster an unnatural environment because the species they promote may not survive without them. Irrigation can be expensive. Fertilizer usually is composed of nitrogen and phosphorus, applied at a rate of 20-30 pounds per acre. Both practices are used in the West. However, their use is not common, and the long-term implication of this practice is not yet known.

Trees and Shrubs

Although many mine sites in the West were originally grassland, shrubs and trees that existed in the pre-mining environment must be replanted. This process involves four kinds of growing techniques: *bare-root* or *tree-spade transplants*, direct seeding, and *containerized transplants*. Bare-root seedlings are transplanted by hand, with no soil around the roots before they are placed in the ground and covered. Unlike bare-root transplants, tree-spade transplants are performed by a machine that lifts the tree, roots, and soil at once and places it in the desired location. Direct seeding and containerized plantings are both usually done by hand, planting and covering either a seed or a

containerized unit of plant, soil and nutrients. In western reclamation, companies sometimes employ more than one technique,
experimenting to achieve success. All four techniques are employed at western mines.

Land Productivity Measures

One of the most common ways to determine the success of
revegetation efforts is to measure productivity using the dry-
weight method. In this process, plants on a selected sample of
land are clipped, and their water is allowed to evaporate. The
clippings are weighed to determine the pounds of *forage* growing
on an acre of land. This method must be employed carefully.
Several factors influence the productivity of a given piece of
land: the season when the measurement is made, stability of the
vegetation, and the plant species measured. Because of the
variability of these factors, measurement of productivity is not
an exact science but changes according to a region's specific
characteristics, the season, and the length of time the vegetative community has existed. The actual requirement of a given
herd of cattle or flock of sheep will also vary from animal to animal. In addition, land productivity figures do not reflect a region's ability to support wildlife. Information about the amount
of forage necessary to support various species of animals, as
well as the density of wildlife in a given area, is often sketchy.

Closing

Before reclamation is complete, disturbed lands at the first
and last pit must be graded. Spoil from the first pit is generally
piled on adjacent undisturbed land (from which topsoil has been
removed and segregated), creating a ridge equal in height to the
depth of the first pit. These spoilpiles are gradually integrated
into the slope of reclaimed land, producing a contour somewhat
higher than that of the pre-mining land at the first pit, which
gradually declines at the following pits. This process leaves
enough spoil to fill the last pit to a level somewhat below that of
its pre-mining contour.

After revegetating the last pit, the dragline, *tipple*, offices,
equipment shed, shop, sediment ponds, and miscellaneous mining
structures are removed. Because of its size, the dragline is
taken apart piece by piece so that it can be transported off the
site. Cleaning and production facilities are removed, and sediment ponds are either refilled with earth and returned to a contour consistent with the surrounding topography, or left as permanent impoundments for watering livestock if allowed by regulatory authorities. Finally, the haul roads are contoured to return them to natural topography.

Once the operator stops maintenance on the land after it has been seeded, mulched, managed for grazing and, if necessary, fertilized and irrigated, the company officially starts its period of "extended responsibility."[22] During this period, which lasts for ten years, OSM prohibits the operator from performing any maintenance on the land other than that which would occur in normal post-mining land use, such as grazing. The agency requires that after ten years "groundcover and productivity" must equal the standard established in the permit application. The land may then be released from bond.[23] At this point the coal company's legal responsibility for the land ends.

Guidelines

Surface coal mining in the West presents both opportunities and problems. Although surface mines disturb grazing and farming, many western lands used for these purposes were severely abused before mining began. The U.S. Bureau of Land Management states that of the "170 million acres of public rangeland in 11 Western states [an area almost the size of Texas], 80 percent of that land is in fair, poor or bad condition."[1] Revegetating disturbed lands with native plants and instituting proper range management practices may even make western mine sites more productive after mining than before. Successful reclamation could allow more water to be trapped in the soil to recharge underground water sources and provide more vegetation for domestic and wild animals. To succeed, however, reclamation must create stable and self-regenerating plant and animal communities.

Surface coal mining in the West may cause serious problems. These operations can easily disrupt the fragile balance of ecosystems in the arid and semi-arid western states. Animal and plant communities that have evolved over many years can be temporarily or permanently lost in months as wildlife habitats and migratory paths disappear. Wind and water erosion may gouge the land and carry sediment onto farmlands and into water supplies. The loss of grazing lands and croplands can alter the lives of the region's inhabitants. Erosion may also reduce the amount of annual forage produced in a given area and its ability to provide sufficient food for livestock or wildlife.

The West is notably short of water. Large surface mines may degrade or diminish many available water sources. Erosion

and sedimentation may pollute surface waters. Mining through
layers of rock and earth that serve as pathways for underground
water (aquifers) may destroy these networks, or reduce the
quality and/or quantity of water flowing through them. The
land's ability to recharge aquifers with water may be imperiled.
Although the replaced earthen layers may allow more water to
reach the lower geologic zones, this water may carry with it increased amounts of minerals or toxic substances, degrading the
water in the aquifer system.

The Coal Picture

    In 1977, the six principal western surface mining states--
North Dakota, Montana, Wyoming, Colorado, Arizona, and New
Mexico--produced 36 percent of the 690 million tons of coal mined
in the United States. Surface mines accounted for 61 percent of
all coal mined that year. The largest U.S. surface coal mine, the
Belle Ayr Mine, located near Gillette, Wyoming, produced more
than 18 million tons of coal in 1978.[2]
Western coal is also important because it is low in sulfur,
making it attractive to utilities faced with strict air quality requirements. In 1974, the United States had 86.5 billion tons of
low-sulfur (1.0 percent by weight or less) coal reserves that
could be mined from the surface. The six states analyzed in
this study contained 69 percent of these reserves.[3] But most
low-sulfur coal has a lower energy or British thermal unit (Btu)
content than a medium- or high-sulfur coal. The National Coal
Policy Project, a group of environmental and industry representatives, has suggested that more low-Btu/low-sulfur coal may
have to be burned to produce the same energy as high-Btu/
high- or medium-sulfur coal. The net effect might be the release
of more sulfur into the atmosphere.[4] The Harvard Business
School Report, *Energy Future*, stated:

> 54 percent of coal by weight--but only 30 percent
> by heat content--is estimated to lie west of the
> Mississippi....[Although] western coals are gen-
> erally less polluting,...they also provide less
> energy per pound. So, on a comparable basis of
> heat value, the difference in the emission pollut-
> ing potential between Eastern and Western coals
> lessens, although it is still significant.[5]

In 1979, U.S. coal production disturbed approximately 1,000
acres of land each week.[6] The U.S. Bureau of Mines projects
that some 1.86 million acres of land in the six western states
INFORM analyzed overlay coal reserves that can be reached by
surface mines.[7]

Successful reclamation of western surface coal mine sites is vital if the United States is to increase its use of coal without permanently compromising the valuable resource that the land itself represents. Yet reclamation is still largely experimental in the West, and the characteristics of specific sites, including soils, climate and water, make it difficult to establish a basic program guaranteeing reclamation in these arid and semi-arid lands. It will take years to determine whether land reclamation efforts to date will succeed. Nature may eventually revegetate lands disturbed by surface mining, but according to a 1979 Environmental Impact Statement prepared by the U.S. Geological Survey and the Montana Department of State Lands, this process may take fifty years, and even then the quality of the vegetation created might not be compatible with planned land uses.[8]

INFORM's Guidelines

INFORM examined information from a variety of sources to determine the basic measures that constitute a successful reclamation program at a surface coal mine. Researchers reviewed comprehensively literature in the fields of land reclamation, vegetative establishment and growth, range management, and semi-arid and arid land ecology; interviewed experts in reclamation technologies and corporate practices, including industry reclamation experts, environmentalists, surface-mine management teams, and environmental scientists from many disciplines; and visited more than 25 mine sites in the midwestern and western United States, as well as in West Germany. We interviewed state and federal regulators and reviewed state enforcement programs to determine similarities and differences between legal requirements and performance in the West. Based on research findings we drafted the series of reclamation program guidelines presented in this chapter. These guidelines were reviewed by coal company officials, federal regulators, environmentalists and others knowledgeable in the field (a list of reviewers is found at the end of this chapter).

The coal industry and those federal agencies involved in land reclamation produced much of the information on which INFORM's guidelines are based. Coal companies have engaged in experimentation with private researchers and universities, sponsored seminars and reported case studies in trade journals. Federal agencies such as the Soil Conservation Service, Bureau of Land Management and Environmental Protection Agency, and federal and state Geological Surveys have also conducted experiments and collected data on land reclamation techniques and practices. Laboratories like the Argonne National Laboratory,

operated by the Department of Energy, also conduct reclamation
experiments.

INFORM's guidelines deal with 5 areas of reclamation:

- earthmoving
- water quality
- seeding and amendments
- wildlife
- vegetation

Had these guidelines been written in the early 1970s, they
would have been unique. However, with increasingly stringent
state regulation (notably Montana) and the passage of the federal
Surface Mining Control and Reclamation Act, the basic reclamation
principles (and many others) have been laid out. INFORM's contribution takes these principles a step further by explaining
clearly in layman's language the thinking behind them, and describing the range of potentially negative effects if these guidelines are not followed. Such description was possible as a result
of the two years of field research, at 15 sites where such effects
could be concretely evaluated. The principles described in IN-
FORM's guidelines for restoring vegetation are neither unique
nor foreign to coal company staff. They are standard agricultural practices that must be applied to lands devastated by surface mining. Other INFORM guidelines address actions that can
be taken during the pre-revegetation period (e.g., grading,
protection of water resources) so that revegetation may be successful and disturbances beyond the mine's boundaries minimized.

INFORM's guidelines are used in this study to evaluate corporate land reclamation performance at a limited number of western mine sites. However, their future applicability lies in several areas: they can be used to assess the basic environmental
soundness of mine and reclamation plans that companies propose
for future western development. Community organizations, public interest groups and Native American governments may find
them a valuable tool. The guidelines may also be used by local,
state and federal agencies and legislators to assess state, local,
or federal legislation and enforcement. For the coal company
executive or investor, the guidelines are a valuable source of
fundamental information on a process that is becoming more and
more important to the daily operations of the coal producer.
They may be used to measure performance at other sites which
INFORM did not study. Finally, in several years they may be
used to reassess the sites which INFORM or others have studied,
to see how much progress has been made in protecting land and
water resources.

These guidelines are based on the premise that, as federal
law requires, coal surface-mine reclamation programs should be

designed to return the land to a productive use at least equal to
that which the land could have supported before coal mining began. This does not mean that the goal is achievable on all lands,
nor that severely over-grazed land should be returned to the
productive levels caused by such mismanagement rather than
levels that the land could support if properly maintained. Federal law requires a mine operator who chooses to return surface-
mined land to a more productive or "higher" land use to justify
the change and demonstrate its prospects for success to the
state or federal regulatory agency with jurisdiction over the
site, as well as to local agencies such as district planning com-
missions.

INFORM's guidelines are used in this study to evaluate the
current state of basic land reclamation practices at 15 surface
coal mine sites in the West as of the summer of 1979. Some IN-
FORM guidelines are used to assess the results of practices that
may have been carried out four years ago, such as the degree of
success in revegetating a mined area, or the elapsed time between mining and grading. Many activities practiced today rely
upon past activities. Grading and topsoil removal, stockpiling
topsoil, and developing reclamation programs continue from year
to year and affect the next year's work. In four years, an initial evaluation can be made of revegetation programs and changes
that occur as experience and new regulations alter reclamation
practices at the mine sites.

Conditions at each mine involve a unique combination of climate, terrain and soil characteristics, and biological and hydrological systems. As a result, certain guidelines permit exceptions where the environment or a mine's design prevents a company from carrying out these practices. However, in these
cases, successful reclamation must include alternatives capable
of fulfilling the aim of a specific guideline.

Many guidelines presented in this chapter highlight experimentation as an important practice in developing reclamation programs for specific mine sites. A report prepared by the Land
Reclamation Program of the Argonne National Laboratory stated:

> If successful reclamation is the desired end for
> surface mining activity, then the reestablishment
> of an ecosystem must be the driving force that
> guides reclamation strategy. Research should
> include studies on the responses of all components
> of the ecosystem to the stresses that occur dur-
> ing the reclamation process. Only by under-
> standing these responses can the correct deci-
> sions be made on timing, species, and soil
> amendments to ensure successful reclamation.[9]

State laws governing land reclamation also recognize this
fact. Colorado, for example, requires that a revegetation plan
"consider environmental factors such as seasonal patterns of
precipitation, temperature and wind; soil texture and fertility,
slope stability, and direction of slope faces."[10] Montana mandates that a mine operator "consider soil, climate, and other relevant factors when planning and/or seeding to provide for the
best seed germination and plant survival."[11] There is no single
way to reclaim a surface mine in the West. Data on land productivity and on plant growth on reclaimed grazing and agricultural
lands are also lacking. Only experimentation can provide this
information. The Draft Environmental Statement for Montana's
Northern Powder River Basin Coal, prepared by the U.S. Geological Survey and the Montana Department of State Lands in
1979, suggests the need for experimentation: "There has not
been enough experience with mined-land reclamation to learn
what procedures and standards are necessary to ensure the
long-term survival of self-sustaining vegetative communities."[12]

These guidelines are one instrument by which to judge reclamation practices at a western mine site. Coupling the guidelines with a knowledge of environmental factors such as climate,
soils and topography, as well as changes in state and federal
laws mandating surface mine reclamation, a picture of the mine's
practices and a company's commitment to reclamation can be
achieved. The guidelines do not tell the mine operator how to
seek a permit. They do indicate practices (such as immediate
use of topsoil) that might affect the design of the mine, and
thus affect reclamation success.

In evaluating reclamation practices at a given mine site, IN-
FORM did not rigidly apply the guidelines established here but
took into account the characteristics of mining and climate at a
specific site. If a company undertook alternative measures that
achieved the goals of the guidelines, the practice was rated
"good." INFORM rated practices that did not achieve the guidelines' goals from "fair" to "poor" based on the effects of these
practices on the mine's total reclamation program. The ultimate
aim of all reclamation practices--returning all lands disturbed by
surface mining and related activities to their former or higher
productivity without compromising the quality of land or water
resources off the mine site--is the paramount criterion. (The
INFORM report does not address air quality.) Finally, innovative practices such as the use of runoff to irrigate transplanted
trees and shrubs were rated "good" to "excellent" on the basis
of their contribution to the success of the total reclamation program.

Analysis of current company practices can establish a basis
for evaluating effects of federal environmental regulation upon

the coal industry in the next few years. It also provides a yard-stick against which to judge coal company initiatives in prevent-ing or repairing the damage caused by mining an increasingly valuable energy resource. Will practice in the industry change? Do current practices warrant modification of federal legislation? Are these practices the result of regulatory requirements or company initiatives? President Carter's call for the use of coal to produce synthetic fuels would increase surface mining in the West tremendously. Before the United States embarks upon this path, it should be sure it will be able to strike an acceptable balance between developing the nation's domestic energy re-sources and protecting western land and water.

EARTHMOVING

Long before seeding and revegetation of a surface mine be-gin, many acres of land are stripped of vegetation and soil. Hundreds of tons of rocks and dirt ("overburden" or "spoil") are removed to expose the coal seam. This material is placed in the huge pit created by the previous operations of the earthmov-ing equipment. The process is then reversed as piles of earth are graded and leveled to create land that can be used as it was before mining began or put to a new use that regulators have approved.

The prompt return of mined land to productive use minimizes the adverse environmental impact that earthmoving operations can create at a surface mine. This strategy includes the protec-tion and reuse of topsoil and subsoils found on the mine site. Also, to encourage the most favorable medium for plants to grow, as well as to protect water resources, any toxic or acidic materi-als exposed in the mining process must be segregated from plant root zones and groundwater. How soon reclamation follows mining dictates how long the mined land remains unproductive. Once the coal has been removed, an operator must shift his emphasis to reclaiming land for other purposes. The guidelines below include essential earthmoving procedures on which establishment of veg-etation and the success of final reclamation depend.

Grading and Contouring

1. To ensure that seeding operations closely follow coal produc-tion, grade within three spoil ridges of an active mine pit, unless pit configuration or contouring needs justify exceptions.

In surface coal mining, when a cut is made in the ground to create the "working pit," the earth and rock that are removed

from the pit are placed in the nearest empty and abandoned pit.
In effect, the last hole dug is filled with the earth and rock, or
overburden, from the pit being mined. (Overburden from the
first hole dug must be placed on the surface, as no other hole
yet exists.) Before the overburden is removed by the mine
operator, it is tightly compacted by the weight of the soil and
rock above it. Removal operations break up the overburden and
increase its volume as air fills in areas between the broken rock.
Thus, the amount of overburden sometimes exceeds the capacity
of the hole to be filled. In areas where a thin layer of overburden lies atop a thick coal seam, the spoil from the active mining
operation may not fill the hole remaining after coal has been removed. The thickness of the coal seam that has been removed
and the amount of expansion of the overburden determine the
size of ridges left by the overburden. These are usually termed
spoil ridges and must be "graded" (or bulldozed) to the reclaimed area's final contour before topsoil is replaced and planting can begin.

Ungraded, unreclaimed land can create serious environmental
problems. Wind and water can erode the spoil ridges and increase sedimentation on and off the mine site. Dust blown from
these piles may degrade air quality. As the mine expands and
more lands are stripped of topsoil, graded areas must be ready
to receive this topsoil or the material must be stockpiled (see
Guidelines for stockpiling). Two spoil ridges are usually necessary for grading to blend the graded lands into one another most
effectively. By grading within three ridges of the active mine
pit, two ridges can be worked while the third remains to keep
grading operations from interfering with mining; thus the grading and subsequent reclamation operations can keep current with
mining. Gary Deveraux of the Northern Energy Resources Company, which operates the Jim Bridger Mine in Wyoming, reports
that "in all our operations we [grade] right up to the mine pit or
to one spoil ridge. We usually leave one spoil ridge for a safety
factor, and also so the next spoil ridge will have a buffer rather
than coming on to [any] reclaimed area."[13] The closer grading
operations are to the active pit, the sooner reclamation planting
can begin, and the better the coordination between coal production and land reclamation. This also makes the continuous handling of topsoil easier and more efficient.

The design of certain mining operations may not always allow
this three-ridge standard to be met. Weather conditions may
also hinder grading. Until grading is completed, however, none
of the other processes necessary for adequate reclamation--
replacement of surface soils, application of mulches or other
amendments (such as fertilizer) and the seeding of plants--can
begin.

Montana law requires that "on lands affected by area strip mining, the grading and backfilling shall not be more than two spoil ridges behind the pit being worked."[14] When a surface mining operator cannot meet this requirement, the operator must seek a variance from the Montana Department of State Lands to be exempted. The federal Office of Surface Mining (OSM) has less stringent requirements: "Rough backfilling and grading shall be completed within 180 days following coal removal and shall not be more than four spoil ridges behind the pit being worked, the spoil from the active pit being considered the first ridge."[15] OSM rejected a two-ridge standard "because of climatological and operational problems."[16] It did not address a three-ridge standard.

2. To ensure adequate drainage of surface water and the reclaimed land's ability to support post-mining use, grade land to approximately its premining contour.

Unless the mine operator can justify an exception, the contour of the land should be no steeper than it was before mining began. (The land may need to be flatter in order to establish vegetation and prevent erosion. However, if land is graded flatter than the surrounding topography, some slopes will be steeper than the surrounding topography. In this case, the benefits of grading to a shallower slope must be weighed against the potential problems from having to revegetate steeper slopes as well as the potential change in water drainage patterns.) The shape of the graded land should complement that of the surrounding lands so that surface water flow and drainage patterns are similar.

Wyoming, Montana, North Dakota and Colorado require, like New Mexico, that grading and backfilling "be carried out so as to produce a gently undulating topography or such other topography as is consistent with the proposed end use of the area."[17] Montana's law states that "the area of land affected shall be restored to the approximate original contour of the land."[18] Wyoming's requires that "all backfilling, grading, and contouring will be done in such a manner so as to preserve the original drainage or provide for approved adequate substitutes."[19] OSM stipulates that "final graded slopes need not be uniform but shall approximate the general nature of the premining topography."[20] In many cases, the contour should not be uniform. In an area of undulating hills, if grading were done to create an average topography the result might be a single uniform slope. Bruce Boyens of the Office of Surface Mining in Tennessee observes that long, uniform slopes tend to be unstable, and "sheet erosion [erosion across the land surface instead of along cuts in the soil] is a large problem."[21] One long slope neither provides the same pattern of drainage for water nor produces the shaded areas a

slope graded to the pre-mining contour and topography would
produce.

*3. To create the maximum opportunity for successful revegeta-
tion, grade land so that seeding and other reclamation machinery
can operate on the contour.*

Because successful reclamation involves re-establishing a
vegetative cover on graded lands once coal has been removed,
the land should be graded so that the best seeding and reclamation techniques can be employed. Unless application of seeds
and amendments is to be done aerially or by methods that do not
require that machinery operate on the slopes, these machines
should be able to operate along the contour of the land. The
grooves created in the soil by machinery tires or treads can
benefit revegetation efforts if they run perpendicular to the
slope (see Guideline 21). However, if these tracks run parallel
to the slope, the potential arises for increased erosion and sedimentation and reduced infiltration of water into soils. Areas
with slopes too steep to support this machinery should not be
mined if they cannot be reclaimed by some other technique.

The Rehabilitation Potential of Western Coal Lands, a report
of the National Academy of Sciences, states, "Maximum vegetative stability cannot be attained on slopes steeper than 33 percent (3:1). Optimum vegetative stability requires slopes of less
than 25 percent (4:1) and use of agriculture machinery may require that slopes be no greater than 20 percent (5:1)."[22] A coal
company wishing to mine lands whose slopes exceed 20 percent
should be able to revegetate an area effectively by methods
equal to or better than those suggested here. In fact, some coal
companies have developed their own reclamation machinery to
cope with the need to apply seed or other amendments along the
contour. Richard Kail, director of reclamation at the Jim Bridger Mine, has developed a mulch-spreading machine for this
purpose.

Topsoil

*4. To provide the best soil for revegetation after mining, re-
move and segregate topsoil from subsoil so that it can be reapplied homogeneously.*

Topsoil is the one medium on the mine site that definitely will
support plant life, as it was doing so before mining began. A
combination of characteristics--including the mineral and water
content as well as porosity, which allows water to percolate to
the roots of the plant--makes topsoil the likely choice as a plant-
growth medium unless other soil layers are shown to be better

suited. Thus topsoil and the soil layers or "horizons" directly beneath it should be removed separately and replaced on top of graded overburden to establish vegetation. (These horizons are usually classified as A-, B- and C-horizons by certain characteristics, such as permeability to air and water and nutrient content, although all horizons are not always present. Topsoil is usually synonymous with the A-horizon.) However, in areas where alluvial soils (of an unconsolidated nature and texture) may exist to a depth of several meters, all the material could be considered topsoil. In some areas, such as the tops of heavily eroded hills, no topsoil may exist. On any given site, opinion may differ on the best material to be used as the plant growth medium on lands being reclaimed. In this case, if the coal mine operator determines not to segregate topsoil and subsoil so that it can be reapplied, the onus falls upon that operator to show the existence of superior substitutes.

In a report published in October 1974 by the Mining Pollution Control Branch of the U.S. Environmental Protection Agency, E. C. Grim and R. D. Hill determined that "the removal and placement of growth supporting material, or 'top soil,' is one of the most beneficial methods for assuring establishment of vegetation."[23] In 1974, at the Second Research and Applied Technology Symposium on Mined-Land Reclamation, D. E. McCormack of the U.S. Department of Agriculture's Soil Conservation Service presented a paper stating:

> Burying the A horizons under many feet of soil during the surface mining operation is certainly not compatible with full restoration or productive potentials. Neither is mixing the A horizon with the next lower horizon--the B...horizon covered by the A horizon....Replacing the A horizon at the surface will greatly favor revegetation.[24]

In 1976, he wrote:

> In most areas, the A horizon of natural soil is vastly superior to any underlying soil horizon or geologic strata. Even if it is only 3 or 4 inches thick, careful handling and return of this horizon to the surface is required for most successful reclamation...including texture, structure, organic matter content, and pH [the measure of acidity or alkalinity].[25]

OSM regulations state:

> All topsoil shall be removed in a separate layer from the areas to be disturbed, unless use of substitute or supplemental materials is approved

by the regulatory authority....The B horizon
and portions of the C horizon, or other underlying layers demonstrated to have qualities for
comparable root development shall be segregated
and replaced as subsoil, if the regulatory authority determines that either of these is necessary or desirable to ensure soil productivity consistent with the approved postmining land use.[26]

The regulation goes on to describe how alternative materials
should be identified.

*5. To protect the quality of topsoil removed for mining, replace
it immediately on graded lands. Stockpile only that topsoil removed to create the first mine pits and seed the stockpile until it
is ready to be used.*

As the topsoil removed in digging the first mine pits cannot
be placed on any graded areas (as none yet exist), this topsoil
is best stockpiled in an area that would not be disturbed
throughout the life of the mine (or until it can be used), and
seeded to vegetate and stabilize the stockpile for that period.
Once grading begins (see Guideline 1), topsoil should be immediately spread over graded land without being placed in a stockpile, then seeded or protected in some other way. This process,
called *direct haulback*, requires less vegetation to be removed
from the topsoil. Organic materials in the topsoil, such as large
roots and stumps, improve its quality and add nutrients to the
soil as they decompose. If topsoil is stockpiled, more of these
roots and stumps must be removed before the soil is stockpiled
because they impede removal of the soil from the stockpile.
Micro-flora and fauna may also be preserved if soils are directly
replaced on graded lands. Immediate respreading of the soil is
also important because land covered with stockpiles--often up to
several acres--cannot be used for other purposes, and because
stockpiled soil can also gather windblown seeds of undesirable
species, and lose nutrients and soil material to wind and water
erosion. Topsoil may be the best supplier of native seeds available at the mine site.

Certain factors may not allow all topsoil to be immediately respread on graded lands. For example, stripping of topsoil ahead
of mining operations may be increased in the fall so that the
company is not forced to attempt this operation in winter. Or,
where grading must be delayed because of the mine's design,
direct haulback may not be possible. However, industry representatives agree that the direct replacement of topsoil is the
best practice. Gary Deveraux writes that "in general, we certainly support the technique of directly applying all topsoil re-

moved because of the costs involved in stockpiling, as well as
the biological factors that [INFORM] would consider.   However
...there are situations where direct application cannot be accomplished."[27] (See Guideline 6.)

Grim and Hill determined that "topsoil can only be stockpiled
for a limited time or it will lose its ability to enhance vegetative
growth."[28] The National Academy of Sciences concluded, "The
values to be derived from adding topsoil are often decreased by
stockpiling the soil, since one advantage of spreading topsoil is
the transplanting of live seeds and plants....If soil is stockpiled,
the revegetative potential may be lost before it is placed on the
spoils."[29] The Draft Environmental Statement on Northern Powder River Basin Coal, Montana, prepared by the U.S. Geological
Survey and the Montana Department of State Lands, states,
"topsoil placed directly without stockpiling would produce a much
greater percentage of growth from salvaged seeds and plant materials."[30] In its recommendations of specific reclamation practices for the arid and semi-arid western United States, the National Coal Policy Project, a coalition of industry and environmental experts on coal, concluded, "Continuous handling of topsoil is desirable in order to preserve nutrient storage equilibria
and soil micro-organisms. Top soil should only be stockpiled
when absolutely necessary."[31]

Wyoming regulations also recognize the value of immediately
replacing topsoil by requiring "reclamation shall follow mining as
soon as is feasible so as to minimize the amount of time topsoil
must be stockpiled."[32] OSM mandates that "topsoil...shall be
stockpiled only when it is impractical to promptly redistribute
such materials on regraded areas."[33] Only the absence of
graded areas would make immediate replacement impossible.

*6.   To create an opportunity for successful revegetation over the
entire mine area, replace topsoil and subsoil to a uniform depth
on graded lands.*

Topsoil may not have been uniformly distributed over an entire mine area before mining began, but it should be replaced in a
uniform fashion to equalize the revegetation potential of all disturbed areas. Natural forces will eventually determine which
areas of a reclaimed surface mine will accumulate more topsoil and
which will lose the replaced soil. Thus, unless a coal-mine operator can show that non-uniform placement will achieve a higher
level of vegetative growth over the whole mine area, topsoil
should be replaced evenly.

Situations may arise where this guideline cannot be fully followed. If an area is mined where topsoil (and subsoil) are
deeper than average, some of these soils may have to be stockpiled for replacement on lands with less of these soils. In these

cases, immediate replacement of topsoil in lieu of stockpiling will
not be achievable for all the soils removed. However, the mine
operator should still strive to replace as much of these soils directly as possible (see Guideline 5).

Wyoming regulations require that "topsoil...shall be evenly
distributed on the surface of all lands affected during the course
of the operation."[34] Colorado mandates that "when growing
media is replaced, it shall be done in as even a manner as possible."[35]

OSM regulations state, "Topsoil and other materials shall be
redistributed in a manner that achieves an approximate uniform,
stable thickness consistent with the approved postmining land
uses, contours, and surface water drainage system."[36]

Topsoil Stockpiles

7. *To prevent erosion of stockpiled topsoil, vegetate the stockpiles as soon as soil is placed in them.*

Vegetation of stockpiles is important in stabilizing the soil to
prevent pollution or degradation from erosion, collection and
germination of wind-blown seeds of undesirable plants, and
death of trapped native species seeds, as well as to prevent sedimentation, which would pollute the nearby area (see Guideline
5). Where stockpiles are required, they should be stabilized.

Even if more materials are to be added to the stockpile later,
the pile should be vegetated. Additions to the stockpile may
cover the vegetation on top of the pile, requiring reseeding, but
the sides of the stockpile will remain stabilized. In Wyoming,
Peter Kiewit Sons, operator of the Rosebud Mine, argued to the
state regulatory authority that because the company might add
topsoil to stockpiles within a year or more, it should not be required to vegetate the stockpiles until then. The Department of
Environmental Quality acceded to this request, according to a
spokesman for the mining company. As a result, stockpiles at
the mine have little protective vegetation (see Rosebud profile).

Grim and Hill write, "Temporary soil stabilization measures
should be established immediately after the stockpile operation is
completed. If the stockpiling is a continuous operation, then
temporary stabilization methods [should be] implemented in
stages as stockpiling progresses" to prevent erosion and sedimentation.[37]

Montana, Wyoming, and Colorado require prompt seeding and
stabilization of topsoil stockpiles to minimize erosion.[38] OSM
regulations state:

> Stockpiled materials shall be...protected from
> wind and water erosion...accomplished either by

an effective cover of nonnoxious, quick-growing
annual and perennial plants, seeded or planted
during the first normal period after removal for
favorable planting conditions; or other methods
demonstrated to and approved by the regulatory
authority to provide equal protection.[39]

Toxic Overburden and Interburden

*8. To preserve surface vegetation, bury toxic material deep
enough to prevent it from harming the vegetation.*

Potentially harmful materials present in the overburden (material under the topsoil and above the coal seam) or interburden (material found between coal seams if more than one seam is being mined) should be buried and isolated beneath enough overburden to prevent them from impairing vegetative growth, or polluting groundwater or surface water systems. The long-term viability of the land is at stake. Some of the most desirable native grasses in the West (needle and thread, western, and thickspike wheatgrasses, and blue grama) have roots that penetrate to depths of five feet or more. Roots of shrubs like the silver sage penetrate as much as eight feet into the soil. In most of the West, the greatest toxicity problem is the overburden's salt content. Salts that migrate into plant root zones can hamper plant growth. Acidic materials may also occur in the interburden between coal seams (see Seneca II profile, Rosebud profile, Seminoe II profile, Indian Head profile).

Grim and Hill state: "A positive preventive method [for keeping toxic materials from oxidating and forming acids] is to cover [them] as soon as possible with earth, which serves as an oxygen barrier....Soil thickness should be designed on the basis of the worst situation...a 'safety factor' should be included to account for soil losses such as erosion. Vegetation not only serves as a barrier, because the pores [in the soil] are filled [with] water and not gases [but] as the vegetation dies, it becomes an oxygen user during the decomposing process and further aids the effectiveness of the barrier."[40] Robert Curry of the University of Montana's Geology Department has recommended that toxic materials be buried at least 10-20 feet to provide a buffer zone between the farthest reaches of the root zone and the area where toxic materials have been placed.[41]

The National Coal Policy Project stated:

Potentially toxic spoils should be handled and
disposed in such a fashion that their ultimate
negative effects upon future site use options are
minimized. This should include burial of such

> materials at depths greater than the deepest ex-
> pected root penetration and greater than depths
> of expected upward migration of toxic materials
> that can be expected to be mobilized through so-
> lution.[42]

Montana requires that "the operator...bury under adequate
fill all toxic materials, shale, mineral, or any other material determined by the Department [of State Lands] to be acid producing, toxic, undesirable or creating a hazard."[43] All other states
in INFORM's study mention the need for proper disposal of toxic
materials and of wastes, but they do not establish a criterion
similar to Montana's. OSM mandates:

> A person who conducts surface mining activities
> shall cover...all exposed coal seams remaining
> after mining, and all acid-forming materials,
> toxic-forming materials, combustible materials, or
> any other materials identified by the regulatory
> authority...[and] if necessary, these materials
> shall be treated to...minimize adverse effects on
> plant growth and land uses.[44]

WATER QUALITY

Water is especially critical to the western United States,
where rainfall is low and evaporation due to heat and hot dry
winds is high. Surface mining can pollute surface and underground water with sediments and toxic minerals, and it can alter
flow and distribution patterns of surface and groundwater upon
which homes and businesses rely.

Until the Surface Mining Control and Reclamation Act was
passed in 1977, the federal government did not tie the regulation
of mining and land reclamation to that of water quality. Federal
water quality legislation was first passed in the 1940s. However,
the Federal Water Pollution Control Act Amendments of 1972
brought a sweeping overhaul of previous water-protection legislation. A major innovation of the 1972 Amendments was the creation of the National Pollutant Discharge Elimination System
(NPDES) which requires that water effluent standards, standards
for the quality of the water discharged, be established and compliance enforced for all facilities that discharge water. The
NPDES program requires that facilities discharging water--pri-
marily factories and utilities--obtain a permit for such activity.
In 1977, the Federal Water Pollution Control Act was further
amended, and its name was changed to the Clean Water Act. In
defining a "direct discharge facility," the Environmental Protec-

tion Agency (EPA) includes those that channel surface runoff
into "point" sources. It is virtually impossible to measure the
quality of water runoff from many acres of land, but quality can
be measured at the point where the runoff collects. This definition subjected actively mined and reclaimed lands, from which
water was channeled into a drainage course or sediment pond, to
the federal regulations.

Under the Surface Mining Control and Reclamation Act,
water quality standards and effluent limitations are coordinated
with the NPDES program. Under a memorandum of understanding between EPA and OSM, jurisdiction over application and distribution of water discharge permits falls to the state or federal
regulatory authority enforcing the SMCRA. This authority varies from state to state. EPA reviews applications for permits but
the regulatory authority under the SMCRA issues the permits
and receives quarterly discharge monitoring reports prepared by
the mine operator. Inspection of mines and water discharges will
be made by the regulatory authority. EPA has jurisdiction to
make independent inspections.

The Office of Surface Mining has established effluent guidelines for surface coal mines related to total iron content, total
manganese content, pH (acidity level), and total suspended solids (sediment). The standards for total suspended solids are
being reviewed by EPA because some receiving waters near coal
mines have levels of total suspended solids above those allowed
for discharge. (The mining industry argued that it should not
be required to clean waters to levels higher than those that receive the discharge [see Glenharold profile].)

Groundwater is a crucial land disturbance issue related to
surface mining in the western United States. Many coal seams
now being mined provide the pathways in which water flows underground. Removing these water-bearing strata (aquifers) will
affect the quality and availability of water flowing from the point
where they have been removed or disturbed. Because refilling
and grading mine pits does not return the subsurface geologic
layers precisely to their pre-mining positions, it is unclear
whether an aquifer can be recreated to perform the original
function once earthmoving operations have been completed.

Despite the importance of protecting groundwater supplies
and systems, there seems to be no consensus among experts as
to how such protection can be assured when mining occurs. In a
1976 study, South Dakota School of Mines and Technology's
Perry H. Rahn addressed the question:

> A major concern is the general effect that in-
> creased mining activities has on [ground] water
> levels and flow direction, ground water recharge
> and discharge and changes in ground water

> chemistry. These parameters have not been suf-
> ficiently studied and consequently the effects of
> these changes are not known.[45]

Effects of mining on groundwater vary from site to site, depending on such factors as soil type and make-up.

Once a mine operator has received a permit to mine through
an aquifer, the aquifer will be destroyed. Most experts believe
even selective placement of certain soils or other materials in refilled pits cannot assure successful re-creation of the aquifer.
Moreover, reclaimed areas next to active mine pits most likely
will not show the effect of efforts to re-create the aquifer, as
the working mine pit will be certain to disrupt an aquifer a short
distance away. Only after mining ends in the area will the results of attempts to reclaim aquifers be measurable (see Belle
Ayr profile).

Not all aquifers exist in the coal seams or in soil layers
above the seams being mined, but many coal seams being removed
by mining are principal aquifers in the West. Presumably, mining
operations above the level of aquifers will not affect them. Yet
experts have not determined whether the changes in surface
areas that recharge aquifers by allowing infiltration of water to
the aquifer zones will have a permanent effect on water quality
and quantity in these underground streams. These lands that
overlie aquifers and allow water to infiltrate the soils and enter
groundwater systems are known as "recharge areas." After
mining, the land in a recharge area will be replaced by mine
spoils, topsoil and vegetation. However, Rahn's study in the
Powder River Basin found that:

> Coal strip mines in the Powder River Basin will
> inadvertantly produce numerous man-made aqui-
> fers....Spoils will contain sufficient porosity and
> permeability which will be equivalent to or
> slightly greater than existing natural shallow
> sandstone aquifers....Water in spoils was found
> to be significantly more highly mineralized than
> natural ground water in terms of total dissolved
> solids, calcium, magnesium, and sulfate. Spoils
> water exceeds the recommended drinking water
> limits in these and other ions, and it is doubtful
> that the water could be used for long-term irri-
> gation.[46]

Thus, the major groundwater issues are the modification of
recharge rates through refilled mine pits, modification of aquifer
flow patterns, and the changes to water quality in that aquifer.

INFORM was not able to suggest guidelines for the protection
or reconstruction of groundwater systems in areas where surface

mining is being conducted. However, Guideline 12 addresses the
need for careful monitoring of groundwater. Also, certain lands
may have to be deemed unsuitable for mining if it would permanently harm valuable groundwater supplies (see Absaloka profile). Russell Boulding of the National Coal Policy Project told
INFORM that the "principles of aquifer reconstruction are fairly
well understood (although untested). It will be complicated and
expensive, however."[47]

Surface Water

9. *To keep sediment from damaging land and water off the mine
site, consolidate all runoff from disturbed and revegetated areas
into sediment ponds.*

Surface mining destroys much of the vegetation on large
areas of land. This vegetation holds soils together and protects
them from eroding and washing away, carrying sediment into
areas through which the water flows. When vegetation is absent,
some system must be established to contain the sediment within
the mine-permit area. A drainage system must not only trap all
water flowing over the permit area but carry it to areas where
the water can be held and monitored for quality before it is discharged into natural drainageways or waterbodies. The drainage
system at surface mines may consist of troughs along roads running through the mine. These troughs act as canals to guide
water into sediment ponds. The drainage system should be
maintained to prevent erosion that would increase the quantity
of materials carried in the water to the sediment pond. On other
mine sites, water flowing over the land may be directed into
strategically placed sediment ponds by the natural contours of
the land. Sediment ponds are also vital to water quality, as
they retain water and allow accumulated solids to settle to the
bottom before water is discharged.

Grim and Hill, in their report for the EPA state:

> The key to minimizing erosion and sediment prob-
> lems is in the control of water flowing into, with-
> in, and from the surface mining area. [The]
> control...objective is to prevent accelerated ero-
> sion and sedimentation....Sediment basins should
> be located on all drainways carrying concentrated
> flows from the disturbed areas...since sediment
> causes more off-site damage than any other as-
> pect of strip mining.[48]

Other methods for cleaning water runoff before it leaves the
mine site have been suggested. One would involve leaving un-

disturbed buffer zones, allowing the natural vegetation to filter
sediment from the water before it leaves the mine site. EPA has
been evaluating this and other alternatives to sediment ponds.
However, Lewis McNay, of the Technical Services and Research
branch of the Office of Surface Mining, told INFORM that sediment ponds were still considered the state of the art for controlling water quality before discharge from a mine site without using chemical additives.[49] (For a discussion of the use of chemical additives, see Gascoyne profile.) If effluent guidelines are
changed to allow discharges from coal mines to exceed those limits now being enforced, sediment ponds may become unnecessary.

Montana regulations state: "the operator...shall...impound,
drain, or treat all runoff or underground waters so as to reduce
soil erosion, damage to grazing and agricultural lands, and pollution of surface and subsurface waters...."[50] OSM regulations
also specify: "All surface drainage from the disturbed area, including disturbed areas that have been graded, seeded, or
planted, shall be passed through a sedimentation pond or a series
of sedimentation ponds before leaving the permit area."[51]

*10. To prevent erosion of the system that consolidates and dis-
charges water runoff, establish vegetative cover along channels
and on the banks of sediment ponds, and use devices to slow the
flow of water.*

Drainage systems capture water and direct it across the mine
area to sediment ponds, but they also increase its potential to
erode as it flows. Upon discharge from the mine permit area,
the water flow is concentrated, thus more likely to erode land.
By establishing vegetation on the ground over which the water
will flow, a mine operator can slow the flow, reduce the water's
scouring action, and filter out some sediment. The flow may also
be slowed by placing rocks or other objects in the water's path
to the sediment pond or point of discharge from the mine site.
Gravel dams have also been used to slow the water flow and to
provide some filtration.

Grim and Hill reported: "A quick growing cover of herbaceous species is necessary to obtain quick stabilization [of the
soil] and initial protection against erosion"[52] when stabilizing
regraded slopes. The same holds true for any area on the mine
site where water running across soils may scour the land and
pick up sediment, leaving behind gullies created by erosion. The
National Coal Policy Project's report, *Where We Agree*, states:
"Properly designed and maintained sediment ponds can control
the movement of sediment into streams....The greatest sediment
transport problem resulting from current mining activities is the
improper construction and maintenance of sediment ponds."[53]
Such maintenance would include keeping the ponds themselves--

and other parts of the water diversion network as well as natural drainages on the mine--from eroding.

Wyoming regulations established in 1975 require:

> Temporary diversion ditches...be built to the
> following standards:...in soils or unconsolidated
> materials, the sides and in ditches carrying in-
> termittent discharges, the bottom shall be seeded
> with approved grasses so as to take advantage of
> the next growing season. Rock riprap [rocks the
> size of bowling balls or larger strewn over the
> ground to slow water flow], concrete, soil cement,
> or other methods shall be used where necessary
> to prevent unnecessary erosion.[54]

OSM mandates "permanent diversions...be constructed with gently sloping banks that are stabilized by vegetation...[and] designed, constructed, and maintained in a manner which prevents additional contributions of suspended solids to stream flow."[55] The agency also requires "discharge from sedimentation ponds, permanent and temporary impoundments...be controlled, by energy dissipators, riprap channels, and other devices,...to prevent deepening or enlargement of stream channels."[56]

11. To preserve surface water and land off the mine site, observe the limits for total suspended solids, pH, iron and manganese as set forth in the mine's National Pollutant Discharge Elimination System permit.

Water discharge permits, formerly under EPA jurisdiction, specify discharge limits for major pollutants found in the water released from the mine site. OSM now establishes water effluent standards for each mine site and administers these regulations under the federal Surface Mining Control and Reclamation Act of 1977.

As noted earlier, regulators are considering revising the limits for levels of total suspended solids (or sediment) allowable in discharged water, as coal companies have challenged current effluent limits for total suspended solid levels in water discharged from a mine site which are lower than the levels of the receiving water bodies. Current EPA studies will produce some changes in these guidelines for coal mines that receive permits under the National Pollutant Discharge Elimination System program.[57] While the EPA is reworking its water effluent guidelines, the agency remains the most complete source of information on water quality in the United States. Since the advent of the Surface Mining Control and Reclamation Act (SMCRA), mine operators have been required to collect data on water quality near their mine sites. The information should help EPA create new guidelines.

Robert Coats, a soils and water specialist in wildlands resource sciences, suggests that the EPA also develop guidelines for sodium, trace mineral and total dissolved solids levels in waters discharged from western mines. With the high level of sodium in many western soils, this might be an important criterion for effluent control.[58]

12. To evaluate the effects of surface mining on the quantity anc quality of surface water and groundwater, monitor water quality and quantity on and off the mine site.

Although some surface mining operations affect only the surface area where that disturbance occurs, water pollution and degradation may take place miles away. Sediment carried from the permit area may be deposited along the banks of streams, doing minimal damage, but because channels have narrowed and become shallower, the next rainstorm may cause floods and sweep sediment into fields or public recreation areas. Although there may be no water users within one mile of a working mine, the loss of water caused by leakage from aquifers could affect landowners miles away. Hence, the effects of the mining operations upon the water systems of the area should be monitored to indicate their impact on the entire environment. Of course, if water discharges are properly controlled, their effect on surface water quality will be minimized.

In a 1974 study of western coal surface-mining rehabilitation potential, the National Academy of Sciences found:

> Improvement of rehabilitation techniques and the reduction of environmental impacts depend critically upon monitoring and evaluation. Therefore, [it] recommend[ed]...continuing observations to monitor the on-site and off-site effects of mining and rehabilitation...and the hydrologic effects of surface mining on groundwater, surface drainage and water quality as affected both on-site and off-site.[59]

The National Coal Policy Project stated: "Disruption of a groundwater supply may be a hardship to local users near a mine; however, contamination of groundwater, once it occurs, is essentially irreversible and more widespread in its effect upon users who are dependent upon it."[60] It also recommended:

> An inventory should be made of groundwater wells within a radius of several miles of an existing or proposed mine site. Monitoring of ground-

water levels and water quality at least one-half
mile from the area of influence down dip from the
mine, and in other directions of possible impact,
should be standard procedure for any mine operation that might affect an aquifer already in use.[61]

Geraghty and Miller, Inc., a consulting group of groundwater
geologists and hydrologists, suggests that:

The major reason for installing monitoring wells
is to provide early warning of ground-water contamination....Its design should be based on demonstrated hydrogeologic principles and site-specific data. [However,] No specific statements
can be made about the distances that contaminants will travel in the ground simply because
hydrogeologic conditions and type of contaminants differ widely.[62]

Thus, although the distance that a company must go to measure
the effects of mining upon groundwater may vary according to
site-specific characteristics, the fact remains that monitoring
should be performed.

Wyoming law states: "Monitoring surface water conditions
may be required during the course of the proposed operation...
[and] the operator may be required to conduct...monitoring in
order to determine the exact depth, quantity and quality of
groundwater in geological formations affected by the mining operations."[63] OSM requires:

Ground water levels, infiltration rates, subsurface flow and storage characteristics, and the
quality of ground water shall be monitored...to
determine the effects of surface mining activities
on the recharge capacity [the ability of the land
to transmit surface water to groundwater streams,
sometimes called aquifers] of reclaimed lands and
on the quantity and quality of water in groundwater systems in the mine plan and adjacent
areas[64]....When surface mining activities may
affect the ground water systems which serve as
aquifers which significantly ensure the hydrologic
balance [the quality and quantity of water flowing
into, held in, and flowing out of a given area]
of water use on or off the mine plan area, ground
water levels and ground water quality shall be
periodically monitored....Surface water monitoring shall...be adequate to measure accurately
and record water quantity and quality of the dis-

charges from the permit area[65]....After disturbed areas have been regraded and stabilized
...the person who conducts surface mining activities shall monitor surface water flow and quality.[66]

Alluvial Valley Floors

*13. To preserve highly productive land, protect alluvial valley
floors.*

Vegetative productivity is much higher on alluvial valley
floors than on surrounding lands. The soils on these lands are
easily penetrated by plant roots. They also allow water from
shallow subsurface natural irrigation networks to percolate upward, providing water to the roots of the plants growing on
them. These characterisitics make alluvial valley floors highly
valuable agricultural lands. In arid or semi-arid areas like the
western United States, alluvial valley floors are small oases of
highly productive lands relative to surrounding lands. It has
yet to be demonstrated either that an alluvial valley floor can be
reconstructed or that the water system that feeds it can be returned once the land is mined, as the mining destroys the soil
layers that transmit the water. The possibility of destroying an
alluvial valley floor would preclude mining an alluvial valley floor
or would prohibit nearby operations that might damage the
floor's hydrogeologic balance and preclude return to pre-mining
levels of productivity.
The National Academy of Sciences stated in 1974:

> It is essential...that alluvial valley floors and
> stream channels be preserved. The unconsoli-
> dated alluvial deposits are highly susceptible to
> erosion....Removal of alluvium from the thalweg
> [the center or middle] of the valley not only
> lowers the water table, but also destroys the
> protective vegetation cover by draining soil mois-
> ture. Rehabilitation...would be a long and ex-
> pensive process, and in the interim these highly
> productive grazing areas would be in disuse.[67]

OSM requires:

> Surface coal mining and reclamation operations
> shall be conducted to preserve, throughout the
> mining and reclamation process, the essential hy-
> drologic functions of alluvial valley floors not
> within an affected area....Operations shall not
> interrupt, discontinue, or preclude farming on
> alluvial valley floors, unless the premining land

A possible alluvial valley floor into which mining may proceed at Westmoreland's Absaloka Mine in Montana.

use is undeveloped rangeland which is not significant to farming; or the area of affected alluvial
valley floor is small and provides or may provide
negligible support for production from one or
more farms.[68]

Erosion

*14. To protect off-site property, disturb no land off the mine-
permit area.*

Companies operating surface mines receive permits from state
authorities and/or OSM to produce coal and reclaim the land disturbed by mining. These permits usually limit mining-related
disturbances to the mine permit area. Mine operators are required to post performance bonds only for the land area within
their permit. Off-site disturbances, if unnoticed by regulatory
authorities or concerned citizens, become the public's responsibility unless the cause for their occurrence can be directly
linked to the operations of one mine or to one company's action.
Sediment carried into streams and rivers can kill fish and cloud
the water, cutting off light to aquatic plants. It can add high
concentrations of toxic substances or minerals, such as iron, and
can generally degrade the water quality, making it unsuitable for
human beings, domestic animals or wildlife. Sediment can also
clog water passages and be deposited in fields or along the
banks of streams, polluting the local environment by rendering
fields unsuitable for grazing or causing flooding by reducing a
stream bed's capacity to transport water. Spoil piles placed near
the mine site's boundary or along streams should be kept from
sliding onto land that has not been permitted. This movement of
dirt or rock onto off-site lands or into these waters may create
sedimentation problems or otherwise interfere with the ordinary
use of the land.
   In *Where We Agree*, a 1978 study conducted by the National
Coal Policy Project, it is stated:

> The impact of mining on water may reach far be-
> yond the area disturbed by mining, as in the
> case of pollution of surface streams...the pollu-
> tion of surface water may severely affect aquatic
> ecosystems as well as a large number of water
> users downstream from a mine....Sediment load-
> ing of streams resulting from erosion of exposed
> mine spoils has been a major surface water qual-
> ity problem associated with surface mining. Sed-
> imentation can have an adverse impact on aquatic
> ecosystems, and in some situations it may change

the morphology and stability of stream channels
and flood plains so as to increase the frequency
and severity of flooding.[69]

The five western states that have regulations dealing with
mining and land reclamation--North Dakota, Colorado, New Mexico, Montana, Wyoming--require that a mine operator obtain a
mining permit before mining begins. The permit restricts mining
and related activities in most cases* to the area covered under
the permit. The state authority then regulates the environmental
disturbances occurring throughout the mine's operation. OSM
regulations require that "all surface coal mining and reclamation
operations [be] conducted only under permits issued in accor-
dance with the requirements of the regulatory program."[70]

*15. To prevent an increase in sedimentation from disturbed
lands, control erosion to the levels of nearby, well-managed
lands similar to the mine area. If erosion does occur, repair it
immediately.*

Water or wind that erodes soils carries the sediment until the
movement of water or wind slows or stops. To prevent sedimentation, it is necessary to prevent erosion. All lands show some
erosion. However, the vegetation on the land and the degree of
erosion are related. The more plants, the more they help hold
soils in place and filter water flowing over the land. While the
soils remain in place, they provide a medium in which the plants
will grow. Even if erosion takes the form of pencil-sized troughs
in the soil surface, these troughs can channel water down the
slope, increasing the size of the trough and carrying away more
and more soil. Thus, immediate repair of erosion-induced
troughs will eliminate serious erosion before it worsens.

The National Coal Policy Project found that "ideally, runoff
from reclaimed spoils should be equal to or less than pre-mine
conditions."[71] Thus, if vegetation is re-established to pre-mine
levels, presumably erosion would also be controlled to similar
levels. The caucus of industrialists and environmentalists also
found that "increased runoff [above pre-mining levels] heightens
the effects of erosion and reduces the water available for plant
growth, which in turn impairs the ability of the associated vegetation to stabilize the spoil surface."[72]

---

*A common disturbance that cannot be limited to the mine site is
the transport of coal from the mine to market. Mine operators
are not responsible for damage produced by the transportation
of this commodity, although some public roads used for hauling
coal must be maintained or rebuilt by coal companies in return
for using them.

In the preamble to its final regulations, OSM states:

> Rills and gullies concentrate runoff water into
> tiny rivulets and small channels and accelerate
> erosion....Sediment derived from rills and gullies
> can be detrimental to water quality and every ef-
> fort should be exercised to prevent such erosion.
> Furthermore, rills and gullies interfere with
> achieving revegetation and postmining land use.[73]

Many state regulations mention erosion prevention as a key
to specific requirements for a successful reclamation program.
Erosion prevention during topsoil stockpiling, grading and revegetation and in the handling of water at the mine site are central to many requirements of Colorado, New Mexico, Montana,
Wyoming, and North Dakota reclamation laws. OSM requires:

> When rills or gullies deeper than 9 inches form in
> areas that have been regraded and topsoiled, the
> rills and gullies shall be filled, graded, or other-
> wise stabilized and the area reseeded or re-
> planted....The regulatory authority shall specify
> that rills and gullies of lesser size be stabilized
> and the area reseeded or replanted if the rills or
> gullies are disruptive to the approved postmining
> land use or may result in additional erosion and
> sedimentation.[74]

OSM justifies the nine-inch reference point as "the maximum
depth that can be stabilized by most grasses, since a large portion of their roots occur in this surface layer."[75] Yet erosion
must be repaired as soon as it is identified to prevent the effects
of sedimentation and further degradation of the soil surface.
Note that even reducing erosion to levels similar to those on
well-managed similar lands may still produce runoff with sediment levels higher than those in the mine's effluent limitations as
established in the permit for water discharge. This raises again
the question of limits for total suspended solids, discussed in
Guideline 11. However, erosion of the land is a key source of
sediment and minerals taken up by water flowing over the land.
Moreover, erosion can accelerate, carving out larger and larger
gullies. The final analysis will require that erosion be no
greater than that which occurs on well-managed lands similar to
the mine, but also that the quality of water that flows from these
lands is also as high.

SEEDING AND AMENDMENTS

Seeding and "amendments" (a technical term referring to fertilizer, supplemental water, or other additives) at a surface mine are the heart of a company's revegetation program. The final land-use and stability of a mined area depend directly upon the amount, type and quality of vegetation growing on the site after it has been reclaimed. Initial stabilization of the site, followed by progressive growth of desirable plant species, is the key to successful reclamation. The guidelines below relate to the choice of a seed mix, the preparation of the land before the seed mix is applied, the equipment that should be used to seed and replant an area, and the relationship of amendments such as mulches, supplemental water, and fertilizers to successful reclamation.

Revegetation Timing and Currency

16. To ensure that reclamation promptly follows mining and to minimize the amount of disturbed land, seed an area equal to that disturbed by mining.

Surface mining disturbs the plants, animals, soils and water systems on the land where it occurs. Reclamation tries to repair the damage by re-establishing vegetation and speeding its growth. Because mining is only a temporary land disturbance, reclaiming the land for its intended use, whether it is a new, higher land use, or the original pre-mining use, should take place as quickly as possible. Disturbed land remains barren until natural forces return the vegetation. Without this vegetation, the soils erode more easily and are deposited in waterways, or on residential property or rangelands. Yards and fields can be covered, making them useless for grazing or preventing the growth of plants, and waterways can become buried or clogged. Grading, replacement of topsoil and seeding accelerate the process by which the land can become revegetated. Planting seeds shows a mine operator's commitment to establishing vegetation that may eventually facilitate productive post-mining land use. By seeding an area approximately equal to the land area disturbed by coal production operations, the mine operator can keep from increasing the ratio of disturbed and unproductive land to that as yet undisturbed or already in the process of reclamation. Although grading land and replacing topsoil are important elements of reclamation, seeding plants actually begins the growing process on otherwise barren soils.
In their 1974 study conducted for the EPA, Grim and Hill state:

> Reclamation should follow closely behind the min-
> ing operation....Backfilling and grading should
> be kept current with the operation....Revegeta-
> tion should follow the grading as soon as possible
> in order to establish a quick protective plant
> cover...because the freshly moved material is
> easier to grade, handle and plant than older,
> compacted material.[76]

In the rush to produce coal, reclamation can easily take a back
seat to a coal company's production priorities. (Coal is usually
sold on a contract basis, which requires a constant rate of production and supply.) John C. Doyle, Jr., of the Environmental
Policy Institute in Washington, D.C., inventoried and analyzed
in 1977 the key statutory provisions of surface mining laws in 28
coal-producing states. He wrote: "During the mining operation,
the operator should be required to...keep reclamation current
with his mining operation, particularly with regard to backfilling
and grading" to speed recovery of the land.[77]

Opening a new coal mine will disturb many acres of land.
Offices and workshops, equipment storage facilities, coal preparation buildings (if needed), rail lines, roads and coal transportation areas such as those where coal is loaded aboard trains
and barges--all of this construction disrupts natural patterns.
When mining begins, and as it continues, sediment ponds and
water diversion networks, more roads and mining pits will be
opened. Some of these land disturbances will remain for the life
of the mine and cannot be reclaimed until all coal production
ceases. However, the amount of land disturbed by these "permanent" facilities usually becomes constant within a few years
after a mine opens. A serious commitment to reclamation requires
a coal producer to seed areas equal to those disturbed by annual
mining operations.

State and federal regulations also require prompt seeding.
Wyoming regulations of 1975 state that "reclamation shall follow
mining as soon as is feasible...."[78] Montana regulations in 1977
require: "As rapidly, completely, and effectively as the most
modern technology and the most advanced state of the art will
allow, each operator...shall reclaim and revegetate the land affected by his operation...."[79] Office of Surface Mining regulations also require that grading and seeding be kept current, the
latter as soon as climate permits.[80]

*17. To revegetate disturbed lands as soon as possible, seed
during the first normal planting period after topsoil is replaced.*

After topsoil is replaced, seeding should begin as soon as
possible to keep erosion to a minimum and to start re-establish-

ing plant cover. The seeding program should ensure that seed is in the ground before germination is most likely to occur. As germination usually depends on the amount of water available, enough must be present to allow seeds that have germinated to continue growing and producing their own seeds, or they will fail to propagate. Thus seeds should not be planted when they might begin to grow but die in the dry season, but should have the benefit of the longest possible growing season. In addition, topsoil left without vegetation is more susceptible to invasion by weeds. Once seeded plants begin to grow, weeds such as Russian thistle (which dries and forms "tumbleweed" after one year) can help seeded plants by trapping snow and moisture, and providing shade and cooler temperatures. In most cases, however, an undesirable plant that invades an area before other vegetation begins growing will monopolize important water and mineral nutrients, leaving little for the desired vegetation for several years.

Grim and Hill contend that "a quick growing cover of herbaceous species is necessary to obtain quick stabilization [of the soil] and initial protection against erosion."[81] *Revegetation Guidelines for Surface Mined Areas*, a 1974 study conducted by C. Wayne Cook and colleagues of the Colorado State University's Range Science Department, said "planting should be employed during the proper season"[82] and "seeding should be done soon after the [top]soil is added to prevent establishment of weeds and loss of topsoil by erosion."[83] Willis G. Vogel, a range scientist with the U.S. Department of Agriculture, concluded in a 1974 study that "seeding immediately after grading takes advantage of a freshly prepared seedbed and improves the chances of establishing cover before erosion patterns are formed. Once erosion has begun, the rills enlarge rapidly and make it increasingly difficult to establish a complete and effective plant cover."[84]

Montana regulations governing surface mine reclamation require that "areas shall be planted or seeded during the first appropriate season following completion of grading, topsoil redistribution and remedial soil treatments."[85] OSM regulations mandate that "seeding and planting of disturbed areas shall be conducted during the first normal period for favorable planting conditions after final preparation."[86]

18. To ensure efficient seeding, choose seeding time on the basis of on-site experimentation or research applicable to the site.

As each mine site has a mix of climatic, geologic, and hydrologic characteristics that makes it unique, an effective reclamation program should be tailored to that site. When seeding does not take place at the best time in a given area or region,

the mine operator must justify the change from this accepted
period. Research on sites where soil and climatic conditions are
similar may be applicable. Experience on the site, too, may be
evidence enough that deviations in the time when seeding is done
are justified. However, planting seeds at other times can jeo-
pardize the entire revegetation program at a mine site and waste
company resources; these changes must be carefully justified.

Although OSM regulations require that seeding and planting
be conducted as soon as possible, they also state that this period
should be "that planting time generally accepted locally for the
type of plant materials selected."[87] Cook et al. believe that, "in
general the best season to seed is just prior to the season that
receives the most dependable precipitation."[88]

Seed Mix

*19. To create and sustain the plant community best adapted to
the mine area, seed a mixture of native plants similar to the pre-
mining community in both diversity and seasonal variety. Use
fast-growing, introduced plant species only to create initial vegetative cover.*

Most experts agree that introduced plant species will grow
faster and produce higher yields sooner than native species.
However, native plant species have become adapted over the
years to the area's climate and topography. Adaptation to the
environment changes the genetic makeup of plant species through
years of natural selection. Introduced species do not have these
advantages, although they may serve to establish a ground cover
to minimize erosion. Reclamation is incomplete if the area is dominated by introduced species whose ability to withstand the environmental pressures of a given region is not proven.

Some introduced plant species can actually be the products
of hybrids of a native plant and an introduced species. These
should be proven as acceptable to achieve reclamation goals before they are included in the seeding program. Russell Boulding, a reclamation consultant who has served with the National
Coal Policy Project, told INFORM that introduced plant species
"also usually require continual maintenance,"[89] not a desired
trait of self-sustaining groundcover. Margaret MacDonald of the
Northern Plains Resource Council reports that, at Western Energy's Rosebud Mine, "the use of introduced species, even in small
proportions to the native seed, caused significant setbacks to the
the native species because of the short-term competitive advantage of exotics in establishing themselves."[90]

Thus, the seed mix of plants to be established in a reclaimed
area should include native plant species, and their growth should

not be threatened by competition from either introduced plants or other plants used initially as a ground cover to protect newly graded and topsoiled areas from erosion. This standard is altered if a land use change is justified (see Introduction to Guidelines).

On the basis of diversity and seasonal variety, two basic types of plant characteristics can be distinguished: annual vs. perennial, and cool- vs. warm-season. Annual plants complete their life cycle and die in one year or less, while the roots of perennial plants continue to sprout new plants. Cool-season plants grow in the fall and spring and die in the summer. Warm-season plants grow primarily in late spring, summer, and early fall. To produce a plant cover in all seasons when snow does not cover the ground, both cool- and warm-season plants must grow on revegetated land. It is important to have more than one type of plant, because conditions fatal to one species may not affect others in the same manner. Few natural vegetative communities contain only one or two types of plants.

Cook *et al.* determined, "It is important to consider [seeding] species that are adapted to the area or site to be revegetated and likewise suited for the uses to be made of the area."[91] Ray W. Brown and Robert S. Johnston, plant physiologist and research hydrologist, respectively, with the U.S. Forest Service's Forestry Sciences Laboratory in Logan, Utah, found, "Native species are apparently better adapted for revegetation of alpine disturbances than are introduced species [because]...native plants are at least climatically adapted and are more capable of surviving periods of environmental stress that may be unique to the area."[92] Because of the range of temperature and precipitation extremes in the western United States, this finding can be extrapolated to any areas affected by a wide range of climatic conditions.

OSM regulations require: "All disturbed land, except water areas and surface areas of roads that are approved as a part of the postmining land use, shall be seeded or planted to achieve a permanent vegetative cover of the same seasonal variety native to the area of disturbed land."[93] They also allow the use of introduced species:

> After appropriate field trials have demonstrated
> that the introduced species are desirable and
> necessary to achieve the approved postmining
> land use; (or) the species are necessary to
> achieve a quick, temporary, and stabilizing cover
> that aids in controlling erosion; and measures to
> establish permanent vegetation are included in
> the approved plan.[94]

In most cases, if introduced species are a major portion of

the seed mix planted, native plant species will take substantially
longer to invade the area than if native plants are successfully
seeded from the start.

*20. To achieve the maximum benefit from seeding operations,
choose the type and amount of seed on the basis of on-site experiments or research applicable to the site.*

Because each mine site is unique and certain soils or slope
aspects will differ even within a site, it becomes important to
vary the seed mix used to create the best combination of plant
types seeded, amount of seed distributed, and the kinds of land
planted. All reclaimed areas are not the same. Southwest-facing
slopes are usually drier sites than north- or east-facing slopes
and may require the seeding of more drought-tolerant plant species. Soils vary greatly, and only a soil analysis using proper
and accepted techniques will determine the specific characteristics of a given site. On-site experiments will best determine
whether changes in the quantity or composition of the seed mix
are necessary to produce maximum success in revegetation. Experience in areas similar to the one being reclaimed will also help
in determining the type and amount of seed to be used.

In *Revegetation Guidelines for Surface Mined Areas*, Cook *et
al.* suggest "more critical sites such as west and south slopes
should have the quantity of seed increased by 50 to 100 percent"
and "as time progresses rehabilitation procedures must be constantly revised to give suitable guidelines for successful revegetation."[95] However, it is possible to place too many seeds on a
given area of land. If all seeds compete for a limited amount of
available water and minerals while germinating, none may be left
when the plants begin to grow. This increases the need for
specific research to determine the best amount of seed to be
used.

No state or federal regulations address this issue specifically.

Seedbed Preparation and Seeding Machinery

*21. To increase the ability of seeded lands to retain water and
thereby prevent erosion, operate reclamation equipment along the
contour.*

By applying seed and amendments, such as mulch or fertilizer, along the contour of a graded slope, all tracks from the vehicles or machines used in this process will also run along the
contour. Thus machinery tracks and wheel ruts will run across
the slope and act as small furrows in the soil to trap water running downhill and snow blowing across the area. This method
also provides shelter for plants from wind. Such furrows would

also keep plants from being washed away by water speeding
down the slope unobstructed (see Absaloka profile). (See Guideline 22 also.)

When lands are graded to slopes too steep for this method,
other techniques must be employed to plant the seeds without
creating potential erosion problems. This might require seeding
with a hydroseeder or from the air. The operation of seeding
machinery on the surface could cause water erosion if the machines travel up and down the slopes instead of across them.
The ruts and grooves from their tires or wheels may actually increase the potential for erosion on the slope, as water would be
channeled along these grooves as it flows downhill. When slopes
are to be graded to contours that would not allow operation of
seeding or other equipment along the contour, alternate methods
for preventing erosion and providing a suitable seedbed should
be proven effective before their use is allowed. Otherwise, the
area may have to be graded to lesser slopes, or not mined at all.

In *Processes, Procedures and Methods to Control Pollution
from Mining Activities*, a report prepared for the U.S. Environmental Protection Agency by the firm of Skelly and Loy, Engineers Consultants, and Penn Environmental Consultants, found:
"surface scarification is an effective runoff control technique....
Scarification is performed by contour plowing, furrow grading,
contour discing, or any other means of abrading [gouging] the
surface parallel to contour...The wheel or tract scars then act as
ridges and valleys."[96] T. R. Verma and J. L. Thames of the
Arizona School of Renewable Natural Resources found that "in
general, contour terracing, contour furrowing, contour trench-
ing, [and]...surface manipulating are [some of] the most commonly used techniques for maximum retention and conservation
of runoff. Most of these mechanical and cultural techniques have
a dual role for erosion control and moisture conservation."[97]

The Wyoming regulations of 1977 require: "Seeding which is
accomplished by mechanical drilling shall be on the contour unless specific situations dictate that other methods of seeding
should be used."[98] Montana regulations state: "All drill seeding
shall be done on the contour."[99] Both regulations apparently
assume that other forms of seeding would be performed without
machinery that must travel on the slopes being seeded--e.g.,
hydroseeding with a spray from the air (aerial broadcasting).
Areas to be reclaimed should be graded so that they are not too
steep for reclamation machinery to travel along the contour (see
Guideline 3).

22. *To provide the greatest opportunity for plants to flourish,
create furrows before seeding.*

The creation of ridges or furrows along the contour of a

graded slope, enhances the chances of successful revegetation
and reclamation. These furrows break up the flow of water down
the slope, reducing the speed and scouring effects of flowing
water. They also help to collect sediment as the water flows
down the slope, and they provide catchbasins for snow, which
then melts into the soil in warmer periods.

In *Range Developments and Improvements*, John F. Vallentine, Professor of Range Science at Brigham Young University,
writes,"Mechanical land treatments such as furrowing, terracing,
ripping...are water-conserving practices...[whose] benefits include increasing the quality and quantity of forage...through
more efficient use of rainfall [and] simultaneous control of erosion...."[100] Cook *et al.* have also determined: "It is necessary
to use small basins or pits constructed on the contour to conserve natural moisture for the seeded species."[101] In 1973, the
EPA found,"Scarification serves to help establish vegetation and
control erosion until the vegetation is established. Scarification
further serves to concentrate water in the low spots [which] is
helpful for establishing vegetation in arid areas."[102] From the
point of view of efficiency and economics, B. T. Wahlquist and
R. L. Dressler, of Westinghouse Electric's Environmental Systems
Department, and W. Sowards of Utah International report that
"surface manipulation seems to offer the greatest increase in
seedling establishment per dollar spent if proper equipment can
be designed to function on a large scale."[103]

Montana regulations mandate: "If necessary, redistributed
topsoil shall be re-conditioned by discing, ripping, or other appropriate methods."[104] Although this process may serve only to
break up soil, the requirements cited in INFORM Guideline 21
(that drill seeding be on the contour) effectively cause these
contoured furrows to be created in the seeding process, as ripping or discing (processes that use blades to break up the soil)
would precede the seeding. OSM allows the use of "other soil
stabilizing practices" in lieu of the use of mulch to control erosion, promote germination, and increase a soil's ability to retain
moisture.[105] The agency does not define these practices because the mine operator must convince the regulatory authority
that the proposed soil stabilizing practice will achieve its goal.

23. *To spread seed accurately and evenly on the land being revegetated, use machinery that opens a groove in the soil, drops
the seed, and covers it in one continuous operation.*

In order to create the best environment in which seeds may
germinate and grow, they must be placed in the soil and covered
to protect them from heat, winds, and foraging wildlife. Seeding
can be accomplished in many ways, not all desirable. Broadcasting seed from an airplane or helicopter, for example, or broad-

Furrows placed along the contour of regraded and topsoiled slopes help retain water and trap sediment flowing downhill. Hardhat indicates scale.

casting it upon the soil surface from a tractor-pulled machine, do
not achieve the goal. Wind can prevent the even distribution
and placement of the seed. Spraying seeds in a mixture of water
and mulch and/or fertilizer--hydroseeding--does not selectively
place seeds in the soil, and it creates other problems: the seed
is immersed in the water-and-fertilizer solution, which can cause
early germination, thus rendering the seed useless if planting is
done in the fall with the intention of spring germination and
growth.

One way to approximate the use of a drill-seeding machine
(which fits these standards) is the use of a broadcasting machine
that is immediately followed by a second machine that tills the
soil onto the seed cast on the surface. The drill-seeder's only
advantage over this method is that since placement of the seed is
more direct with a drill-seeder, less of the seed mixture must be
used, and distribution may be more even.

Cook *et al.* divide seeding into two types: broadcast and
drill. They define them as follows: "Any method that scatters
the seed directly on the soil surface without soil coverage is
termed broadcasting....The seed, however spread, must be
covered with soil in some way if it is to germinate and become
established."[106] Drilling fulfills these requirements. Vallentine
argues that if broadcasting must be done, it is best that it
"should follow such operations as deep plowing, [and/or] disk-
ing,...but precede shallower treatments such as railing, chain-
ing, and pipe harrowing,"[107] which would open grooves in the
soil and then cover the seed again. But Vallentine also states:
"Drilling provides the best method of obtaining uniform distribution and depth of seeding."[108] Grim and Hill, in their report
for the EPA, write, "Whenever possible, grasses and legumes
[plants that can fix gaseous nitrogen into organic nitrogen]
should be drilled...and [drilling] is best accomplished by using
a drill equipped with depth bands [to set the depth at which the
seed will be planted] and packer wheels [to smooth the soil after
seeding]."[109] Finally, Paul E. Packer and Early F. Aldon of the
Intermountain and Rocky Mountain U.S. Forest Service Forest
and Range Experiment Stations in Utah and New Mexico state
that "whenever possible, grass and legume seeds should be
drilled....In general, aerial seedings have not been satisfactory
on mine spoils in the semiarid West. If spoil surfaces are rough
enough for wind or rain action to cover the seed, satisfactory
stands might be obtained, but the uncertainties are great."[110]

Montana regulations established in 1977 require that "an operator shall establish a permanent diverse vegetative cover...by
drill seeding...."[111] Problems with drill seeding usually occur
when a slope is too steep to be traversed by a drill seeder on
the contour. It is possible that on slopes that are too steep to

be traversed on the contour, drill seeders could be driven at an
angle across the slope in two directions to insure even and adequate distribution of seed over the graded area, but this requires more seed (see Guideline 3).

Amendments and Maintenance

*24. To use mulch most effectively, choose its type and quantity
on the basis of on-site experimentation or research applicable to
the site.*

Mulching can aid germination and growth of vegetation by
decreasing erosion, holding water, adding nutrients to the soil,
reducing evaporation and reducing soil surface temperatures, all
of which further the growth of new plants. Many types of
mulches exist; the most common are hay, straw, and wood chips.
Mulch can be created by seeding annual grains and mowing them
before they establish seed heads. Hay mulches are composed of
mown grasses and forbs (any herb-like plant that is not a
grass), which are spread over the soil of the graded area. These
grasses and forbs can be any combination of introduced, native,
or leguminous (nitrogen-fixing) plant species. Many times the
hay mulch will contain seeds from its constituent grasses and
forbs. Tests should establish the effect the growth of these
seeds may have on the revegation efforts. Grain mulches, either
straw or seeded and mowed, may yield seeds as well. Seeds from
the plants used as cover crops may use up water and soil nutrients needed by the newly planted seeds, undercutting the revegetation program. Mulching can work in the company's favor,
if the seeds contained in the mulch are the same species as the
plants seeded. In this case, the mulch provides an excellent
source of extra seeds on the land to be revegetated.

Skelly and Loy's report for the EPA, Vogel of the U.S. Department of Agriculture, and others have agreed that mulching
can facilitate germination and growth of vegetation on seeded
areas. In *Revegetation Techniques for Dry Regions*, Packer and
Aldon state,"Surface mulches of various kinds effectively modify
environmental factors to benefit plant growth. Organic surface
mulches conserve moisture, reduce temperatures, prevent erosion, and supply organic acids and essential plant nutrients."[112]

OSM regulations require that mulches be used in revegetation programs but permit alternate techniques, such as contour
furrowing without mulch, to achieve the same ends if approved
by the regulatory authority.[113]

*25. To keep mulch on the land and protect soil from flowing wa-
ter, anchor mulch to the soil immediately upon application.*

Crimping or tacking mulch by using rollers to punch it into the surface of graded and topsoiled lands, or using chemicals to hold it in place, assures that the mulch will not be blown or washed away, which would defeat its purpose. Most companies in the West use either a straw or hay mulch, which is blown or placed on the surface of land to be reclaimed. When applied this way, mulch can be carried off by wind and water, or eaten by wildlife. By spreading the mulch and immediately crimping or tacking it to the soil, the material will be distributed evenly and can stabilize the soil and aid in re-establishing vegetation.

An alternate method is the spreading of nets, made from paper or other materials, over the mulch to hold it in place. This may not be the best practice in windy areas of the West. At the Rosebud Mine in Wyoming the operators found that the matting used to cover mulch at the site blew around too much to afford adequate protection. When annual grains are seeded as a mulch, crimping and tacking are unnecessary.

Grim and Hill determined: "Mulch, necessary for good seedling establishment, required some means of holding it in place."[114] Also, Packer and Aldon wrote "effective mulching requires...a means for keeping the mulch in place...[and that on] slopes of less than 3 to 1 [for every one foot on the vertical, three feet on the horizontal], straw and hay can be tacked down by cutting them into the soil with a crimper."[115] Barbara West, of the Office of Surface Mining in Denver, told INFORM that on steep slopes where contour furrowing is practiced, there may be no need to crimp or tack the mulch to the ground. Wind patterns are such on these slopes that the mulch material will be caught in the furrows and not be lost. However, this would still leave certain lands between the furrows bare of the material.[116]

OSM regulations leave the question of anchoring mulch to the soil to the regulatory agency overseeing reclamation. "When required by the regulatory authority, mulches shall be mechanically or chemically anchored to the soil surface to assure effective protection of the soil and vegetation."[117] OSM allows this latitude because, under some conditions, crimping is not necessary for some mulches.[118] For example, a mulch such as wood chips cannot be tacked to the soil (see Seminoe II profile). Chemical glues that could serve this purpose are expensive and seldom used.

26. *To ensure that revegetation efforts produce a self-sustaining plant community, fertilize or irrigate land only on the basis of on-site experiments or research applicable to the site.*

Fertilizers supply nutrients such as nitrogen, potassium and phosphates to soils which may lack these compounds vital to plant growth. Irrigation adds water to an area being vegetated, and is needed especially in dry seasons when plants lose water to heat and cold.

Both practices alter the plants' natural environment. Adding fertilizers and water to an area being revegetated risks raising plants that grow well under this type of maintenance but may not be able to survive without it. Most pasture and grazing land in the western United States is not regularly fertilized or irrigated. Experiments and tests on the mine site (see Introduction to Guidelines) should determine the specific need for fertilizers and water before they are added. The experiments should determine not only whether fertilizer and additional water are necessary, but the effect of discontinuing their use. The company that studies the need for fertilizer and water before adding them to large areas of land may save money and time if it determines that the practice would not aid long-term reclamation.

INFORM found no specific data on the use of irrigation in reclamation, and according to some experts little is known in the area. In their paper entitled "Use of Irrigation in Reclamation in Dry Regions," R. E. Ries and A. D. Day of the U.S. Department of Agriculture's Agricultural Research Service state:

> Data are limited on the amount and frequency of water needed as precipitation or supplemental irrigation for many grass and forb species used in revegetation work. Even less is known about water requirements for plant establishment, as compared with production and quality. Few data are available on the length of time vegetation should be irrigated for establishment.[119]

They also found that:

> Special consideration will need to be given to water infiltration and subsurface movement when irrigating disturbed lands. The rates of application of irrigation water for each area will need to be considered individually.[120]

Fertilizers, however, have been tested on lands disturbed by surface mining. Cook *et al.* state,"Any fertilizer application on mined land should be based on intensive soil testing results. Fertilizers, consisting of nitrogen, phosphorous, or potassium alone and in combination, when applied at the time of planting when topsoil is replaced have shown only modest benefits to seedling establishment."[121] Vallentine agrees: "In semiarid regions the use of starter fertilizers has often failed to be advantageous except on irrigated lands and very deficient soils. The benefits of starter fertilizer have been more consistent with legumes than with grasses."[122]

Wooded draw -- common in parts of North Dakota, these natural areas provide excellent wildlife habitats.

The OSM requires:

> Nutrients and soil amendments in the amounts de-
> termined by soil tests shall be applied to the re-
> distributed surface soil layer, so that it supports
> the approved postmining land use and meets the
> revegetation requirements....All soil tests shall
> be performed by a qualified laboratory using
> standard methods approved by the regulatory
> authority.[123]

Trees and Shrubs

*27. To recreate the pre-mining plant community, plant trees
and shrubs similar to those that were present on the site before
mining began.*

Trees and shrubs represent a higher order of plant in the
development of a vegetative community. Plant communities are
not static: they grow and reach the limits their environment will
sustain. A mature plant community generally consists of a mixture of grasses, shrubs and trees. Many western rangelands in
the United States are heavily populated by shrubs. Shrubs and
trees shelter plants from wind, sun and excess water, protect
soil from heavy rains and hailstorms, and provide habitats and
forage for wild and domestic grazing animals. Moreover, in the
arid and semiarid West, where much precipitation falls in the
form of snow, trees and shrubs catch and retain the snow, allowing it to irrigate soils near the plants as it melts.

In a 1978 article, Neil C. Frischknecht of the Intermountain
Forest and Range Experiment Station Shrub Science Laboratory
found that "approximately three-fourths of the western rangelands of the continental United States, where mining is or will be
important, consist of ecosystems dominated by shrubs...[and]
that shrubs and trees should be used extensively in revegetating
mine spoils within these ecosystems."[124] In his book, *Range
Ecology*, Robert R. Humphrey found that shade provided by
trees reduces the evaporation of water from plants and the soil
due to the heat of the sun, and the annual fall of leaves creates
a nourishing mulch for the soils in the immediate area. Branches
and leaves also reduce erosion from wind and rain.[125] Trees on
grazing lands also provide a direct economic benefit. Humphrey
cites studies that have shown that "animals...made significantly
greater [weight] gains in the shade than in the sun. Difference
in rate of gain by the cows was roughly proportional to the
amount of shade [available]."[126]

OSM regulations require,"Each person who conducts surface
mining activities shall establish on all affected land a diverse,

effective, and permanent vegetative cover of the same seasonal variety native to the area of disturbed land or species that supports the approved postmining land use."[127]

28. To ensure success in establishing trees and shrubs, use containerized plantings or tree spade transplants, unless on-site experimentation or research relevant to the mine site shows that bare-root transplants and direct seeding of trees and shrubs will be equally successful.

Seeding shrubs and transplanting trees and shrubs whose roots are exposed should be thoroughly investigated on a site-by-site basis (see Introduction to Guidelines) to determine whether these methods will be successful. Planting trees or shrubs in soil-filled containers or transplanting trees with a machine that lifts the tree, soil, and roots at once (tree-spade transplants), ensures that a layer of soil surrounds the roots of the plant and protects them from a harsh external environment. This can save transplanted species from shock and even death.

Ingvard B. Jensen and Richard L. Hodder of the Reclamation Research Unit at the Montana State University's Montana Agricultural Experiment Station performed numerous experiments with tree spade, containerized plant, and bare-root transplants of trees and shrubs from 1970 through 1977. These showed that the success of bare-root planting techniques varied greatly from site to site.[128] Jensen and Hodder also found:

> One of the greatest challenges reclamation specialists are facing is the rapid and efficient reestablishment of native shrubs and trees. In order to comply with new reclamation standards improved methods of establishing individual trees or shrubs must be developed....Data show that up to three times more plants can be established with tubeling stock as compared to bare-rooted stock. The results show that generally the more sensitive the plant to transplanting shock and an arid environment, the more beneficial the tubeling treatment.[129]

A study by E. J. DePuit, J. G. Coenenberg, and W. H. Willmuth of the same unit found that bare-root transplants failed to establish Ponderosa pine because of competition from already established plant species, and confirmed the findings of earlier studies: containerized seedlings increased initial survival and/or vigor when transplanted, when compared to bare-root seedlings.[130]

Reclamation plans that include consideration of wildlife needs for food and shelter recognize the interdependence between animals and plants. Small animals eat insects and pests, which can damage the plant community if not controlled. Small animals in turn are the prey of larger animals, which contribute to the overall food chain and complete the web of relationships among plant, animal, and climate. Removing one animal probably will not affect an area's ecosystem. But removing many or all animals from a site during mining must be stopped if reclamation is to be complete. The post-mine plant community will evolve into a new form if the animals who feed on it disappear.

In their study of the *Rehabilitation Potential of Western Coal Lands*, the National Academy of Sciences determined:

> Rehabilitating disturbed areas for animal habitats is of great importance and should be a primary consideration in designing and reshaping the soil material and in selecting plant species to meet habitat requirements. In general, the best habitat for native animals is one similar to the habitat in which the species evolved. Therefore, the more closely that conditions prior to mining are approximated, the more suitable will be the results for native animals that were on the site previously.[131]

29. To enable wildlife to return to reclaimed areas (if this is a proposed land use), seed plants both palatable and non-palatable to these animals.

When either the primary or secondary proposed post-mining land use includes wildlife habitat, the vegetation that is established on these reclaimed lands should be palatable to a wide range of animal species. This should include plants eaten by both game and non-game animals, as a wildlife habitat is not necessarily the same as a hunting preserve. Creation of a diverse animal habitat should include plant species that will attract a range of animal species similar to those that lived in the area before mining began.

The incorporation of non-palatable plants ensures that when vegetation is grazed by wildlife, certain plants will remain to provide shelter as well as to protect the land surface from erosion. Thus, extremely palatable plants should be offset by extremely untasty plants, so that there is a range of vegetation left on the land after animals on the site have fed.

These rocks were piled by the mining company to form wildlife habitat on grazing lands. Though suitable for small animals, it is inadequate for a wide range of wildlife.

The National Coal Policy Project, in *Where We Agree*, recommended: "When mined land is reclaimed to create wildlife habitat and food, research on the palatability, digestability and nutrient content of plant species should be expanded to determine what species are most beneficial [for planting]."[132]

The National Academy of Sciences study reported: "The plant and animal communities in the arid West represent a successful evolutionary attempt by species to accommodate themselves to harsh sites. Large scale disruption of the habitats of endangered species...may lead to their eventual extinction with the [even greater] irreparable loss of genetic diversity."[133]

Wyoming and Colorado both require:

> If compatible with the subsequent beneficial use of the land, the proposed reclamation plan shall provide for protection, rehabilitation, or improvement of wildlife habitat. Habitat management and creation...shall be directed toward encouraging the diversity of both game and non-game species.[134] Operators are required to restore wildlife habitat, whenever possible, on affected land in a manner commensurate with or superior to habitat conditions which existed before that land became affected.[135]

OSM regulations stipulate:

> If fish and wildlife habitat is to be a primary or secondary postmining land use, the operator shall ...select plant species to be used on reclaimed areas, based on...their proven nutritional value for fish and wildlife, their uses as cover for fish and wildlife, and their ability to support and enhance fish and wildlife habitat after release of [reclamation] bonds.[136]

30. To encourage animals to return to reclaimed areas, provide wildlife habitats.

Animals need places to live, and many natural rock outcroppings and small variations in surface topography that existed before mining often are not restored. Construction of wildlife habitats could include building rock piles for small animals or saving a rock wall or cliff area for birds and small rodents.

Cook *et al.* suggest:

> Any rehabilitated mining area may have important wildlife values....The rehabilitation planning should consider both shelter and food requirements for wildlife that is desired in the area....

Reshaping mine spoil materials to encourage the
return of wildlife species that formerly inhabited
the area should be of primary concern.[137]

Colorado requires "protection, rehabilitation, or improvement
of wildlife habitat."[138] OSM mandates,"Each person who conducts
surface mining activities shall, to the extent possible using the
best technology currently available, restore, enhance where
practicable or avoid disturbance to habitats of unusually high
value for fish and wildlife."[139]

VEGETATION

The final measure of a reclamation program's success is vegetation. Unless the mine operator can prove that a change in
land use is desirable, and regulatory authorities approve the
change, the land must be returned to its former use and level of
productivity. The final results should prevent further environmental degradation through erosion and sedimentation, should
approximate the vegetative community that existed before mining
began, and should return the mine site to a use comparable to
the land's pre-mining use.
To do this a successful reclamation program must prevent
erosion. Soil erosion not only harms plants, which lose the medium in which their roots grow, but degrades the air by creating
wind-blown dust,and the water and land by depositing sediment.
Although erosion occurs naturally on lands exposed to wind and
water, stabilizing the land remains necessary (see Guideline 15).
The existence of erosion before mining began does not absolve
the coal mining company of keeping future erosion to the level
found on undisturbed land in the area.
The laws of all states that regulate reclamation of surface coal
mines include provisions to prevent erosion above the natural
levels in the area where the mine is located and require a mine
operator to establish plant cover on land disturbed by mining.
OSM has also stated,"One of the principal effects of vegetation is
to stabilize the soil surface with respect to erosion"[140] (see
Guidelines 14 and 15).
Reclamation is based primarily on revegetation. Before major
land disturbances occur, such as those that surface mining can
produce, vegetative communities contain wide ranges of plant
types. A simple grassland can have more than 100 plant species.
Pre-mining vegetative types best suit the former land use, as
they grew on the site before mining began and have adapted to
it over time. Many of the lands mined in the West served as
grazing lands for cattle and sheep before mining. One cannot
expect to restore a vegetative community plant by plant, but di-

verse species are crucial to the community's stability. Successful reclamation requires that a combination of cool- and warm-
season plants be established on reclaimed areas. Land dominated
by plants that grow only in cool seasons will have little vegetation in summer and will not provide forage for wild or domestic
animals.

The National Coal Policy Project recommended that in the
arid and semi-arid portions of the United States:

> Reclamation of prairies and grassland ecosystems
> should reestablish vegetation species composition,
> diversity, and agricultural potential equal to or
> greater than pre-mining ecosystems. The re-
> claimed ecosystems must be able to withstand the
> stress of drought and reasonable agricultural
> use, just as that which existed before mining.[141]

In *Range Developments and Improvements*, John F. Vallentine writes,"Seeded stands [a group of plants in a continuous
area] on range should not be grazed until after the second full
growing season following seeding. If dry years prevail during
establishment of range seedings, an additional year or more may
be required to allow seedlings to become firmly rooted."[142] Because this study focuses primarily on 1977-1979, lands seeded in
1976 will have had the benefit of as much as three years growth
before evaluation. Return to native diversity indicates at least
initially the level of success of a company's reclamation program.

Finally, once a mine operator has established vegetation and
is controlling erosion, the land's productivity becomes crucial to
its planned post-mining use. In surface coal mining, the vegetative productivity of the land temporarily takes a back seat to
its mineral resource productivity. However, once the coal has
been removed, federal law requires that the land be returned to
its former use and regain its productivity, unless the regulatory
authorities allow another use. Productivity is measured by forage production and ground cover. It assumes a pre-mining
grazing intensity, haycutting, or cultivation and fertilization and
irrigation consistent with the normal land use in the area. When
grazing is the dominant land use, data on weight gain based on
trial grazing of animals should help evaluate the reclaimed area's
productivity.

The mineral content of the vegetation should also be analyzed. Because the soils in the overburden have been mixed and
may contribute some minerals to the topsoil and subsoil, eventually reaching the plants growing on the land, the identification
of trace minerals that may harm grazing animals should be attempted.

The environmentalists and industrialists involved in the National Coal Policy Project said:

> The ultimate goal of reclamation in the [arid and
> semi-arid West]...should be to establish a viable
> progressive, self-regenerating ecosystem. Rec-
> lamation of prairie and grassland ecosystems
> should reestablish vegetation species composition,
> diversity, and agricultural potential equal to or
> greater than premining ecosystems.[143]

They added:

> Where a return to equal or greater productivity
> is a reclamation goal...[this equal productivity]
> should be interpreted to mean that a burden is
> not imposed upon future land managers which is
> greater than that for other lands treated similarly
> in all respect except mining....The goal should
> be to provide for sustainable productivity
> through whatever natural perturbations may af-
> fect the site.[144]

The Draft Environmental Statement for Northern Powder
River Basin Coal, in Montana, prepared by the U.S. Geological
Survey and the Montana Department of State Lands, found that
"under sound management, postmining productivity may exceed
levels actually achieved before disturbance perhaps by 50 percent or more. Such increases are sometimes due in part to heavy
grazing on the minesites before disturbance, reducing the natural productivity."[145]

OSM mandates that when a coal company is operating in areas
of under 26.0 inches average annual precipitation, it is responsible for a ten-year period following the establishment of a
ground cover which equals the OSM approved standard, and must
show in the last two years of this ten-year period that:

> Both ground cover and productivity of the re-
> vegetated area shall be considered equal if they
> are at least 90 percent of the ground cover and
> productivity of the reference area. When the
> approved postmining land use is range or pasture
> land, the reclaimed land shall be used for live-
> stock grazing at a grazing capacity approved by
> the regulatory authority approximately equal to
> that for similar non-mined lands, for at least the
> last two full years of liability.[146]

INDIVIDUALS WHO REVIEWED AND COMMENTED
ON INFORM'S DRAFT GUIDELINES

Genevieve Atwood, staff officer for the National Academy of Sciences study, *Rehabilitation Potential of Western Coal Lands*; and a member of the Utah House of Representatives

J. LeRoy Balzer, Director of Environmental Quality, Utah International Inc.

Russell Boulding, member of the Mining Task Force of the National Coal Policy Project and former consultant to the Environmental Defense Fund on coal issues

Bruce Boyens, Office of Surface Mining, Region II

James A. Brown, Director of Environmental Control, Western Division, North American Coal Corporation

James Cannon, Director of Citizens for a Better Environment, and author of *Leased and Lost*, an examination of the federal coal lands leasing program

Jon Cassady, Director of Regulatory Affairs, AMAX

Robert Coats, Environmental Defense Fund

Gary Deveraux, Manager of Mined Land Reclamation, Northern Energy Resources Corporation

George Dials, Director of Technical Services, Mining and Reclamation Council of America

Edward Dobson, Montana representative for Friends of the Earth

Sheridan A. Glen, Director of Environmental Affairs, Arch Mineral Corporation

Sarah Gorin, Director of the Powder River Basin Resource Council

Alten F. Grandt, Director of Reclamation, Western Division, Peabody Coal Company

Donald Hawes, Energy Consultant, Investment Operations, Equitable Life Assurance Society

James R. Jones, Vice President, Environmental Affairs, Peabody Coal Company

Margaret MacDonald, Northern Plains Resource Council

Gary Melvin, Environmental Quality Manager, Peabody Coal Company

Steve Shapiro, United Mine Workers of America

David W. Simpson, Environmental Administrator, Westmoreland Resources

Gary E. Slagel, Manager of Reclamation, CONSOL

Stewart L. Udall, former Secretary of the Department of the Interior

Barbara West, Office of Surface Mining, Region V

John D. Wiener, Attorney at Law, consultant to the National Coal Policy Project, Wyoming representative for the Friends of the Earth

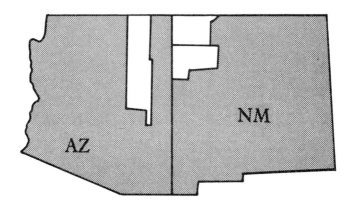

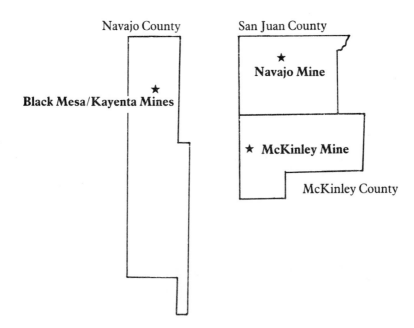

Navajo County

San Juan County

★
Navajo Mine

Black Mesa/Kayenta Mines ★

★ **McKinley Mine**

McKinley County

The Southwest

The arid Southwest adds many factors to the already complex issue of energy development and land reclamation. Huge deposits of coal, gas, oil and uranium await the eager developer, and much of it lies under lands owned by numerous Indian tribes, including the largest--the Navajo Nation. With many poorly developed soils and wildly fluctuating climatic conditions, the region is poorly suited for successful land reclamation and revegetation. Mix in the clashing cultures of the Native American with the Anglo, and land reclamation plays only one part in the maze of problems that beset coal production. Regulatory activity by the states is minimal; regulations concerning reclamation on Indian lands have not been established (much less enforced), and little environmentalist activity is to be found related directly to land reclamation and surface mining. The major environmental and resource group is the Southwest Research and Information Center in Albuquerque, but it has recently concentrated its efforts on nuclear waste disposal and uranium mining issues.

Black Mesa-Kayenta/Peabody/AZ

Mine: Black Mesa/Kayenta

Location: Kayenta (population 2,740), Arizona, in Navajo County (population 71,365)

Operating company: Peabody Coal Company

Parent company: Peabody Holding Company*

Parent company revenues: *

Parent company net income: *

First year of production: 1966

Coal production in 1978: Kayenta, 6,771,768 tons; Black Mesa, 2,515,820 tons

Acres for which a mining permit has been issued: NA

State regulatory authority: None

Reclamation bond posted: None†

Reclamation cost: reclamation cost per acre, $12,000-$18,000; reclamation cost per ton of coal, $1.34

Expansion plans at mine: None

Additional acres for which the company will seek a permit: Approximately 4,400 over the life of the mine for the next 20 years

Parent company U.S. coal reserves: Owns or leases 8.9 billion tons of coal reserves

Development and expansion plans for surface coal mines in western United States: At present, Peabody Coal operates five surface mines in the western United States--Big Sky Mine (2,064,886 tons in 1978) in Montana, Nucla Mine (102,393 tons in 1978) and Seneca II Mine (1,372,251 tons in 1978) in Colorado, Black Mesa and Kayenta. Peabody plans to produce 5 million tons of coal per year at the Big Sky Mine by 1985. The company is also developing three mines in Wyoming to produce a total of 21 million tons per year by 1986.

NA = Not available

*Peabody Holding Company is a consortium of six companies including: Newmont Mining Corporation (27.5 percent ownership), Williams Companies (27.5 percent), Bechtel Corporation (15 percent), Boeing Company (15 percent), Fluor Corporation (10 percent), and Equitable Life Assurance Society (5 percent).

†However, a $15,000 security collateral and $105,000 collective bond to ensure lease provisions have been filed with the U.S. Geological Survey.

Sources: Peabody Coal Company; 1977 and 1979 Keystone Coal Industry Manual; "New Coal Mine Development Survey 1978-1987," Coal Age, February 1979; Rand McNally Road Atlas, 54th Edition; Navajo Research and Statistic Center; OSM Inspection August 17, 1978

SITE DESCRIPTION

The Black Mesa dominates the landscape in northeastern Arizona, in the Navajo Nation. Inhabited by the Navajo and Hopi Indians--whose reservation lies in the midst of the Navajo Nation --Black Mesa is a 2.1 million-acre highland, 7,500 feet above sea level at its northern end.[1] It is on this end that the Black Mesa/ Kayenta mining complex is located, about 20 miles southwest of the town of Kayenta, which is off the Mesa. Although Peabody prefers to call the mining pits that provide coal to the Page electrical station in Page, Arizona, the Kayenta Mine, and the other pits the Black Mesa Mine, the two can be considered as one large mining complex, administered by common offices on the site and in Flagstaff, 120 miles southwest.

The topography is one of rolling slopes, some as steep as 45 degrees or more.[2] Small valleys can be found where several water drainages--or "washes"--cut through the land.

The climate is highly variable and unpredictable. The precipitation in northeastern Arizona ranged from 8-22 inches annually between 1940 and 1970.[3] The mesa receives an average of 12 inches of precipitation yearly, but toward the northern end--

near the mines--the average is from 15-17 inches.[4] Yet, much of
it falls in a relatively short time. Approximately 25 percent of
this moisture falls as snow in the winter and 50 percent--as rain
--falls between July and September.[5] The growing season lasts
from about mid-May through mid-October,[6] and high winds are
frequent throughout the year.

The land in the area is primarily used by Navajo herdsmen
for grazing sheep, cattle, goats and horses. Although some
agriculture is possible in the valleys, little is practiced. None of
the land can be considered prime agricultural land, although
there is some question whether washes to the south of the existing mine area could be classified as alluvial valley floors.

OVERVIEW

It took passage and vigorous enforcement of a new federal
law--the Surface Mining Control and Reclamation Act--to get
Peabody Coal Company to improve its reclamation practices at the
Black Mesa/Kayenta mining complex after years of shoddy performance. Peabody is generally characterized as a company that
has usually met the demands of the law once these demands have
been enforced. Owing to the lack of a strong regulatory program
in Arizona and the rather meager program established by the
Navajo tribe, Peabody has had a virtually free hand to do as it
pleased. Now the picture is changing. The new federal law, as
well as divestment of Peabody Coal from Kennecott Copper Corporation and its subsequent purchase by the six-company consortium that comprises Peabody Holding Company, has led to a
sharp reversal of inadequate reclamation efforts of the past at
this site, one of the most expansive operations in the entire
country. Today, new programs and a new management structure
and staff are combining to improve considerably the company's
approach to reclaiming the land. Still, the results of past poor
reclamation planning and practices plague the company--steeply
graded slopes, no segregation and saving of topsoil, poor experimentation, etc.--and even in its new more diligent efforts, there
is much room for increasing their utility.

Improvements still need to be made. A better, more diverse
seed mixture could be developed. The establishment of trees
and shrubs could also use some further attention, and experimentation on the site--more closely related to reclamation problems encountered by Peabody--could be undertaken.

In addition to Black Mesa's operational problems, an equally
pressing problem is human. And no federal laws or regulations
seem to affect the way Peabody has ignored the human beings
most directly affected by its mining operations: the Navajo tribal
members who have lived on and worked the land for generations.

At no other mine in INFORM's sample was the direct impact of
mining on humankind so apparent and dramatic. In 1980, when
the nation has at long last turned its attention to the plight of
its Native Americans, Peabody has allowed a vast communications
gulf to develop between the company and the Navajo on Black
Mesa. The Navajo are bitter about it. Strong company initiative
to improve dialogue, understanding and joint reclamation planning could make a valuable difference here. But, unfortunately,
these things have not yet occurred.

In 1978 the complex produced 9,287,588 tons of coal--the
third highest production rate of all U.S. mines.[7] Before 1979
revegetation efforts at the complex were fair; the results remain
poor today. However, new techniques for seeding, water control
and surface preparation--including the replacement of topsoil--
promise to increase the company's potential for success in the
future.

Major changes on the site include the segregation and replacement of topsoil--some by the direct-haulback method--a good
practice. Grading and contouring have been improved; the company has regraded slopes that were graded too steeply in the
past, and a more gently rolling topography has resulted. Soil
surveys (which should have been conducted before the mine
opened) have finally begun, and much of the baseline data on
soils, as well as other features of the mine area, are now being
collected.

Seeding is now accomplished with the use of drill-seeding
machines, and slopes are furrowed along the contour to allow
moisture to penetrate the soil and to reduce the potential for
erosion--all good practices. But the company's seed mixture still
leaves much to be desired, as introduced grasses and legumes
dominate. The company has done little to replace trees and
shrubs, and efforts to date have met with total failure. According to Peabody officials, the company plans to try again, but at
present this aspect of its reclamation program is in a holding
pattern.[8]

Water control is achieved by creating water impoundments
and holding water on the site instead of releasing it and being
forced to comply with federal water-discharge quality standards.
Peabody plans to leave many of these water impoundments permanently for use by livestock and wildlife. Erosion is not a major
problem, and undisturbed lands exhibit at least as much erosion
as mined and graded lands. Only the final results of revegetation efforts will determine exactly how well the land will be protected from erosion in the future.

To date, the vegetation on lands reclaimed on the Mesa is
dominated by Russian thistle--which turns to tumbleweed when
dry--and introduced grasses and legumes. Productivity figures

for the site are incomplete, as the long-term productivity of reclaimed lands cannot yet be measured, many of them having been
seeded for the first time, or completely regraded and seeded in
1979.

Peabody has done little to benefit the wildlife population of
the Mesa.

One of the great difficulties faced by Peabody--largely the
result of inaction on the company's part--is its lack of communication with the Navajo tribal members living on and using the
land that is being disturbed by mining. Dealings between the
top administrators of Peabody Coal and leaders of the Navajo Nation have resulted in little direct communication with the Indians
actually living on the Mesa and the mine site. Numerous examples
of misunderstandings and poor interaction between the company
and the Navajo herdsmen came to light during INFORM's investigation. Faced with a language barrier--many Navajo do not speak
English, or do not understand the language sufficiently to deal
with the company--and a fundamentally different social structure
and value system, Peabody has relied on intermediaries such as
the Black Mesa Review Board, Navajo Environmental Protection
Commission, the Office of Surface Mining's (OSM) staff members
working on issues fundamental to the development of Native
American resources, and others to relay exactly what is taking
place on the Indians' land. Gary Melvin, environmental quality
manager for Peabody's Arizona office, expressed a sentiment that
is surely shared by all parties involved when he said the tribe
and the company do not communicate, and this proves "very
frustrating."[9]

Although Peabody is improving its reclamation operations,
the company must also improve its community relations with those
most directly affected by the mines. Homes are moved, and
herdsmen are prevented from using the land they have used for
many generations. More persons are directly affected by mining
on the Mesa than at any other mine studied by INFORM--and
possibly more than at any other mine in the country, except for
the eastern coal provinces of Appalachia. It is up to the company
that invaded the homes and lives of the Navajo to show that mineral resource development can be conducted in an environmentally and socially acceptable manner.

EARTHMOVING

Grading

Mining is conducted in many areas at the Black Mesa/Kay-
enta complex. Multiple seams of coal exist in patches of deposits
that are mineable from the surface, and in February 1980 there

were seven open mining pits.[10] Grading has been catching up to
disturbance since late 1977.[11] By February 1980, of the 4,707
acres disturbed by mining and related operations 2,699 acres
had been graded, more than 57 percent. Because the mine complex is so large and diffuse, about 446 acres of disturbed, ungraded land consist of roads alone.[12] (Other lands yet to be
graded are topsoil stockpile areas, the active mining pits themselves, and the ramps leading into those pits.) OSM said in January 1979 it felt that Peabody had constructed too many roads at
the mine.[13] Peabody officials told INFORM these were access
roads, and the company does not feel the OSM criticism was
valid.[14]

Grading has been conducted up to one spoil ridge from all
the active mining pits--a vast improvement over previous practices. As late as January 1979 some mine pits had up to eight
spoil ridges behind them.[15] Wayne Hilgedick, the mine's reclamation manager, told INFORM part of the reason for this extremely close grading is to control fires in the mine pits by covering combustible materials.[16] (See Toxic Material.) Peabody
would prefer not to have to grade so close to these pits because
it believes that better contours can be created when two spoil
ridges exist for grading.[17] However, adhering to the mine's fire
control plan has priority.

Graded areas on the Mesa now show gently rolling hills with
no slopes greater than a 5:1 angle, according to Hilgedick.[18]
However, Peabody has had to do extensive regrading in the past
year of areas where slopes were in excess of 3:1.[19] The amount
of grading and regrading is reflected in the amount of land
seeded by Peabody in 1979 (see Seeding Procedures).

Peabody readily admits that it is not replicating the pre-mining topography,[20] but the company believes that the greater
number of gentle slopes will be easier to reclaim and should provide more quality rangeland than existed previously.[21]

Another change in the contour of lands on Black Mesa results from the construction of water holding ponds, which Peabody plans to leave as permanent water impoundments. Although
the Navajo generally react favorably to these impoundments--
which will provide more watering holes for their livestock--a
problem develops if the placement of these impoundments is not
coordinated with the individual Navajo who use the land for
grazing. According to a U.S. Geological Survey (USGS) report
in April 1978, because there are no fixed boundaries separating
the grazing areas of one herdsman from those of another, tribal
officials are concerned that disputes might arise between individual herdsmen, and that if water impoundments are too close to
one another, lands nearest to them will be severely overgrazed,
as animals tend to stay near sources of water.[22] INFORM does

not know how Peabody plans to decide on which impoundments will be left on the Mesa, but it should be watched carefully. Judging from comments made by Gary Melvin, there has been little communication between the company and the individual land users. Most of the dealings between the tribe and the company are conducted between the head of Peabody's Arizona office and the tribal leaders.[23]

The changes in the nature of the land on the Mesa are disturbing to some of the Navajo interviewed by INFORM. Their culture is based on the natural harmony between the sky--that rains on, and nourishes the land--and the earth, which provides grazing and sustenance for the tribe. Changes in the contour of the land, as well as its initial disturbance, are difficult for the Navajos to accept. For example, they want at least *some* flat areas for building their homes (or hogans), but little is being created by grading. Peabody conceded to INFORM that it has created rolling land even in the few areas where flat land existed before.[24] This is only one issue that points up the poor communication that perpetuates the dislike and distrust felt by many Navajo for Peabody.

The contours of the graded lands on the Mesa are all traversable by reclamation machinery. In September 1979 Peabody submitted a map of the final contours for the mine to the OSM.[25] Before the 1979 regrading, many slopes at the mines would not have been traversable.[26]

Topsoil

During the first decade of mine operations at Black Mesa, what little topsoil did exist on the site was never saved by Peabody until it was required by federal regulation. Only in November 1977 was the first salvage of topsoil at the mining complex undertaken,[27] just before the December 13 deadline established by the OSM.[28] Subsoils are not differentiated from topsoil, and Wayne Hilgedick reports that all materials on the surface of land to be mined are removed to the depth of rock, or clayey or shaley overburden.[29] Although the topsoil contains little organic matter, Hilgedick concedes that the soil provides a better medium for plant rooting than graded overburden.[30] The company is only now conducting an extensive survey to ascertain the quality of the different soils and the quantity of soil available for salvage and re-use.[31] Alten Grandt, Peabody's environmental affairs director in St. Louis, told INFORM the company expects to be able to replace 6-8 inches of soil over the land after grading. The soil survey, which was expected to be completed by the summer of 1980, will be more accurate.[32] Some areas at the mine may have less than 8 inches of topsoil on them. Some areas may have virtually none, while in some washes or drainages topsoil materials exist to much greater depths.[33]

When topsoil removal began, all materials were stockpiled, and a portion of the topsoil already in stockpiles will remain until it can be used later. In some parts of the mine, all topsoil must continue to be stockpiled and no direct haulback can be accomplished. But now, according to Gary Melvin and Wayne Hilgedick, there are areas on which 100 percent of the topsoil can be directly hauled back. Topsoil stripped ahead of the active mining pit is thus immediately replaced on lands graded behind the pit. The controlling factor is the extent to which grading and topsoil removal are accomplished simultaneously. Hilgedick could not estimate the ratio of topsoil that can be directly replaced to that which must be stockpiled. He said the company would like to conduct as much direct haulback as possible, as it greatly reduces topsoil handling costs.[34] Topsoil is replaced to an even depth over all graded lands, even though some areas may not have had much topsoil before mining.[35] However, of all the land that has been seeded at the mine, topsoil has been replaced on less than 20 percent.[36]

Topsoil stockpiles observed by INFORM in February 1980 showed their vegetation to consist primarily of Russian thistle (a weed), although snow prevented observation of other vegetation that may exist. However, an August 1978 inspection by OSM found that stockpiles lacked vegetation, and many had not been seeded. OSM also noted that wind erosion could cause the loss of much of the stockpiled material and should be kept to a minimum. OSM also noted some small areas of water erosion.[37] According to Peabody, recent additions of topsoil to some of these stockpiles accounted for the lack of vegetation.[38] Yet a USGS inspection in November 1978 found that "the lack of vegetative cover on all storage areas was still apparent....Recent stockpiling activity can only explain the lack of cover on only a portion of the storage areas. Apparently, Peabody has no plans in the near future to correct this situation."[39] James Jones, vice president for environmental affairs at Peabody, says stockpiles are now seeded with crested wheatgrass--an introduced grass--and mulched with straw which should help considerably.[40]

Toxic Material

Fires in the mine pits and spoil piles at the Black Mesa/Kayenta complex, which increase air pollution, as well as affect surface vegetation, are the most prevalent of any mine in the INFORM sample. Numerous inspection reports and articles on the mines note that fires are common on the site, and an OSM inspection in January 1979 resulted in a requirement that Peabody develop and institute a fire prevention and abatement program.[41] (The results of this plan can be seen at the mine, as grading is

within one spoil pile almost everywhere on the site, thus covering many of the combustible materials.) However, smoke from
fires could still be seen coming from the open pits and ungraded
lands, during INFORM's visit to the mine in February 1980. The
effects of these fires upon vegetation could be significant as
heated spoils may present conditions unsuitable to the germination and growth of plants. However, it remains to be seen
whether fires will cause patches of graded ground to remain bare
while lands surrounding them continue to produce vegetation.
Fires occur naturally in the area of the Black Mesa where coal
seams surface, according to Peabody's James Jones,[42] but the
increased amount of coal that is exposed to the air during mining
increases the fire potential.

With the mining of up to six different coal seams in any given
mine pit, many layers of interburden--materials lying between
the coal seams, sometimes called "the parting"--must be removed.
According to Wayne Hilgedick and Gary Melvin, the interburden
found on the mesa is generally not toxic, but its high sodium
level could cause vegetation problems if it is not buried in the
pits before grading. Some of the overburden removed during
mining may also be high in sodium salts.[43] Hilgedick notes that
until Peabody's extensive soil survey is completed in the summer
of 1980, he cannot be certain what measures the company may
take, but it may have to replace 18-24 inches of topsoil on
graded spoils that exhibit high salt levels.[44]

Another issue involving toxic materials at the mine is less
clear than that of the quality of the overburden or interburden.
Piecing together stories from Navajos and Peabody, INFORM
found that a number of sheep being grazed near the mine area
died after drinking water from an impoundment on the site. According to a Navajo who lives only a few hundred yards from the
mine, when his sheep drank from this impoundment 13 died immediately and another nine died that night.[45] According to both
Peabody and the man, he received $40 for each of the sheep
killed. Peabody reports that the sheep died from nitrate poison-
ing because high levels of nitrates in an area where explosives--
commonly composed of ammonium nitrates--are stored, had
washed into the water impoundment.[46] Whether the nitrates that
poisoned the water should have been allowed to contaminate the
pond in the first place is a question that INFORM had neither
the time nor the resources to pursue. However, there is an additional question as to whether Peabody fenced off the pond from
which the sheep drank. Peabody reports that it did, but the
Navajo whose sheep were killed asked INFORM why the company
had not fenced the area to prevent such a thing from happen-
ing.[47] Again, it seems that poor communication, if not a poor

basic practice, was evident between the company and the people
who use the land around the mine.

WATER QUALITY

Surface Water and Groundwater

Much of the water that flows over the surface of lands at the
mine complex drains naturally into sediment (or holding) ponds
constructed by Peabody. As of February 1980, 11 ponds already
existed on the site, but had not been designed to physical-engi-
neering standards. An additional 17 ponds are planned, and
were to be completed by the summer of 1980.[48] Mine officials say
that more than 80 percent of the water draining from spoils and
graded lands remains in ponds on the site.[49] According to studies by researchers from the University of Arizona in 1975, water
falling on graded lands will be more apt to sink into the soils
than to run off the area[50]; however, the water that does run
from the graded land will have higher concentrations of sediment
and salt than water running from undisturbed areas.[51] Peabody
is moving to correct deficiencies in its pond system. OSM inspectors in the past have found ponds designed without spillways
(with the risk of erosion when excess water leaves the ponds
during storms), and spillways and impoundments unprotected by
either vegetation or rip-rap.[52] After a January 1979 inspection
and notice of violation levied by OSM, the company took measures
to correct the problems. OSM found these measures satisfactory
and in June 1979 terminated the violation notice.[53]

Because Peabody plans to discharge no water from the mine
site, except in cases of extremely heavy precipitation, it does
not have a water discharge permit under the National Pollutant
Discharge Elimination System (NPDES) administered by the Environmental Protection Agency (EPA).[54] Thus no discharge monitoring reports are filed with the EPA.[55] According to Peabody,
newly planned sediment ponds (as previously described) will be
engineered to retain waters from the most severe precipitation.[56]

Water monitoring by Peabody has been virtually non-existent
on Black Mesa. According to Gary Melvin, since the company
was never required by law to monitor groundwater or surface
water, it never has.[57] Yet since 1973 the U.S. Geological
Survey has had 10 surface water and 12 groundwater monitoring
stations and wells for monitoring water on a regional basis
around the Mesa, as part of a program funded by Peabody and
the Navajo and Hopi tribes.[58] However, these wells and monitoring stations exist over a larger area than that encompassed by
the mining complex--providing baseline data against which Peabody will measure its mine-specific data--and the company has
proposed an extensive monitoring system to be installed and

operating by mid-1980.[59] This program will include the installation and monitoring of 29 groundwater wells and 8 surface water flumes (concrete structures placed in water drainage areas to measure water flow and quality).[60]

Results of this monitoring will be critical in determining the quality and quantity of surface water running from the mine, as much more surface water is infiltrating soils and not flowing as it did before mining, thus affecting water availability off the site. Monitoring also will assess the changes in the quality of the water that does flow from reclaimed areas.

Although mining does not appear to be interrupting aquifers on the Mesa, the increased infiltration of water into soils on disturbed lands, and the possible effects upon groundwater systems will also be important for ascertaining the full impact which mining has on water quantity and quality in the area. (See also Toxic Material for surface water quality.)

Although not directly related to land reclamation on Black Mesa, Peabody's operations are affecting groundwater. Aquifers below the level of the coal seams being mined are tapped to provide water to transport coal through a slurry pipeline from the mine to the Mojava power plant near Bullhead City, Arizona. The company has also provided local Navajo with some of the water from its wells, but the full impact of this use has not been quantified. A 1973 USGS report found that:

> Although present withdrawals seem small compared to the extent of the aquifer, an artesian aquifer is a pressure system and any significant withdrawal will reduce the artesian pressure over a wide area....It should be stressed, however, that development of the water supply will result in interference between wells or well fields, especially in the artesian parts of the aquifer. The Navajo Sandstone is the principal fresh-water aquifer in the Black Mesa area; therefore, future development should be evaluated carefully with respect to the effects on the existing water supplies.[61]

Edward Evans, the project manager for a study by the Williams Brothers Engineering Company of Tulsa, Oklahoma--to be finished in the spring of 1980--contends that even though major regional impacts cannot be expected from the tapping of this aquifer, Peabody should stop using this high-quality water and tap inferior water from an aquifer nearer the surface. Evans believes the lower-quality water would be sufficient for the company's needs, and that every gallon of potable water from the Navajo

Sandstone aquifer that is used for coal transport is wasted if not used for domestic purposes.[62]

Alluvial Valley Floors

Peabody believes no areas on the mine site qualify as alluvial valley floors under OSM criteria,[63] and the company is now studying water and soils in an effort to prove its belief. However, certain lands south of the mine may indeed be AVFs according to studies conducted by researchers from the University of Oklahoma.[64] In this case, Peabody would have to assess any impacts which its mining operations might have upon these alluvial valley floors. It is doubtful that immediate concern is necessary, but if further mining was planned for the Mesa, greater consideration would have to be given this issue. Again, groundwater monitoring programs should help define the impacts of mining upon water resources in the area--some of which may be feeding shallow limited aquifers that underlie potential AVFs. The areas most likely to exhibit AVF characteristics are the natural water drainages--or washes--where alluvial soils have collected over many years.

Erosion

A 1978 OSM inspection found that erosion did not appear to be a major problem on graded slopes--but sedimentation was a problem. This is due in part to poor stabilization of water holding areas at the mine complex.[65] (See Surface Water and Groundwater.) Water impoundments and haul roads have been found to erode and release sediment into water drainages, and spoils placed in drainage areas have contributed to sedimentation of lands downstream.[66] In June 1979 OSM found that Peabody was making an increased effort to rectify erosion and sedimentation problems noted in previous inspections.[67]

Because of the dry climate, which hardens some surface soils, and the occurrence of rain in sudden, intense storms, erosion is common to all lands lacking adequate vegetation in this part of Arizona. University of Arizona researchers conclude that vegetation is the best means of controlling erosion in the area.[68] As noted above, reclaimed lands will generally allow less water to run from them, but the water that does run off will contain higher concentrations of sediment. In this dry climate, high winds--common to the region--increase soil erosion on disturbed and undisturbed lands alike. Because soils are loosened in mining, more erosion can occur on disturbed lands when winds sweep the Mesa. OSM has noted wind erosion to be very evident in these periods.[69] But Hall Susie, district mining supervisor for the U.S. Geological Survey, says the wind erosion affects

undisturbed areas as much as it does mined lands.[70] Again, good
vegetative cover, which Peabody is attempting to create, helps
to hold soils together best.

SEEDING AND AMENDMENTS

Seeding Procedures

Seeding at the Black Mesa/Kayenta complex has increased
dramatically in the last two years. Owing in part to a dry year
and a dearth of graded land, very little seeding was done in
1977. In 1978 more seeding was accomplished.[71] However, in
1979 Peabody seeded a huge amount of land--957 acres at Kayenta and 1,153 at Black Mesa.[72] Although about 622 of these
acres were reseeded to remedy poor vegetative establishment in
the past, the remaining 1,488 acres were either lands seeded for
the first time, or lands that had once been seeded but were regraded in 1979.[73] (See Grading.) However, about 30 percent of
the seeding was done from the air, a poor practice which the
company says will not be continued but which was used in Peabody's effort to seed an enormous amount of land in one year.[74]
(See below.) As mentioned earlier, topsoil has been replaced on
less than 20 percent of all the land seeded to date, which is
poor.[75] About 45 percent--above average for INFORM's sample--
of all the lands disturbed by mining and related activities, and
almost 80 percent of all graded lands, had been seeded as of
February 1980.

Peabody must always attempt to outguess the extremely
fickle weather in order to plant its seeds just before the heaviest
rainfall. To illustrate the uncertainty: INFORM found that two
suggestions by the same member of the Navajo Environmental
Protection Commission's (NEPC) staff differed as to the optimum
planting time. According to a March 1979 memo by Leonard Robbins, an NEPC inspector, the best time for seeding on the Mesa
is from mid-May to mid-June.[76] But in an interview with INFORM
only three months later, he said the proper time for seeding is
June and early July,[77] showing the problems that the climate can
create for reclamation success.

Peabody's Wayne Hilgedick and Gary Melvin said the company
was to attempt in 1980 to pinpoint the best time for seeding. Because so much seeding took place over seven months in 1979, on-
the-ground results in 1980 should show which seeding periods
yielded the best (and worst) germination and growth.[78] Still,
the unpredictability of the rains leaves a large margin for error,
leaving irrigation as perhaps the only way for Peabody to provide adequate water (see Amendments and Maintenance).

The 1979 seed mixture is only fair for establishment of a native, diverse plant community which will be suitable for grazing.

TABLE A

1979 SEED MIXTURE
FOR LANDS BEING RECLAIMED AT BLACK MESA

Plant Species	Introduced/ Native	Cool/ Warm	Pounds Applied Per Acre
Grasses			
Western wheatgrass	N	C	3.0
Crested wheatgrass	I	C	3.0
Pubescent wheatgrass	I	C	1.0
Indian ricegrass	N	C	1.0
Blue grama	N	W	3.0
Alkali sacaton	N	W	0.5
Sand dropseed	N	W	0.5
Russian wildrye	I	C	1.0
Bromegrass (Smooth brome)	I	C	3.0
			16.0 Pure live seed
Legumes			
Alfalfa	I	C/W	5.0
Yellow sweetclover	I	W	5.0
			10.0 Pure live seed
Forbs			
Fourwing saltbush	N	W	0.5
			0.5 Pure live seed

Source: Peabody Coal Company

It consisted of six introduced grass species seeded at a rate of
18 pounds per acre and five native grass species seeded at a
rate of 8 pounds per acre. A half-pound of fourwing saltbush--
a forb--was also used. However, the seed mixture relies heavily
on introduced legumes--almost 40 percent by weight--a common
Peabody practice at other mines (see Seneca II profile). Three
times as much cool-season seeds (by weight) is being used as
warm-season seed, also a poor but fairly common practice.
  Peabody says the mixture is determined year-to-year, based
on the results of the past year's seeding. However, the number
of species seeded has actually been reduced, and the ratio of introduced to native species has increased, both poor practices.
Officials at the mine say the company also factors in research by
the University of Arizona on test plots at the mine. They say
the company would like to establish more forbs on the site for
grazing purposes, but the seed mixture being used to date does
not reflect that desire.[79]

Amendments and Maintenance

Because of the 1979 regrading and Peabody's policy to grade no slopes steeper than 5:1, all reclamation machinery can be operated along the contour.[80] The tracks along the contour retain water and prevent erosion. In 1979 Peabody began to create furrows along the contour of all lands being seeded. Before that year's extensive seeding the company made such furrows 14 inches deep.[81] Frank Farnsworth, engineering director for Peabody's Arizona division, calls this one of the most significant changes in the seeding program.[82] Gary Melvin told INFORM the practice was instituted after a visit to the Pittsburg & Midway McKinley Mine.[83] (See McKinley profile.)

Peabody told INFORM it formerly seeded both from the air and on the ground (using a drill seeder) but found that drill seeding was superior; henceforth, officials said, Peabody will use drill seeding exclusively as another major improvement in its program.[84] According to one Navajo interviewed by INFORM, whenever the company seeded from the air, flocks of birds would follow, feeding on the seed being dropped, leaving little to grow.[85] But in 1979, during the company's massive effort to seed 2,110 acres, almost one-third of the seeding was still done from the air.[86] When drill seeding, Peabody reports, the seed is covered through the use of bars dragged behind the seeding machine.[87]

Mulching has been employed for the past two or three years, says Gary Melvin.[88] The company uses a straw mulch of wheat, which is blown onto the ground and crimped into place by discing. However, at the time of an OSM inspection in August 1978, John Kyle of Peabody told the inspectors that mulch had never been used at the mine, although plans for its use were being implemented.[89] It seems reasonable that Melvin's mistake is attributable to his relatively new position with Peabody, and mulching is a fairly new practice on the site. Peabody officials told INFORM in June 1979 that they felt Russian thistle (a weed that grows profusely in the West, and when dry is known as tumbleweed) to be adequate mulch while growing on the land.[90] This is highly dubious, as once Russian thistle becomes established it is very difficult to get much else to grow. This is especially apparent on both disturbed and undisturbed lands on Black Mesa and surrounding areas. Again, Peabody's policy is difficult to ascertain, although the installation of the new reclamation staff at the mine may be changing the company's previous posture. On a visit to the mine in February 1980 INFORM suggested the use of a standing mulch--such as that used at many mines in the West. Peabody officials said they were considering the practice,

although they fear that the growing mulch might take moisture
away from grasses and other plants the company is trying to establish. In that event, Peabody might consider establishing the
growing mulch in the first season following replacement of topsoil, then seeding the desired plant mixture in the next season.[91]
Further experimentation is warranted.

According to Hilgedick and Melvin, Peabody fertilizes to enhance vegetative establishment by using nitrogen and phosphate
at levels determined by soil tests of graded lands. On occasion,
potassium may also be used.[92] The use of fertilizer is new to the
Peabody reclamation program; evaluation of its use cannot yet be
made.[93] Although irrigation is not practiced, it may become necessary for the establishment of desired plants. However, the
picture is confused. According to a USGS inspection conducted
in March 1978, Peabody said it could not irrigate the whole mine
even if it wanted to, as water rights leased by Peabody restrict
water use to mining activities only. USGS questions this inter-
pretation.[94] On a previous visit to the mine in November 1977,
USGS inspectors reported that in the irrigation test plots observed, the irrigated portion of the plots was the "only area that
showed any noticeable amount of vegetation."[95] To add to the
confusion, Hall Susie told INFORM in March 1979 that the Navajo
tribe did not want Peabody to irrigate.[96] INFORM could not cor-
roborate this assertion. Further experimentation seems war-
ranted--as does full consultation with the Navajo.

A major controversy surrounds the fencing of seeded lands.
Fences are erected to prevent local herdsmen from allowing their
animals to graze on lands where the company is trying to establish vegetation. The controversy arises--once again--because
Peabody has had little or no communication with local land users
about any of its operations, including the use of fences. Navajo
grazing areas are delineated only by long-accepted imaginary
lines based on traditional uses. Informal though these boundaries may be, Navajo family units do have fairly precise notions
of where their grazing areas begin and end. In at least one area
of the mine, fences installed by Peabody have been cut through
to allow animals to graze on the newly reclaimed site.

Peabody officials believe the issue can be resolved only by
seeking out each individual herdsman and explaining why he
must not cut the fences or use the seeded land. They also suggest that the herdsman be compensated for lost grazing land by
the Navajo tribe, which receives a royalty on every ton of coal
produced on the Mesa.[97] In discussions with local residents on
the Mesa, in the nearby town of Kayenta, and Peabody employees, INFORM was told that compensation by the tribal government has been meager at best. Royalties paid to the tribe are
not used to benefit the people most affected by mining, but the

tribe as a whole.

Again, for Peabody, the lack of understanding of historical Navajo social structure or land-use traditions is critical. On the issue of reclamation, the company failed to consult the heads of the family units whose overall grazing area is well-delineated. Working through them, Peabody might well have developed an understanding among the family unit members as to which lands in their traditional use area would be fenced and unavailable for grazing. Then the family units could have either partitioned the remaining land equitably, or found other ways of compensating the individual herdsmen most severely affected by the loss of land. The family units also would have greater strength in dealing with the Navajo tribal government.

A body called the Black Mesa Review Board--comprised of five tribal members--has been established to deal with controversies between the company and individual Navajo living on the mine site. The Board has held approximately 50 meetings with the residents since it was established in 1975.[98] Apparently, communication between the Black Mesa Review Board and Peabody has improved greatly since a strike by Navajo miners shut down the complex in February 1979. The Board and the company now meet every two months, and are beginning to become more involved with issues of compensation for land and homes (the issue from which it was born), miner safety, social impacts of mining, and land reclamation.[99] However, it is unclear how well the Board is relating to the residents. Many conflicts are resolved between Peabody and residents directly, and this raises further the question of why both Navajo in the area, and Peabody expressed to INFORM the frustration that communication was so difficult. Yet, as Andrew Bennallie, the chairman of the Black Mesa Review Board, told INFORM, "There are lots of problems we can't solve as long as there is mining on Black Mesa. It's not just Peabody, but any mining is hard on the people."[100]

Ideally, a mining company in Peabody's position should have made significant efforts as mining plans first were made, to understand the social--as well as environmental--context into which it was moving. However, there is room still for Peabody to improve both its understanding and relationships with the Navajo. Gary Melvin and Frank Farnsworth believe communication will develop with time and experience.[101] However, history shows this passive approach to have been unavailing. Peabody will have to explore fresh initiatives in dealing with family units and the Board--if it is to resolve such specific questions as fencing and grazing rights--and in establishing broader and more sensitive channels of communication overall.

Trees and Shrubs

Peabody has done little to re-establish shrub and tree growth on lands disturbed by mining. What little planting has been done has not survived. Bare-root stock--inferior to containerized or other forms of transplants--has been used, and has died.[102] In one inept attempt to transplant trees, the company failed to ball up the roots of the trees while they still had soil on them; total failure resulted. The company has no present plans to try further tree or shrub transplants,[103] although Navajo residents interviewed by INFORM said such vegetation was desirable.[104]

VEGETATION

Peabody's vegetation record at Black Mesa has always been poor. It may be improving.

Vegetation on most of the land being disturbed by mining on Black Mesa consists of big sagebrush and pinyon-juniper vegetation.[105] Trees, shrubs, forbs, and grasses are common on most of this land.[106] Grazing productivity is among the lowest found on any mine site studied by INFORM. According to both the Bureau of Indian Affairs and Peabody, to graze one animal unit for one month on this land (an animal unit is equivalent to five sheep, or one cow and calf) requires about 40-45 acres of land.[107] However, the information used to determine these stocking rates is old, and Peabody--as required by OSM regulations--has recently established reference areas on unmined land to provide up-to-date productivity measurements.[108]

According to a 1975 study by EPA and the Argonne National Laboratory, and a 1977 survey of the mine by the U.S. Fish and Wildlife Service, undisturbed lands have more ground cover and many more shrubs than revegetated lands, which are mostly covered by grasses and legumes.[109] Drier south- and west-facing slopes were found to have more introduced grasses than the cooler and more moist north and east slopes, which showed some native grass growth.[110] Studies by the U.S. Forest Service showed that undisturbed lands around the mine produced from 118.4-432.5 pounds of vegetation per acre. Data from revegetated areas showed productivity to be from 4-135 pounds of vegetation per acre.[111]

More recently, Thomas Ehmett, an inspector for the Denver OSM office, told INFORM that in the summer of 1978 he had been unable to find any germinated seeds on reclaimed areas that he toured.[112] A Peabody study that summer indicated that from 1,796-2,652 pounds of vegetation were found to be growing on lands seeded in 1972 and 1974.[113] However, much of the sampled vegetation included Russian thistle (a weed), crested wheatgrass (an introduced grass), alfalfa (a legume--also introduced), and some native wheatgrasses. The study took place, moreover, at peak productivity, according to Peabody officials.[114] According to an inspection report by Ehmett after his visit to the mining complex in August 1978:

> Revegetation has pretty much been a failure, previously, at these mines. According to [Peabody officials] the best revegetation results were obtained on the J-3 area of the Black Mesa Mine. This area was inspected, and was noted to be very sparse, in cover, with desirable species, although Russian Thistle and other volunteer weed was abundant. The N-1 area had been the most recent revegetation effort (spring, 1978). This area was inspected, and no plants were present.[115]

USGS inspectors who accompanied Ehmett on this inspection wrote that the J-3 area:

> has been reseeded numerous times since about 1972. At best, the revegetation has had very limited success. Native and introduced species of shrubs and grasses were sparse, but volunteer weed species were abundant, especially Russian Thistle. This result is due, in part, to the lack of summer moisture this year.[116]

After a follow-up inspection in November 1978, USGS wrote: "The major reclaimed area, J-3, is still very sparse, and most vegetation, especially the Russian Thistle, appeared to be either dead or dormant."[117]

OSM inspected the mining complex in June 1979, as the company was conducting its massive seeding and contour furrowing program (described above). It reported:

> It is hoped that the use of topsoil will greatly improve the revegetation program....All seeding and mulching activities noted were taking place on the contour, [and] it was evident...that Peabody is making an increased effort in the reclamation phase of the mining operations.[118]

With the advent of Peabody's increased and improved reclamation efforts, it remains to be seen whether the results of work done in 1979 will also improve. Erosion, as noted, is not a major problem on the site. Only time will tell whether the original diversity and productivity of the land can be achieved, or even increased.

The Black Mesa/Kayenta complex has been the scene of many experiments and reclamation studies by the University of Arizona and other researchers. However, some of the reclamation studies have been at least irrelevant (and at most, ridiculous), and although Peabody's reclamation staff in St. Louis continues to use the studies as proof that it experiments on the site, the new reclamation staff in Flagstaff and on the mine site considers most of the work "academic exercises with no practical applicability" and would prefer not to waste the company's time or money supporting further work that will not enhance actual reclamation practices.[119] Two such experiments have included one involving the culturing of African fish in a holding pond (see Wildlife) and another seeking to grow fruit trees and vegetables in graded mine spoils--termed an enhanced land-use project--launched in 1979. The University of Arizona had a 20-acre, bowl-shaped area graded into some mine spoils, then paved 5 acres of this land with gravel and asphalt. This area collects water, directing it to fruit trees and vegetables planted at the base of the bowl. Although Alten Grandt, Peabody's environmental affairs director in St. Louis, supported this project, telling INFORM that water harvesting was important because it had been practiced "since the time of the Israelis, Hebrews, and Iranians,"[120] Gary Melvin told INFORM in February 1980 that not only were the Navajo on the Mesa unreceptive to this form of land use, but it had no bearing on the company's reclamation efforts, was expensive, and Peabody would do better to aid the university in more relevant work in the future.[121] Experiments done by the University of Arizona on water runoff and erosion have been better received. Peabody will have to play a greater role in the development and implementation of on-site experiments that will be of real use to its reclamation program; it should have done so long ago.

WILDLIFE

Though sparse wildlife is to be found on Black Mesa, Peabody has not conducted extensive experiments that would benefit the wildlife that does inhabit it. Tree and shrub transplants have failed (see Trees and Shrubs), and the value of the grasses being seeded is only as good as the amount of forage produced. The company has begun a survey of all the wildlife to learn whether more than rabbits, assorted rodents, snakes and small birds are using the area.[122] According to residents, bears, foxes, coyotes and badgers are some of the other animals found on the Mesa.[123] Peabody told INFORM that if regulators insist, the company will construct rock shelters for wildlife.[124]

The water impoundments have provided some unexpected benefit to wildlife. According to both the company and local residents, migratory water fowl use these impoundments frequently, especially ducks and geese.[125] However, as noted earlier, the company will have to pay more attention to preventing contamination of these impoundments (see Toxic Material).

In 1975 the University of Arizona brought 110 warm-water fish from Africa--at the suggestion of the editor of Peabody's public relations magazine[126]--to study the potential for fish production on the site. Allowed to swim in a 4-cubic-foot cage lowered into a small impoundment on the site, the fish were fed a diet of cat food twice daily to supplement the diet of plankton found in the pond water. After 78 days, from July 1 to mid-September, the surviving 97 fish were removed, measured and weighed.[127] Although the experiment itself could be considered only an academic exercise--as these warm-water fish would never survive the winters on Black Mesa--Peabody followed up the experiment with an article suggesting that this work could lead to the creation of fishing ponds, providing local residents with the resources to enjoy America's third most popular pastime.[128]

RECLAMATION COSTS

TABLE B*

RECLAMATION COSTS AT BLACK MESA

Operation	Cost Per Acre	Percentage of Total Cost
Earthmoving		99%
Topsoil operations	$ 4,000-$ 6,000	
Grading	8,000- 12,000	
Revegetation	65- 85	negligible
Total Cost:	$12,065-$16,085	

*These costs and percentages are not strictly comparable to those of the same operation or activity at other mines. See the Cost of Reclamation chapter for a detailed explanation.

Source: Peabody Coal

McKinley/Pittsburg & Midway/NM

Mine: McKinley Mine

Location: Gallup (population 19,500), New Mexico, in McKinley County (population 55,000)

Operating company: Pittsburg & Midway Coal Mining Company

Parent company: Gulf Oil Corporation

Parent company revenues: 1977, $19.8 billion; 1978, $20.1 billion

Parent company net income: 1977, $468 million; 1978, $513 million

First year of production: 1959

Coal production in 1978: 2,993,000 tons

Acres for which a mining permit has been issued: NA

State regulatory authority: New Mexico Coal Surfacemining Commission, and Navajo Environmental Protection Commission

Reclamation bond posted: NA

Reclamation cost: reclamation cost per acre, NA; *reclamation cost per ton of coal,* NA

Expansion plans at mine: To reach full production of 5 million tons per year in 1979

Additional acres for which the company will seek a permit: More than 10,000 acres of land have been leased in the area. INFORM does not know the acreage for which the company will seek a mining permit.

Parent company U.S. coal reserves: Owns or leases 2.6 billion tons of coal reserves

*Development and expansion plans for surface coal mines in west-
ern United States:* Besides operating the Edna Mine in Routt
County, Colorado, Pittsburg & Midway is planning a new mine in
Campbell County, Wyoming, to produce 15 million tons of coal
per year by 1987.

NA = Not available

Sources: Gulf Oil Form 10-K, 1978; 1977 and 1979 Keystone Coal
Industry Manual; "New Coal Mine Development Survey 1978-1987,"
Coal Age, *February 1979; James S. Cannon,* Mine Control: West-
ern Coal Leasing and Development *(New York: The Council on
Economic Priorities, 1978)*

AUTHOR'S NOTE: Pittsburg & Midway, a subsidiary of Gulf Oil
Company, would not cooperate with INFORM on any aspect of our
research into the reclamation program at its McKinley Mine.
Pittsburg & Midway was the only one of the thirteen coal companies included in INFORM's sample which refused any discussion
of its program (see Methodology). It did so despite the fact that
its parent company, Gulf Oil, has broadcast the merits of this
program nationally. "Reclamation [at McKinley]," an ad in the
Wall Street Journal read, "will return the land to its original
productivity or better."

INFORM's report on the company is based on use of information
from state and federal government agencies, as well as interviews
with individuals familiar with the Pittsburg & Midway operations.
The format of this profile differs from others, due to our interest
in most clearly presenting and assessing the information that is
available and most clearly defining the many aspects of company
impact and performance which cannot be understood from this
scanty public data. The critical questions which would need to
be answered, if Pittsburg & Midway's reclamation progress is to
be given an accurate public evaluation, are listed.

SITE DESCRIPTION

    Located near the western border of New Mexico, in McKinley
County, and just east of Window Rock, Arizona, Pittsburg &
Midway (P&M)'s McKinley Mine is producing nearly 3 million tons
of coal annually from the rolling and rough topography common
in this part of the Southwest. Stands of sage and pine occur
along the rocky breaks, and sandstone outcroppings are common.
Highway 264, which runs through the middle of the mine, is the
only "access" for individuals not allowed onto the site itself.
    Although the growing season lasts 120-150 days each year,[1]

This is reclaimed land. The coal is out, and soon the sheep will be grazing here again.

"We need the land as much as we need the coal. We found a way to have both."

"For centuries, the Navajo have used this land for grazing their sheep," says Ben Sorrell, a Gulf Land Reclamation Supervisor. "Their whole livelihood depends on it, as it has for centuries.

Riches below

"But this land is some of the best coal-producing country in America. So the Navajo nation, which owns the land, leased part of it to Gulf's subsidiary, The Pittsburg & Midway Coal Mining Co.

"Now it's one of P&M's most productive mines. It's producing three million tons of coal a year,

Gulf Oil Corporation.

and it's being expanded to five million.

Riches above

"It's my job to put things back pretty much the way they were before mining started.

"As much as possible, we try to restore the character of the land, the general contours, and espe-

This is where I work: the McKinley Mine, near Gallup, New Mexico.

cially the drainage patterns.

"When we're finished, it's as good grazing land as it ever was — sometimes better.

"It's a real challenge, getting out the coal we need, without destroying the land, which we need just as much. I'm a Navajo myself, and I'm proud of the way Gulf is meeting that challenge."

Gulf people: meeting the challenge

from about the middle of May to October,[2] precipitation falls only
in certain periods of this growing season. Average precipitation
in this arid area is only about 14 inches, with 11 inches falling
as rain; however, the year-to-year variation from the average is
enormous[3] in both timing and amount.[4] Most rain generally falls
in July.[5] Though this is a high level of precipitation for the
Southwest, the problems of predicting when it will fall are difficult. This means that seeding just prior to the rainy season--
desirable for plant germination and growth--is difficult.

The land is primarily used by the Navajo Indians for the
grazing of sheep and as a general wildlife habitat. Much of the
area has been overgrazed in the past.

OVERVIEW

In 1978, the McKinley Mine produced almost 3 million tons of
coal, making it one of the 20 most productive mines in the United
States.[6]

According to examiners of the McKinley site, revegetation
has been somewhat successful. The responsible regulators (the
federal Office of Surface Mining, the Navajo Environmental Protection Commission, the U.S. Geological Survey, and the New
Mexico Coal Surfacemining Commission) report that a cover of
grasses exists and that they are generally satisfied with the
amount of growth.

According to 1977 and 1978 state inspection reports, the
company had an extensive shrub, tree, and grass planting program. However, New Mexico State University and U.S. Fish and
Wildlife studies indicate that plant diversity is low. The company
may not have tried to restore more than a handful (based on its
seed mixture) of the many species of native plants found in the
region (more than 100, according to a regional environmental impact statement by the Bureau of Land Management). The seeding
and amendment program is only fair. In 1978 company information
described its primary seed mix as almost exclusively grasses,
half of the plants being introduced species, a poor selection.
The U.S. Bureau of Land Management reports that research has
determined native plants are better for revegetation. The company is placing furrows along the contour of graded slopes to
help control erosion and to aid in retaining water for establishing
vegetation. Fertilizer and water for supplemental irrigation, as
well as straw mulch, are used in places at the mine; however,
INFORM does not know the extent of these programs. The U.S.
Department of Agriculture and New Mexico State University are
conducting research on the site, but the extent to which P&M incorporates the results of this work is not known.

Rill erosion (which is common in this region, and usually of minor consequence) is a recurring event, according to state and federal inspectors. Because of this, New Mexico and United States Geological Survey (USGS) inspectors believe the erosion and sediment pond provisions of the federal reclamation law are unnecessary in this region as rain is too infrequent to justify the level of control of other regions. It is hard to judge without actually seeing the extent to which erosion and sedimentation are affecting lands and drainages in the area.

According to both Office of Surface Mining (OSM) and state inspectors, P&M is salvaging and immediately replacing topsoil on graded lands, which is a good practice.

INFORM could find no reports of either major problems or major innovations in the areas of earth handling, wildlife, or day-to-day seeding operations.

A consulting firm in California--Earth Science Associates-- hired by the state regulatory agency, the New Mexico State Surfacemining Commission, has conducted sporadic visits to the site. USGS inspectors have also visited McKinley and have noted on different occasions good practices (soil preparation and early revegetative success) and bad (leaving steep slopes), but no enforcement actions have been taken. The Navajo Environmental Protection Commission has generally viewed favorably the reclamation done at McKinley, and OSM has offered little criticism.

EARTHMOVING

Grading

Latest figures available to INFORM indicate that approximately 1,500 acres of land had been disturbed by mining and related activities at McKinley by April 1978--average for a mine producing 3 million tons of coal annually in the West.[7] Plans called for disturbance to increase to 300 acres annually beginning in 1979 or 1980.[8]

Grading apparently follows mining fairly closely. Jack Reynolds, an inspector for the New Mexico State Surfacemining Commission (SSC), told INFORM in May 1979 that the company has kept current within the three-spoil-ridge standard required by the state.[9] Grading at the mine is generally good, although reports have noted that rather steep slopes are left on some lands. USGS has reported slopes of 35-38 degrees on the outsides of boxcut spoil piles,[10] but the New Mexico SSC has been fairly satisfied with the grading. It has also noted slopes steeper than 18 degrees,[11] which may make operation of reclamation machinery on the contour difficult, if not impossible. OSM has evaluated

grading on the site as acceptable. In May 1979 it reported that
no problems exist.[12]

Questions

  •How much land has been disturbed at the mine site since
   mining began (including lands disturbed by roads, buildings, sediment ponds, stockpile areas, etc.)?

  •How much land has been graded?

  •Will boxcut spoils be graded to more shallow slopes or left as
   they are? If they are not to be graded, how will this affect
   the pattern of water flow and drainage on and off the mine
   site?

  •Are all graded slopes traversable by reclamation machinery?

Topsoil

     There is little topsoil--good organic soil--to be found in the
Southwest, but surface materials suitable for plant growth are
salvaged at McKinley. When found in the gullies and water drainages on the mine, topsoil is taken and redistributed over graded
lands.[13] However, a misunderstanding between OSM and P&M
resulted in the company's failing to recover all topsoil from lands
being disturbed, removing instead large deposits of the material
from arroyos and alluvial fans in the valleys on the site while
neglecting to take soils from other lands.[14] This situation must
be resolved between the company and regulator. A report by
state inspectors in April 1978 indicated that much of the topsoil
being salvaged is directly replaced on graded lands without being stockpiled.[15] On OSM's first visit to the mine in May 1978, it
also found that topsoil is directly hauled back, not stockpiled--a
good practice.[16]
     Although topsoil is being redistributed on graded lands at
McKinley, the extent of the effort is not clear. However, an inspection by state regulators in October 1977 found that more
than 12 inches of topsoil was being respread.[17] Because topsoil
is apparently being hauled directly onto graded lands, no information on stockpiles, if any, has been found.

Questions

  •How much topsoil--or suitable plant-growth material--is being salvaged, and from where?

  •What percentage of this material is being replaced immediately on graded lands, and how much is stockpiled for later
   use?

•If extra soils are found in water drainage areas, is all the
material used, or is some stockpiled? If all is used, will
there be a deficit in soil material when the final graded lands
are prepared for revegetation as the mine closes?

•What measures are taken to protect stockpiled materials from
erosion and loss? How long is material kept in stockpiles?

Toxic Material

INFORM found no information on the presence or absence of
either acidic or sodic materials on the site. However, it is fairly
common in the West for earthen materials to have high levels of
salts. The extent to which this presents a problem at McKinley
is unclear.

Questions

•What acid or sodic materials are found on the site, or in the
overburden? What measures are taken to prevent these materials from harming water supplies, or vegetation?

•Do coal fires occur in ungraded spoils, or in open mine pits?
How are these situations addressed?

WATER QUALITY

Surface Water and Groundwater

Although P&M has some sediment or holding ponds at McKinley, according to state officials it does not discharge water from
these ponds and has no permit to do so under the National Pollutant Discharge Elimination System (NPDES) requirements.[18]
Water monitoring is being done with seven groundwater
wells, and three surface water stations have been established.[19]
Although Jack Reynolds at the New Mexico SSC reports that
groundwater is not found to be entering mine pits from subsurface aquifers, information is insufficient on which to base an
analysis of the effects of mining upon water resources in the
area. Harris Arthur, formerly of the Shiprock Research Center
and now with the Bureau of Indian Affairs, says the site has
four small aquifers, which are tapped by local residents for domestic and stock uses.[20]

Questions

•Is any water now being discharged from the mine?

• Do sediment ponds consolidate all the water running on the mine before discharge?

• How many sediment ponds have been built on the site?

• Does P&M plan ever to discharge water from these ponds?

• If no discharge is planned, what effect will this have on lands downstream that may rely on natural runoff for vegetative growth or wildlife use?

• What are the results of water monitoring conducted in the area?

• What effect are mining and reclamation having upon shallow aquifers used by local residents?

Alluvial Valley Floors

The Bureau of Land Management reports no alluvial valley floors exist on the site.[21]

Questions

• Do any of the small water-drainage or wash areas constitute lands which might qualify as alluvial valley floors?

Erosion

As in many parts of the arid Southwest, erosion is a problem, as hard, dry soils do not allow water to infiltrate as fast and vegetation is not lush. Water runoff tends to scour the surface, eroding it and carrying sediment downstream. Although Jack Reynolds told INFORM erosion was not a problem--by his definition--at McKinley,[22] his view does not accord with other reports. Rill erosion is common on the site,[23] as it is on undisturbed lands. But excessive gullying was noted by USGS inspectors in April 1978[24] and state regulators in the same month.

The state noted that gullies 12-18 inches deep were found on lands that had been graded and seeded only one year earlier.[25] Al Taradash, an attorney for People's Legal Services in Window Rock, reports significant erosion on the hillsides at the mine,[26] and OSM has reported related sedimentation problems as recently as May 1979.[27]

Questions

• How serious is erosion on the site?

• Is sediment carried off the mine?

•What measures are taken to reduce erosion, and how quickly are they implemented when needed?

•How effective is erosion control as practiced?

SEEDING AND AMENDMENTS

Seeding Procedures

By February 1978 P&M had seeded approximately 70 percent of all the lands disturbed at McKinley, a good percentage. Yet much of this land is being re-mined and will have to be revegetated from scratch once mining is complete.[28] However, topsoil had not been replaced on much of this land, which suggests that it would not have shown good productivity in the future.[29]

Reports from April 1978 indicate that planting is generally done in early and mid-summer to take advantage of late summer rains.[30]

The 1978 seed mixture located by INFORM shows that P&M used a poor mixture of introduced and native grasses with neither forbs nor shrubs being seeded.

It is noteworthy that McKinley relies heavily on introduced species, in spite of a 1978 Bureau of Land Management report, based on extensive research in the Southwest, indicating that native plant species offer the best hope for revegetation in the area.[31] INFORM does not know how much of the research that apparently is taking place on-site is incorporated into the mine's reclamation programs. The U.S. Department of Agriculture and New Mexico State University are both conducting reclamation research on the site.

Questions

•How much land has been seeded in total?

•How much of the land that has been seeded has had topsoil replaced upon it?

•Are experiments conducted on the site to determine the exact time when seeding should be done? How does P&M plan its seeding, given the unpredictable weather conditions?

•What is the composition of the seed mixture now in use?

•How has the composition of P&M's seed mixture changed?

•What experiments are used to modify or supplement the basic seed mixture?

TABLE A

1978 SEED MIXTURE
FOR LANDS BEING RECLAIMED AT McKINLEY

Plant Species	Introduced/ Native	Cool/ Warm	Pounds Applied Per Acre
Grasses			
Crested wheatgrass	I	C	3.0
Slender wheatgrass	N	C	2.5
Siberian wheatgrass	I	C	2.5
Tall wheatgrass	I	C	2.5
Western wheatgrass	N	C	3.0
Smooth brome	I	C	3.0
Pubescent wheatgrass	I	C	3.0
Mountain muhly	N	W	4.0
Sand dropseed	N	W	4.0
Streambank wheatgrass	N	C	4.0
Spike muhly	N	W	4.0
			35.5 Pure live seed
Legumes			
Yellow sweetclover	I	W	2.5
			2.5 Pure live seed

Source: Annual Reclamation Report Permit #3, April 27, 1978

Amendments and Maintenance

P&M is placing furrows on the slopes along the contour to retain water and to slow erosion.[32] Leonard Robbins of the Navajo Environmental Protection Commission praises the contour - furrowing, observing that the furrows allow seeds to germinate - better and that they eventually smooth out as they are filled with sediment.[33]

Reports by P&M in April 1978 indicate that the company uses both drill seeders and less precise broadcast seeders.[34] The company reported in 1978 it has a regular program of reseeding and interseeding lands one to two years after their initial seeding.[35]

Barley straw was used as mulch on the site in 1977 and 1978 and was applied at a rate of 2.5 tons per acre after the completion of seeding.[36] However, David Scholl of the Rocky Mountain Forest and Range Experiment Station wrote P&M in December 1977 that the station had found that wood chips produced better growth in test plots than either straw mulch or no mulch. The chips, he wrote, also apparently allowed for better moisture retention.[37] When mulching with straw, the company tacks it to the ground.

Irrigation and fertilization are being practiced at McKinley.[38]

Questions

- Is reclamation machinery operated along the contour of slopes?

- How much drill seeding and how much broadcast seeding is done at McKinley?

- Is the soil disced prior to seeding in all cases?

- Are seeds covered with soil after their placement?

- Is drill seeding used only for reseeding and interseeding?

- What are the results of regular reseeding and interseeding upon long-term growth? Is it done because initial growth is always poor? Is a different seed mixture used in these cases?

- Are wood chips used for mulch, based on research conducted in 1977?

- Are any new mulches being used? Have experiments been conducted to determine the best mulching technique? Has P&M ever hydro-mulched?

- How much fertilizer is applied to the ground, how often, and what type? Why is fertilizer used, and what experiments or tests have been done to determine that fertilization is necessary?

- How much irrigation is conducted? Have any experiments been done to prove the need for this treatment, and the effects upon long-term growth and plant community stability once the treatment is ended?

- When is most of the water applied?

- What is the effect upon plant communities when irrigation or fertilization treatments are terminated?

Trees and Shrubs

Although tree, shrub and grass transplants have been conducted at McKinley, very little is known about the success of this work, and reports are conflicting. Containerized transplants have been used at McKinley, according to reports from 1977 and 1978.[39]

The importance of successful tree and shrub transplants is great, as these plants are necessary to stabilize the land and to provide grazing forage when winter snows cover the ground.[40]

Questions

- What is the extent of tree, shrub and grass transplants at McKinley?

- What have been the results of tree and other transplants on the mine? What species are planted, and which survive best?

- What are the cost implications of extensive transplantation of shrubs, grasses and trees?

VEGETATION

OSM, the Navajo Environmental Protection Commission and the state regulators told INFORM that P&M has achieved adequate plant cover and growth on most reclaimed lands at McKinley.[41] From the vantage point of a public road bisecting the mine, IN- FORM saw both a dense cover of grasses on parts of the mine, and weed growth. OSM and USGS attribute recent reclamation success in part to exceptionally wet weather in 1978 and 1979.[42] Yet an October 1977 vegetation survey performed for P&M showed that surrounding undisturbed lands had twice the cover of re- claimed lands, and that productivity on reclaimed lands was only two-thirds as good. The cover was found to be 97 percent grass (by weight), whereas undisturbed lands are balanced by grass, shrub and forb vegetation.[43] This dominance by grasses does not provide as good forage as undisturbed lands covered with a more diverse plant community. These findings are further cor- roborated by the U.S. Fish and Wildlife Service, which found that dominant plants included crested, western and slender wheatgrasses.[44]

INFORM could locate no more information on the success or failure of reclamation at McKinley. However, Gulf Oil's full-page advertisements in the *Wall Street Journal* and other newspapers and magazines assert that reclaimed land is as "good grazing land as it ever was--sometimes better."[45] From our information, this is highly misleading.

Questions

- Are further studies on productivity being done?

- Are desirable plants that have not been seeded invading re- claimed lands?

• What is the productivity on lands reclaimed long ago?

• What are the long-term results of irrigation and fertilization upon productivity once the treatments are stopped?

• Are there any plans for trial grazing of reclaimed lands?

WILDLIFE

INFORM has no information on P&M programs (or lack thereof) for returning wildlife to the site. Numerous mammals, birds and reptiles are known to live in the region.[46] Plants being grown on the site are generally good for grazing livestock.

Questions

• What work is P&M doing to benefit wildlife?

• Are water holding ponds to be used for wildlife?

• Will these ponds contain water throughout even the driest portions of the year?

• Are any wildlife shelters being built?

RECLAMATION COSTS

William Beck, of the Bureau of Indian Affairs in Window Rock, Arizona, told INFORM that P&M spends about $5,000 per acre on reclamation.[47]

INFORM has no other information on costs at McKinley.

Questions

• What is the cost per acre for the following operations:

--Topsoil removal
--Topsoil replacement
--Grading to approximate original contour
--Seeding
--Creating contour furrows
--Fertilizing, mulching and irrigating
--Reseeding and interseeding

• Are there other reclamation expenses?

• What is the total cost of reclamation per acre?

• What is the total cost of reclamation per ton of coal produced?

Navajo/Utah International/NM

Mine: Navajo Mine

Location: Farmington (population 40,500), New Mexico, in San Juan County (population 78,000)

Operating company: Utah International, Inc.

Parent company: General Electric Co.

Parent company revenues: 1977, $17.9 billion; 1978, $20.1 billion

Parent company net income: 1977, $1.1 billion; 1978, $1.2 billion

First year of production: 1963

Coal production in 1978: 6.2 million tons

Acres for which a mining permit has been issued: 31,416

State regulatory authority: New Mexico Coal Surfacemining Commission

Reclamation bond posted: None (not required)

Reclamation cost: reclamation cost per acre, $5,030; *reclamation cost per ton of coal,* NA

Expansion plans at mine: none

Additional acres for which the company will seek a permit: none

Parent company U.S. coal reserves: Owns or leases 1.3 billion tons of coal reserves

Development and expansion plans for surface coal mines in western United States: Utah International operates the San Juan Mine in New Mexico (2,613,030 tons in 1978) for the Western Coal Company, and the Trapper Mine in Colorado.

NA = Not available

Sources: Utah International, Inc.; General Electric Form 10-K, *1978;* 1979 Keystone Coal Industry Manual; *"New Coal Mine Development Survey 1978-1987,"* Coal Age, *February 1979; U.S. Department of the Interior, Bureau of Reclamation,* Proposed Modification to the Four Corners Powerplant and Navajo Mine, New Mexico--Final Environmental Statement, *1976, pp. 1.32-1.33*

SITE DESCRIPTION

The Navajo Mine of Utah International occupies a site in San Juan County in northwest New Mexico. It stretches over 20 miles in length and one mile in width, south from Fruitland, New Mexico, between Shiprock and Farmington.

Navajo is in the San Juan basin, a circular area 5,300 feet above sea level, ringed by mountain ranges and uplands. Forty miles to the west, in Arizona, dense pine forests cover the western slopes of the Chuska, Carrizo, and Defiance ranges. But the mine area lies to the east of the peaks, which create a rain shadow. The pines thrive on water denied to the San Juan basin.

The climate at the Navajo Mine is the least hospitable to reclamation of any in the INFORM sample. An environmental statement by the Bureau of Reclamation characterizes the site as a "dry cold desert area, with evaporation exceeding precipitation. Twelve-year precipitation records show average precipitation of 6.13 inches, median 5.77, minimum 2.25." Rains are irregular. The year 1972 saw seven consecutive rainless months.[1] Half the precipitation usually falls in thundershowers in August-October; one fourth as winter snow.[2] But in 1978 no rain fell from May 15 to September 15.[3] Precipitation is somewhat less at the south end of the leasehold.[4]

The San Juan River passes to the north of the mine. A narrow band of lush irrigated farmland borders the river. But where the farmland ends and the Utah International leasehold begins, the contrast is stark. To the west and parallel to the mine runs the Chaco River, an intermittent stream that is dry 90% of the time. Several ephemeral arroyos run from east to west into the Chaco through the leasehold. They are dry except during rainstorms or snowmelts.[5] No alluvial valleys or major aquifers exist on the permit area.[6]

The on-site terrain is generally level to gently sloping westward. Many sandstone outcrops, shale flats, eroded washes and arroyos, and rock terraces created by continual wind erosion exist in the area.[7]

The land appears barren to untutored outsiders, but it pro-

vides sustenance to Navajo Indians, who raise sheep. It is heav
heavily overgrazed, especially in the northern part of the leasehold.[8]

OVERVIEW

One of the oldest of the large surface mines in the West, Navajo crystallizes the most difficult issues surrounding coal development on Indian lands. With its poor soils and minimal rainfall,
it has proven to be one of the most difficult sites in the United
States to reclaim--perhaps even more difficult than Peabody's
Black Mesa-Kayenta mine (see Black Mesa-Kayenta profile). The
nearby Four Corners power plant, which operates on coal mined
from Navajo pits, is one of the largest--and was one of the dirti-
est--generating stations in the world.

Reclamation at Navajo did not begin until the early 1970s
when New Mexico passed a law requiring it--and Utah International is only now in the slow process of trying to reclaim old
spoil areas. To date, the company has achieved moderately successful growth of some plants on reclaimed lands--but only with
heavy maintenance by the company, which could bode ill for the
post-reclamation picture.

Moreover, Utah International has been forced to develop its
reclamation policies in the dark; it had no precedents to follow
for such a difficult region when it began operations in New Mexico. The company has a continuing research program to seek answers to such plaguing questions as the best irrigation strategy,
plant mixes, and soil chemistry and micro-organisms. It continues to revise its program as research results dictate.

But even with continuing research, the overriding question
about reclamation at Navajo--to an even greater degree than at
other western mines--is whether it will stand the test of the future. Heavily maintained lands, irrigated and held back from
grazing, do not accurately forecast the state of the land once
normal land use has resumed.

Apart from the harsh constraints of nature, the company's
reclamation program itself has major shortcomings. Utah International has much to do before it can claim to be getting the most
from the land for the Navajo who will use it after reclamation is
deemed complete.

Lack of plant diversity on reclaimed lands is a central problem. The company is seeding a limited range of species, and two
or three of these have grown much more successfully than others.
Sterling Grogan, the company's senior environmental engineer,
suggests that it begin experiments in interseeding (adding additional seed one or more years after initial seeding). It might
also consider experimenting with higher seed application rates

than the 12 pounds per acre it now employs.

Utah International has found it necessary to use a major irrigation program for initial plant establishment. Much of the irrigated land has been weaned from the water for up to several years--and although post-irrigation cover has decreased, there has been no evidence of major diebacks. Cover is roughly equivalent to that on unmined land. But grazing will be the ultimate test, and there remain many years before the reclamation program can be judged a success or failure.

The lack of rain diminishes the importance of erosion and other water problems at Navajo. Erosion occurs during the infrequent thunderstorms, but regulatory agencies familiar with the site do not see this as a serious problem. The company has no water diversion or sedimentation system; it relies on internal drainage and the lack of moisture to prevent off-site consequences.

Current mining shows few recorded grading or topsoiling problems. But Utah International is still trying to grade many old mine spoils from earlier operations; these spoils contain little topsoil, if any. Indeed, it is questionable even to label as "topsoil" the surface soil found in the region, as it is low in organic matter and there is so little of it. Much of the poor topsoil was lost in the first decade of mining, and it is unlikely that the land will benefit much from the little that remains.

Utah International has not made extensive efforts to promote wildlife, which exists but does not abound in the area. Early studies show that wildlife activity is greater on reclaimed land than on unmined land.

Relatively meager inspection activity has been conducted by the state of New Mexico and the Navajo tribe at the mine, and increased oversight by the federal Office of Surface Mining (OSM) will be helpful. In the meantime, OSM has noted its own concerns that long-term reclamation success is still not guaranteed on the site.

The company may show, with much experimentation and innovation, that it can make the desert bloom. But it is just as likely that wretched lands long ago disturbed and rendered useless by coal mining will be impossible to reclaim.

EARTHMOVING

Grading

Utah International did not grade spoils at all from the opening of the mine in 1963 until New Mexico passed its first reclamation law in 1973. The company has graded about 100 acres more than it disturbed in each of the last three years.[9] Currently active pits are graded to within one spoil ridge, and the company plans to finish the backlog of grading at inactive pits by 1985--

much later than the company's original plans to catch up by
1977.[10] By January 1980, 66 percent of the 5,573 acres of land
disturbed by mining operations had been graded.[11] That record,
though better than the average for mines in the INFORM sample,
still leaves more than 1,850 acres of land in a disturbed state.

OSM inspectors found that Utah International's final grading
is being done along the contour, except in areas where slopes
are too steep for the equipment.[12] Outslopes of spoils are graded
to a steep 3:1, interior slopes to a much gentler 10:1--travers-
able by reclamation equipment.[13] The topography is gently rolling, with drainage to the west as it was before mining.[14]

Utah International has created more than a dozen water impoundments, although only one was reported to contain more
than one-half acre-foot of water in 1978.[15] Orlando Estrada,
Utah International's acting senior environmental engineer, told
INFORM three or four water holding ponds will remain on the site
after mining; the Navajo want them for watering their sheep.[16]
But Sterling Grogan told INFORM earlier that OSM was reluctant
to approve this arrangement,[17] and further study is warranted
(see Black Mesa-Kayenta profile).

Utah International told INFORM no subsidence has occurred
on areas graded since 1976,[18] but graded areas have subsided
and created small depressions in some areas of the mine. The
U.S. Geological Survey (USGS) noted depressions greater than
one cubic yard on two inspection trips in 1978.[19] Earth Science
Associates, which inspected Navajo for the New Mexico Coal Surfacemining Commission for several years, noted uneven settling
in one of the mine pits, with many cracks and open fissures, in
1977.[20] These depressions and cracks--though small at the out-
set, could enlarge and create a greater obstacle to final reclamation success and land stability.

Topsoil

Owing to the combined lack of state regulation and company
initiative, topsoil was not saved in the first 10 years of mining,
before enactment of New Mexico's surface mining law. Richard
Hughes, an attorney for the Navajo People's Legal Services
(DNA), said some of the best topsoil on the leasehold was lost in
these years.[21] Thus little material will be available for spreading on lands now being graded from old mining operations. Although the remaining material is not the highly productive topsoil found in parts of the Midwest and is even less productive
than most soils in the West, it is the best medium found so far
for plant growth. This failure to retain topsoil is probably the
company's greatest reclamation error--with the exception, perhaps, of the decision to mine here in the first place.

What little good topsoil existed at Navajo has eroded away

after years of overgrazing. The remaining material is often
sand.[22] Because surface soils have the consistency of dirt or
sand, the distinction between an upper layer of topsoil and a
subsoil layer is vague. Utah International currently removes the
upper layer of material--anywhere from 9 inches to 4 feet--down
to the depth of stratified calcium carbonate layers.[23] Company
scientists have found that this "upper mantle material" is suitable
for reclamation.[24] Utah replaces it upon reclaimed areas in a
uniform thickness.[25]

Utah International generally hauls the upper mantle material
directly from mined areas to reclaimed ones, a good practice.[26]
Sometimes this material is stored in fenced areas prior to replacement. OSM has reported that most of this soil is used soon
enough for seeding not to be required.[27]

Toxic Material

OSM and New Mexico inspectors say no acid or toxic materials
are present on the Navajo site, although some areas are alkaline
or saline.[28] It is still unclear what effect this salty material will
have on vegetative growth once irrigation is ended and vegetation is left to grow unaided. Different theories abound as to
whether continued irrigation only washes the salts down through
the soil layers until it is terminated, which then allows the salts
to rise to the surface, or whether the salts will become fixed.
Yet, because the artificial climate created by irrigation is so different from that which prevails, it may be a long time before the
effect of this combination of factors can be adequately quantified
(see Surface Water and Groundwater).

Coal fires are common. INFORM observed smoke from at
least one such fire on a June 1979 site visit.[29] Utah International
does not mine all of the eight coal seams on the leasehold, and a
good deal of waste coal finds its way into the mine spoils. Here,
as at the Peabody mine at Black Mesa, fires inevitably result.
The company says it has tried to stop them with burial and with
chemicals.[30] This has not eliminated the problem. Bare spots
also have been found on reclaimed land; Utah International says
they are caused by waste coal smoldering beneath the surface
and preventing plant growth.[31] There may not be a solution to
this problem, in which case continued mining may create additional problems.

WATER QUALITY

Surface Water and Groundwater

The low amount of rain and the flat topography make water
handling issues less important at Navajo. Drainage is largely in-

ternal; reclaimed areas slope west into the mine pits, and water does not run off the site. The U.S. Geological Survey reported in February 1978 that "Because of the reverse slopes, Utah International does not feel that sedimentation ponds are required."[32] New Mexico does not require the mine to meet the 10-year 24-hour sediment pond criteria of OSM regulations; as state inspector Jack Reynolds said, "We just really do not have a whole lot of water."[33]

The site contains a number of small impoundments. None of them discharges off the mine site; when water collects, it is used to water roads to hold down dust, and only one stream channel diversion has been constructed, in Chinde Wash.[34] Utah International did, however, obtain a discharge permit from the Environmental Protection Agency in March 1977.[35] The company discharged water from the northern area of the mine into the reservoir of the Four Corners power plant for several months that year. Discharge levels were in a range of 500-4,500 milligrams per liter (mg/liter) suspended solids, far above the permit limit of 70 mg/liter. The company stopped the discharges and, in August 1977, told EPA it was considering installing a treatment system,[36] which it did not pursue further.

Scarce water is a major issue for the Navajo in the area surrounding the Four Corners power plant and the Navajo Mine. Congress established in 1964 a program whereby the Navajo tribe can irrigate 110,000 acres of land with water from the San Juan river, which runs through the area. At present, approximately 25,000 acres is being irrigated for the production of grasses, vegetables and grains. Congress granted the Indians 508,000 acre-feet of water for the irrigation project, while New Mexico may draw only 727,000 acre-feet. Moreover, the Four Corners power plant uses about 20,000 acre-feet of water annually, but it wants to double its present consumption for additional cooling and use in wet-scrubber (flue-gas desulfurization) systems; a coal-gasification plant has also been planned.[37] Thus the careful monitoring of water use and disturbance is essential. (Irrigation at the mine is negligible in comparison to these uses.)

The USGS reported in 1978 that Utah International had a system of 38 wells and 13 surface runoff stations[38] to monitor groundwater and surface water. However, the company discontinued its program of monitoring in May 1978. Sterling Grogan says New Mexico inspectors told the company to discontinue its program.[39] But LeRoy Balzer of the company's headquarters planning staff wrote that "Utah International told New Mexico that UI would discontinue its monitoring program until the permanent regulations for Indian Land were finalized."[40] In any case, in May 1979 OSM issued a notice of violation to Utah International for the discontinuance.[41] Utah International told IN-

FORM that OSM cited the wrong regulations when it issued the notice.[42] INFORM could obtain no results from monitoring conducted during the period when the program did operate, and the issue of monitoring the effects of mining upon water resources in the area should be pursued.

Alluvial Valley Floors

According to OSM, no alluvial valley floors exist on the mine site.[43]

Erosion

Jack Reynolds of the New Mexico Coal Surfacemining Commission says offsite sedimentation is no problem at Navajo.[44] Reclamation director Sterling Grogan told INFORM minor erosion may occur where the boxcut spoils are adjacent to the permit boundary. However, he believes sediment would not reach nearby waters.[45]

Erosion probably occurs during rainstorms. The minesite soil is unconsolidated, and vegetative cover is sparse. But circumstances at Navajo are far different from those at, for example, Powder River Basin mines, where rainfall is 20-50 percent greater. Erosion of unmined land in the region is also quite likely high during thunderstorms. Streams in the region are known to carry heavy sediment loads after storms. As for repairing erosion, the USGS reported in February 1978 that "Utah International now removes all significant rills and gullies."[46]

SEEDING AND AMENDMENTS

Seeding Procedures

The dominant characteristic of the revegetation program at Navajo is the degree of experimentation and irrigation being conducted. Utah International has undertaken a wide variety of research projects to find how best to reclaim its very difficult mine site. Much room remains for improvement in its program. Too much fourwing saltbush appears in the seed mixture and on reclaimed areas; mixtures could be varied by soil type; and the number of species seeded is limited. But generally, the company has made progress in devising a program to suit this unique site. Of course, the long-term results remain unknown (see Vegetation).

The company began a 10-year revegetation "acceleration" program in 1975 to reclaim the spoils untouched in the first 10 years of operation, before New Mexico passed its surface mining law. Utah International seeds approximately 540 acres per year,

100-200 more than it disturbs.[47] However, in 1979 only 322 acres
of land was seeded.[48] By January 1980, 2,496 (45 percent) of
the 5,573 acres of land disturbed at Navajo had been seeded.
Yet topsoil has been replaced on only 598 acres of this land.[49]
Utah International conducted its seeding in the spring several
years ago, then changed to July after experiments showed that
this corresponded better with the timing of natural precipitation.[50]

The current seed mix is native to the United States. Sterling Grogan said this was a practical choice: in early planting
trials in 1972, the company found that introduced species germinated quickly but were soon outcompeted by native plants.[51]
This contrasts with the experience of mines in damper regions,
where introduced wheatgrasses and legumes are common and
flourish readily. Wheatgrasses, alfalfa and sweetclover do not
grow well in this region, although the company tried to plant
wheatgrasses in the past.

In 1977 the company used three different seed mixes, according to soil type and texture.[52] It no longer follows this practice,
although surface material varies from sandy to shaley.[53] Utah
International feels such variations are unnecessary and not "efficient."[54] Slope and moisture do not vary in reclaimed areas,

TABLE A

1979 SEED MIXTURE
FOR LANDS BEING RECLAIMED AT NAVAJO

Plant Species	Introduced/ Native	Cool/ Warm	Pounds Applied Per Acre	
Grasses				
Galleta	N	W	1.0	
Indian ricegrass	N	C	1.0	
Alkali sacaton	N	W	1.0	
Sand dropseed	N	W	0.5	
Giant dropseed	N	W	0.25	
			3.75	Bulk seed
Forbs and Shrubs				
Fourwing saltbush	N	W	4.0	
Shadscale saltbush	N	W	1.0	
Rubber rabbitbrush	N	C	1.0	
Mormon tea	N	W	1.0	
Winterfat	N	C	1.0	
			8.0	Bulk seed

Source: Utah International

and the company told INFORM its current mix is adequate for all
of the soil types encountered, as separate mixes have been combined.[55]

Fourwing saltbush accounts for fully four of the almost 12
pounds of seed mix. This may be excessive; OSM has criticized
the company for the dominance of this species on reclaimed lands.
The overall seeding rate is lower than at most mines in the IN-
FORM study. Utah International should consider experimenting
with higher application rates. In 1978 the company was developing its own stock of rare plant seeds.[56] Sterling Grogan told
INFORM that despite past difficulty getting the right seeds, they
are now available despite tight specifications Utah International
places on suppliers.[57]

Utah International found soon after it began reclamation that
existing literature was generally inapplicable to the mine's particularly harsh climate. The company began limited revegetation
research in 1969, and major test plots--the first in the Southwest, it claims--in 1972.[58] A variety of research projects is under way. The 1976 environmental statement for Navajo lists 15
projects.[59] The 1978 annual reclamation report lists seven
more.[60] During the INFORM site tour, Grogan frequently cited
research results when asked why the company followed certain
practices. The research program thus appears to be both elaborate and intergrated with the revegetation program.

Amendments and Maintenance

The company told INFORM it seeds all lands at the mine on
the contour except on the outslopes of graded areas where slopes
are 3:1.[61] Dike-like structures are also created on slopes to
prevent erosion and to trap water from the irrigation system.[62]

The company uses a rangeland drill and a farm drill in series.
The latter drill plants smaller seeds that a range drill does not
plant well. Both have been modified so that they are, in effect,
broadcast seeders. The company found that seeds planted less
than an inch deep germinate better than those planted at drill
depths; it made the modification, and chains are dragged after
the seeders to cover the seeds with soil. Finally, galleta is
hand-seeded.[63] Utah International notes that seeding application
rates are "erratic," with sometimes more than $1\frac{1}{2}$ times the desired amount applied to a given area of a revegetation plot, owing
to the inaccuracy of the machines.[64]

The company mulches with straw at one ton per acre, but
loses half of the mulch to the wind, even though the mulch is
crimped into place.[65] Utah International told INFORM that on-
the-ground results show mulch does not improve reclamation results, and is continued only to appease regulators.[66] However,
a 1978 company report to the New Mexico Coal Surfacemining

Commission said preliminary study results showed mulch greatly
increased plant growth.[67] Tests of the effectiveness of mulch
continue today.[68] Soil tests showed that fertilizer was unnecessary. The company has tried to use coal as a fertilizer, but
found it had no effect on plant growth.[69]

Utah International has a major irrigation program, which
could prove to be either the blessing or the bane of its total reclamation effort. Revegetation tests in the early 1970s showed a
need to irrigate if revegetation were to succeed. Early revegetated areas were irrigated with 12 inches of water in the spring,
and $2\frac{1}{2}$ inches more the following year. The company has since
decided that irrigation should approximate the average amount of
annual precipitation, that spring irrigation promoted growth of
tumbleweed (Russian Thistle), and that warm-season species
needed more help to germinate. Irrigation is now applied at 8
inches during the July following seeding, immediately before
summer rains. An additional $2\frac{1}{2}$ inches is added the following
spring.[70] Still, the amount of water made available to the plants
remains more than double that which will be available when the
treatment is ended.

The company fences reclaimed areas as part of its long-term
maintenance plans. It believes adequate long-term productivity
can be attained only if the land is fenced into grazing management units. These fences have been controversial. Sterling
Grogan told INFORM in June 1979 that in establishing such a
system, "the cutting edge has to be the feelings of the local residents." The company has held a series of meetings with grazing
permit-holders to explain its program and the need to prevent
overgrazing. Officials say that vandalism of the fences and pre-
mature grazing of reclaimed areas have been minimal--although a
1978 Geological Survey inspection report includes a photo of
sheep grazing on reclaimed land.[71] It attributes this success in
part to the company's education program, and in part to the
shape of the mine: Navajo is long and narrow, and has taken
only a small part of the grazing permit area of each affected
rancher. This experience contrasts with the fencing battles between Peabody, at its Black Mesa-Kayenta complex, and the Navajo of Arizona--although the shape of the Navajo Mine is probably due more to the geology of the coal deposit than to fore-
sighted planning.

Trees and Shrubs

The Navajo leasehold had no trees and a few shrubs (less
than 5 percent of plant canopy cover) before mining.[72] The company has experimented with sage-hybrid transplants, but found
they did not grow well without the deep soils they need. It had
limited success with containerized plantings of fourwing saltbush

--but direct seeding of this plant has produced it in overabun-
dance. No regular transplant programs are conducted, and the
company's reclamation program is poorer for it.[73]

VEGETATION

Vegetative cover on reclaimed lands is moderately good,
though mixed depending on the site. Vegetation has not completely died once irrigation has terminated, but OSM's regional
director called the mine a "biculture" of fourwing saltbush and
alkali sacaton.[74] Utah International has had more success with
these species than with others. Species diversity is on the low
side, though unmined land was less diverse than at other IN-
FORM mines. The company has not had a problem of overabun-
dance of cool-season wheatgrasses and legumes, as have most
other mines, because these plants do not grow as profusely in
the arid conditions of the region.

To determine the type of vegetation that existed on the site
before mining, Utah International engaged consultants to study
16 reference areas from 1973 – 1976. It found that plant cover
ranged greatly, from 1 percent to more than 40 percent--and,
with additional ground litter, sometimes as high as 47 percent.
Vegetation is sparse and grasses dominate, although a variety of
forbs and shrubs is found in the area.[75]

According to Utah International, studies by New Mexico State
University at Las Cruces found that the best land on the site
could support one animal unit (one cow and calf or five sheep)
for one month on between 8 and 22 acres. For the least productive land, 35-118 acres are needed to graze an animal unit.[76]
Thus all but the best land at Navajo is less productive than most
of the worst lands in the Northern Plains. Nevertheless, even
this poor land is critically important to the local inhabitants.

Reports by Earth Science Associates (ESA), a consulting
firm that inspected Navajo for the state for several years, record
a mix of fair to poor revegetation success on seeded lands. ESA's
1977 and 1978 reports on lands seeded in 1973-1976 vary. Earlier
seeded vegetation died off in a prolonged drought in 1975-76 but
came back with a cover of fourwing saltbush and Indian ricegrass. Fourwing saltbush, alkali sacaton, sand dropseed, western
wheatgrass, and Russian thistle (tumbleweed) were reported on
1975 seeded lands. Covers were poorer on heavily textured spoil
sections, better in medium and light spoils. Lands that had been
planted only the year before ESA's observations generally had
scant plant covers, other than the weedy Russian thistle,[77] although one area seeded in 1977 had a cover of fourwing saltbush,
alkali sacaton, and some galleta and Indian ricegrass by the
spring of 1978.[78]

The U.S. Geological Survey and Fish and Wildlife Service observations from 1977 are similar. The USGS observed of 1976 seeding that "The best vegetal growth occurs on medium spoil without topsoil and bottom ash [used as fertilizer]. Fourwing saltbush, rubber rabbitbrush, and alkali sacaton are the harti-est species of vegetal growth. Western wheatgrass, alta fescue, and Indian ricegrass have been abandoned."[79] The Fish and Wildlife Service recorded: "Fourwing saltbush has become established (cover 10-15%). The ground layer is sparsely populated by alkali sacaton and Russian thistle. Groundlayer coverage [is] less than 5%."[80] Cover, in short, is poor.

A 1978 OSM inspection report stated that on land seeded before 1977: "Revegetation efforts have been very successful.... Species coming up in revegetated areas include saltbush, alkali sacaton, greasewood, and to a smaller extent a variety of other grasses and weeds." Plant vigor and ground coverage was two or three times that on adjacent undisturbed land. However, this was attributed to irrigation, barley straw mulch, and the fencing out of livestock.[81] Jack Reynolds said some plots had been successful and others less so. He claimed there had been reseeding, but Utah International denies this.[82]

INFORM's site observations are similar. A 1975-seeded area near Ramp 6 had a substantial amount of fourwing saltbush and alkali sacaton, and some Russian thistle. Sterling Grogan noted that the company had seeded Indian ricegrass, sand dropseed, western wheatgrass, and alta fescue, but that none survived. Grogan planned to selectively disc and interseed this and adjoining areas to alter the species mix and to allow for fewer but healthier plants. A 1977-seeded area had less fourwing saltbush and more galleta, plus barley and rye that grew from the mulch. An adjacent nonmulched area had more native plants, according to Grogan. A 1978-seeded area had the first successful winterfat, but also substantial weed growth.

Plant vigor is greatest in the first years, after irrigation, as one might expect. Reynolds said some of the several thousand acres where irrigation is no longer used had not been fully successful.[83] No massive dieback has occurred, but Grogan said it is too early to tell how permanent growth will compare with irrigated growth. Further, Utah International says long-term growth will spring from dormant seed, and second-generation growth from seeded species on reclaimed lands.[84]

Utah International vigorously disputes the contention of Donald Crane, OSM western regional director, that Navajo is a "biculture" of alkali sacaton and fourwing saltbush, and that it does not meet OSM's criteria for bond release.[85] It says the OSM conclusion is based on a report by Dr. Charles Bonham of Colorado State University, and that Bonham's data collection was a

"casual survey" of only 40 acres. Utah International contends
that the following species are found on reclaimed land, in addition to the two Crane cited: greasewood, Indian ricegrass, galleta, sand dropseed, giant dropseed, mormon tea, globemallow,
yucca, winterfat, snakeweed, and tamarix.[86] Obviously, more
data collection is needed.

The operator has conducted only one grazing trial to date,
sending sheep onto land with a substantial tumbleweed cover in
1976. Grogan said the sheep ate mostly plants other than the
tumbleweed.[87] Utah International also told INFORM that, "As of
September 1979, the experimental grazing plots of 1976 have a
greater diversity of native species and less mature Russian thistle than adjacent seeded areas. Each year these areas have improved as a result of natural invasion of native species."[88] The
company plans to begin two to three years of rigidly controlled
grazing experiments this year or next.[89]

Harris Arthur of the Bureau of Indian Affairs says Utah International has been accused of manipulating its revegetation
data.[90] The company denies the charge.[91] However, no vegetative data were provided to INFORM because no recent data exist.[92]

In sum, Navajo has had mixed reclamation success compared
with other mines in the INFORM sample, and much of the success
can be attributed to irrigation. The company has shown that it
can establish some type of cover that will persist after irrigation
stops, but its quality is uncertain. Species diversity is low. A
diverse environment is more difficult to establish here than at
other sites, and the company would do well to alter its program
to try to improve its record. An experimental program of interseeding would be profitable.

One observation on "absentee landlords": Brant Caulkins,
the Sierra Club's southwest regional director, told INFORM that
Utah International shows a marked difference in its reclamation
practices at the Navajo Mine, which it owns, and the San Juan
Mine--12 miles distant--which it operates for the New Mexico
Public Service Commission (PSC). He believes the regulatory
and political climate in the state make the PSC more responsive
to the wishes of citizens. Thus the PSC has required Utah International to practice better reclamation and take more care with
its work. But Utah International's parent, General Electric,
does not have the same sense of community.[93] INFORM was not
able to visit the San Juan Mine, but comparisons between San
Juan and Navajo would be a worthwhile venture in the future.

WILDLIFE

Wildlife densities are not particularly high, according to the Bureau of Reclamation Environmental Impact Statement. Several kinds of snakes, two lizard species, and low densities of squirrels, rabbits and mice inhabit the area. Prairie dogs died from a plague in 1975. Occasional larger predators, such a coyotes and foxes, are found.[94] There are numerous birds, but most live near Morgan Lake, an artificial body of water used for cooling the Four Corners power plant.[95]

Utah International's wildlife program consists of piling rocks on reclaimed land to create small habitats (similar to many mines in the INFORM sample) and leaving power line poles for nesting. The company said a wildlife survey showed more activity on reclaimed areas: in trap experiments, it caught two mammals in 900 "trap nights" on unmined land, and nine in 600 trap-nights on revegetated land.[96] But the lack of diversified shrubs being established is evidence of minimal concern for wildlife needs.

RECLAMATION COSTS

TABLE B*

RECLAMATION COSTS AT NAVAJO

Operation	Cost Per Acre	Percentage of Total Cost
Earthmoving		74%
Grading	$1,580	
Topsoiling	$2,160	
Revegetation		26%
Seeding, amendments, and irrigation	$1,290	
Total Cost:	$5,030	

*These costs and percentages are not strictly comparable to those of the same operation or activity at other mines. See the Cost of Reclamation chapter for a detailed explanation.

Source: Utah International, Inc.

Colorado

Even though the Routt County region has fairly steep slopes, with coal seams that follow the general contours of the land, the climate in this semi-arid/steppe--with an average annual precipitation of about 14-20 inches--favors revegetation. Used principally for grazing, the land is well covered with plant life of many types, and the challenge of reclamation is not as great as in southern Wyoming or the Southwest. Reclamation is regulated by the Colorado Mined Land Reclamation Board--a fair state regulatory group; local environmental watchdogs include the Public Lands Institute and Colorado Open Space Council, in Denver, and the Western Colorado Resource Council, based in Hotchkiss.

Energy Fuels/Energy Fuels/CO

Mine: Energy Fuels Mines 1, 2 and 3

Location: Milner (population 195), Colorado in Routt County (population 12,200)

Operating company: Energy Fuels Corporation

Parent company: Energy Fuels Corporation

Parent company revenues: 1977, NA; 1978, NA

Parent company net income: 1977, NA; 1978, NA

First year of production: 1962 by Energy Coal Company, purchased and name changed to Energy Fuels Corporation in 1972

Coal production in 1978: 3,770,000 tons

Acres for which a mining permit has been issued: 3,500

State regulatory authority: Colorado Department of Natural Resources, Mined Land Reclamation Board

Reclamation bond posted: $1,700,000

Reclamation cost: reclamation cost per acre, $1,985-4,630; reclamation cost per ton of coal, NA

Expansion plans at mine: Plan to expand Mine 1 onto approximately 1,100 additional acres

Additional acres for which the company will seek a permit: None

Parent company U.S. coal reserves: *

Development and expansion plans for surface coal mines in western United States: Although Energy Fuels Corporation does not

plan to expand its surface mining operations in the West, the
company does plan to open six deep coal mines in Routt County
by 1983. Total coal production from all six mines should reach
5.54 million tons annually by 1986.

NA = Not available

*Energy Fuels Corporation considers this proprietary informa-
tion.*

*Sources: Energy Fuels Corporation; "New Coal Mine Develop-
ment Survey 1978-1987," Coal Age, February 1979; Rand McNally
Road Atlas, 54th Edition; Communication with officials of Energy
Fuels Corporation, June 20, 1979, October 5, 1979, November
26, 1979; Interview with inspectors from Colorado Mined Land
Reclamation Board, January 9, 1980*

SITE DESCRIPTION

Energy Fuels Mines 1, 2 and 3 are located in the mountainous
region of northwestern Colorado's Routt County about 9 miles
southeast of Peabody's Seneca II Mine. The mines are approximately 10 miles south of Milner, the nearest town, and 25 miles
southwest of Steamboat Springs. The Energy Coal Company,
which was purchased by Energy Fuels in 1972, began mining at
this site in 1962. Energy Fuels estimates that Mine 1 will be
fully mined in about 10 years; Mine 2 is to be closed by the
first quarter of 1980, and Mine 3 will cease producing coal by
1981 unless the regulatory authority issues a permit allowing
the company to mine new lands. Such a decision would extend
operations 10-12 years.[1]

Mining at Energy Fuels takes place on moderate slopes in
the foothills of the Rocky Mountains, approximately 7,000 feet
above sea level. Slopes at the mines are generally less steep
(6 - 10%, with 15% slopes near Mine 3[2]) than those of the neighboring Seneca II Mine.[3] Sagebrush communities, mountain
shrub communities and aspen plants dominate the vegetation in
areas surrounding the Energy Fuels site.[4] These areas also
include some wheatgrass species, bromes, blue and needle-
grasses, shrubs and trees.[5] The Mine 3 site contains some
wheatlands, but the topography does not generally lend itself to
farming.

Although the company will be returning approximately 78
acres of land to wheat production, Kent Crofts, chief environmental coordinator for Energy Fuels, reports that the company
does not know how many acres were actually used for crop production before mining began. He says ranchers in the area

have periodically planted wheat for one or two years on grazing lands and then returned the land to grazing use; thus, no accurate figures are available.[6] Soils in the wheatland areas are very rocky,[7] making operation of equipment and farming difficult.

Over the last 30 years, annual precipitation has ranged from 11.5-24 inches,[8] an average of 16 inches a year.[9] Although precipitation in 1977 was only 12 inches, it was above average in 1978 and 1979; this should help vegetative establishment.[10] The heaviest rains fall in April, May and October, averaging 1½ inches per month, and the heaviest snowfalls occur in December, January and February, averaging almost 2.5 inches per month.[11] The company reports that the growing season at the mine is approximately 102 days long, from late spring to early fall.[12]

Most land in the area is used as pastureland for livestock,[13] but grasses are grown for hay production along river bottoms and at the bases of hills in the region. Some wheat farming has been done on permitted lands at Mine 2, but this has not been the rule historically.[14] Energy Fuels will be returning most reclaimed lands to rangeland and pastureland, with a small portion at Mine 2 (78.1 acres) returned to cropland.[15]

Under Office of Surface Mining (OSM) regulations, Energy Fuels must determine whether lands being reclaimed will be returned to primarily rangeland use, where maintenance is achieved by controlling the amount of grazing, or to pastureland which will either be grazed or hayed, and will be seeded with a different seed mixture. Through 1980, the lands are generally being reclaimed to rangeland use. More land will be returned to pastureland and crop productivity in the future.[16] Post-mining land use will be identical to pre-mining use.

OVERVIEW

Energy Fuels is conducting one of the most environmentally sound reclamation operations in the INFORM sample. Practices at Energy Fuels Mines 1, 2 and 3 demonstrate the company's broad commitment to successful land reclamation. Coupled with a climate less harsh than many found in the West, these mines incorporate good management and reclamation techniques with enough innovation to increase the potential for good reclamation on surface-mined areas. Water resources are monitored constantly, and control of discharges from the site has kept the company within the limits of its permit under the National Pollutant Discharge Elimination System (NPDES).

The company is also conducting a good seeding and amendment program at its three Colorado mines. Seeding is done in the fall. It follows land disturbance fairly closely. Energy Fuels seeds a mixture of grasses, forbs and shrubs, of which about

half are native plant species. Machinery operates along the contour, and furrows are made on slopes to control erosion and
enhance water retention (see Erosion). Aside from the use of
these furrows, annual grains are seeded to create a living mulch
on lands being reclaimed. Energy Fuels has had particular success with tree transplants at the site, owing in part to its use of
water running off the mine for irrigating newly transplanted
trees. Seeding and reclamation maintenance (e.g., cleaning of
furrows) is done by a local rancher under contract to Energy
Fuels.

Energy Fuels effectively protects water resources at these
mines. The company has allocated one staff member full-time to
monitor water quality and quantity. This is more than many
companies in the INFORM sample are doing. (For example, hydrologists for the Jim Bridger Mine in Wyoming are located in the
Portland, Oregon offices of the parent company.) Energy Fuels
has not exceeded the Environmental Protection Agency (EPA)
limits for water discharged from the mine under its permit. It
has also avoided mining alluvial valley floors, although some
question remains whether lands near present mining operations
can be so designated. Although the company contends that its
mining operations will actually enhance the area's ability to recharge aquifer systems by loosening overburden, thereby allowing water from the surface to reach into deeper ground layers,
it has not addressed the question whether the quality of the
water in these aquifers will change if more minerals are transported into groundwater systems.

Except for two minor instances--when Energy Fuels failed to
salvage topsoil in a small area, and when spoil piles from the site
slid over a ridge--the company's earthmoving practices have
been sound. After reprimands from state and federal regulatory
authorities, the company improved its topsoil removal practices.
In 1979 Energy Fuels purchased new equipment to improve topsoil removal and handling at the site. Moreover, Energy Fuels,
like the neighboring Seneca II Mine, attempts to use direct haulback as much as possible, instead of stockpiling material for
later use. Grading is done well, and in the instance when spoils
slid onto neighboring ranchland, the conflict was resolved when
Energy Fuels bought the ranch.

Energy Fuels has so far successfully established vegetation,
prevented erosion, and provided food and shelter for wild animals at its mine site. However, the productivity of reclaimed
lands depends heavily upon introduced plants that have been
seeded in the past. Although levels of forage productivity have
sometimes tripled within four to five years after seeding, the
company has yet to establish a diverse vegetative cover of native
plant species. It also remains for the company to demonstrate

that long-term pressures from grazing and normal levels of range
management will not cause this land to become less productive.

Regulators have been relatively satisfied with practices at the
Energy Fuels mines, so enforcement activities have been limited.
The Colorado Mined Land Reclamation Board (MLRB) is the state
authority with jurisdiction over state mining and reclamation
practices. The MLRB inspects the Energy Fuels mines quarterly.
In July 1978 a Board inspector wrote: "Over-all, the operation is
exemplary in terms of both mining and reclamation operations."[17]
The Office of Surface Mining has inspected these mines only
twice--in November 1978 and October 1979--since its Denver regional office was established. Inspections by state and federal
regulators have usually helped identify any problems discovered
on the mine site, and have encouraged Energy Fuels to take
steps to prevent their recurrence.

Grading

Energy Fuels' grading operations are good. Grading generally follows within three pits of mining (though the Mined Land
Reclamation Board reports that the company does occasionally fall
behind), and by late 1979 the company had voluntarily leveled
almost 700 acres of old spoil ridges left by mining operations of
the former owner, the Energy Coal Company. Grading approximates both the surrounding topography and pre-mining topography.[18]

In 1977 Energy Fuels disturbed 101.6 acres and graded 108.3
acres.[19] In 1978, 272 acres of land were mined and an additional
195 acres disturbed by roads, stockpiles, and spoil piles, and
related mining activities. However, by the end of 1978, 55 percent of the 2,516 acres that had been disturbed at the mine had
already been planted.[20] Company officials say that at any one
time, 450-500 acres of land remain disturbed at the mine as a result of mining, spoil ridges and areas where topsoil has not been
replaced.[21] This situation is common for mines with this amount
of coal output.

In one instance, grading operations may have caused off-site
damage to a neighboring ranch. In a February 1977 letter to the
Mined Land Reclamation Board, the neighbor wrote that blasting
at the mine had weakened the foundations of her house and that
Energy Fuels' dumping of overburden on a hill near her home
had caused mud and boulders to slide down roads and onto her
property.[22] A photograph in the Environmental Statement for
Northwest Colorado Coal, prepared by the U.S. Bureau of Land
Management, shows earth slides from spoil piles that were placed
on the crest of a steep ridge near Energy Fuels Mine 1.[23]

In October 1979 Kent Crofts, chief environmental coordinator
for the company at the mine, confirmed that the picture shows

the ridge from which boulders did roll onto the ranch property.
Crofts said no mud from the ridge had transgressed onto the
farmhouse itself; nor had the house's foundations been damaged.
And he denied the allegations of blasting damage, noting that
the ranch is three-quarters of a mile from the ridge where mining was taking place.[24] In any case, one month after issuance
of the complaint, it was withdrawn when Energy Fuels bought
the ranch.

Topsoil

Topsoil removal and replacement practices at the site are
good. Energy Fuels says topsoil on the mine site averages 24
inches in depth.[25] The U.S. Geological Survey has found that in
some areas where surface water flow has increased sediment deposits, topsoil can be 5-8 feet deep.[26] Energy Fuels does not
often face the problem of removing topsoil and subsoil separately
because such an insignificant amount of subsoil exists there.
When subsoil is found, it is removed separately.[27] The company
removes all topsoil and respreads it to a depth of 15-18 inches on
graded lands, providing a high-quality soil where plants have a
good opportunity to grow.[28]

Kent Crofts told INFORM in June 1979 that *only 35-40 percent* of topsoil was replaced immediately.[29] In late 1979 the company reported that on *most* of the 1,500 acres of land where topsoil had been replaced, the operation had immediately followed
the stripping of that topsoil.[30] Company officials at the mine
have agreed that topsoil is the best medium for retaining water
and seed that can be used in reclamation, and that immediate replacement of topsoil on graded lands should be a priority. Yet it
is not clear how much is actually being done.[31]

Procedures for removing topsoil at the mine are now generally adequate. Although no fines have been levied, Energy
Fuels has experienced small problems with these operations in
the past. In July 1978 the Mined Land Reclamation Board suggested that the company remove topsoil from a wider area ahead
of the working dragline and blasting operations in order to allow
more room between these operations.[32] In November 1978 an
OSM inspection found an area of less than one acre where spoil
had been placed upon topsoil during mining operations.[33] In July
1979 the MLRB cited the company for failing to remove topsoil on
one-tenth of an acre of mined land.[34] Since then, Energy Fuels
says it has purchased two new topsoil removal machines to keep
this from recurring. In 1979 the company reported that topsoil
removal usually precedes mining by 75 yards, or the width of
two mine pits.[35]

Energy Fuels carefully protects stockpiled topsoil from erosion by covering it with vegetation. In July 1979 the Mined Land

Reclamation Board reported the company had no problems with topsoil stockpiles at these mines.[36] The stockpiles vary in size from 0.6-6.5 acres.[37]

Toxic Material

Energy Fuels has experienced no difficulty with toxic substances or salts at the mine site as neither is being exposed.

WATER QUALITY

Surface Water and Groundwater

Energy Fuels carefully controls surface runoff. The company has dug drainage ditches along roads to carry water to sedimentation ponds or to intermittent streams. Water carried to intermittent streams, such as Foidel Creek, is filtered through dams built of rocks before it enters sediment ponds further downstream. Other streams on the site include Fish and Middle Creeks, which are tributaries of Trout Creek, itself a tributary of the Yampa River. Energy Fuels is innovative in using water runoff to irrigate transplanted trees and shrubs. Ditches are dug to channel runoff from along mine roads to an area where trees and shrubs have been planted. The channel directing water to the plants is lined with rock and usually vegetated with grasses to reduce erosion. The use of this system may account for the success of tree plantings at the site.

The company operates a good water control system. From January 1977 to January 1979, none of the mines exceeded the limits for sediment levels established for water discharged under the National Pollutant Discharge Elimination System (NPDES).[38] The mine's water-control system is largely responsible for this achievement. In April 1979 discharges from the mine did exceed Total Suspended Solids (TSS) and Total Iron limits as set forth in the mine's NPDES permit. However, EPA determined that these events resulted from melting snow, which produced more water than would be expected from a 10-year, 24-hour precipitation event.[39] (The NPDES permit for the mine describes a 1.8-inch rain in 24 hours as a 10-year precipitation event.[40]) When this occurs, the company is not held responsible for exceeding the prescribed effluent limitations.

The Office of Surface Mining has described sediment-control structures, such as the rock filter dams in Foidel Creek, as "highly successful at retaining sedimentation from the mining operations within the permit area."[41] Energy Fuels has proposed sedimentation ponds with a minimum volume of 500 cubic feet (equaling one foot of water on 50 acres of land) in its latest application for a mining permit. This would be larger than the

ponds now on the site. The application was to be acted upon by
January 1980.[42] Vegetation generally covers the sides of all
sediment ponds now on the site and prevents erosion.

The company has experienced only small problems with its
water-control system. Visiting the mine in June 1979, INFORM
found that a bulldozer had accidentally filled with dirt a discharge channel leading from a sediment pond in the southeast
corner of one of the mine areas. Water was leaking through the
dirt, with the possibility that a good deal of sediment would be
carried from the mine site to neighboring lands and waters.
Energy Fuels hydrologist Geoffrey Saunders said the company
had discovered the problem one week before INFORM's visit, but
that the channel would not be repaired for "one month or so."[43]
Failure to act quickly to eliminate this problem is not in keeping
with Energy Fuels' generally meticulous reclamation practices.
Energy Fuels reports that it regularly cleans sediment ponds to
maintain their capacity to hold water.[44]

A detailed groundwater monitoring program is conducted by
Energy Fuels.[45] It monitors 47 wells in the mining area. Energy
Fuels' chief geologist, Gary Myers, says studies by the company
show that mining has not degraded groundwater resources.[46]
The mine site is a recharge area for groundwater aquifers--
where precipitation permeates the ground and supplies water to
aquifer systems. Energy Fuels says mining operations will improve the land's ability to serve as a recharge area because
"blasting, excavation and replacement of the overburden in the
recharge area will make the overburden more porous and permeable."[47] But further monitoring will be necessary to determine
whether the water that is infiltrating and recharging aquifers is
carrying additional minerals, exposed by mining, that could degrade water quality. Although the Environmental Impact Statement for Northwest Colorado Coal reports that surface mining
will destroy several springs and wells in the area of the mines,[48]
Energy Fuels contends that its operations have destroyed only
monitoring wells on the site, not wells that supply water to
neighbors.[49] The company says no springs exist in the area,
and that in only a few places does water seep to the surface in
wet weather.[50] (Given the extensive water monitoring on the
site, this analysis is probably sound.)

Alluvial Valley Floors

Although the existence of alluvial valleys in the area of the
three Energy Fuels mines has not been conclusively proven, the
company has taken steps to preserve any that may exist. OSM
has cited Fish Creek as a potential alluvial valley, but by June
1979 it had not made a final determination. Energy Fuels reports
that at OSM's request it will not mine the Fish Creek area, and

that it is building channels to divert runoff from the Fish Creek floodplain to prevent sedimentation of the land.[51] Although OSM has not identified Trout Creek near Mine 3 as an alluvial valley,[52] Geoffrey Saunders, the mine's hydrologist, says it is "without a doubt"[53] such a valley, and he says the company has no plans to mine coal in that area.[54]

Erosion

State and federal regulators confirm that Energy Fuels has successfully minimized erosion at the mine site. A March 1978 inspection by the Colorado Mined Land Reclamation Board found that erosion control at the mines works for the most part.[55] Later that same year, OSM inspectors found:

> Rilling, on areas which have received topsoil replacement, has been controlled to an extremely high degree through the combined practice of planting annual vegetation for rapid vegetation establishment, constructing furrows 12 to 18 inches deep along the contour with burms [soil built up on the downhill side to hold water in] or dikes within the furrows approximately every 10 feet to insure the surface water and whatever sedimentation it is carrying, is retained in place rather than being diverted by the furrows, and the planting of perennial vegetation.[56]

These comments confirmed a July 1978 state inspection report, which said that cross ditches or furrows used for erosion control "are doing a good job and should be encouraged."[57]

In June 1979 Kent Crofts said "surface manipulation is the key to retaining water and preventing erosion."[58] He also feels that furrowing will help plants become established to create a stable plant community, the best protection for soil and the most effective method of preventing water from carrying sediment from the site.[59]

Erosion has occurred on one to two acres of land mined in 1974 and 1975, which Energy Fuels did not furrow. In 1979 the company placed furrows along the contour in an attempt to halt this erosion. Most gullies on this land are from 5-12 inches deep[60] but are vegetated in many areas. Mine staff are now filling them with dirt; bales of straw in the larger gouges are being used to trap sediment and slow the flow of water. As of June 1979 no harmful amounts of sediment had been carried from the mine site to neighboring lands.

SEEDING AND AMENDMENTS

Seeding Procedures

Seeding at the mines generally follows mining closely. Although less than 100 acres was seeded in 1977,[61] 272 acres were
disturbed by mining and an additional 195 acres disturbed for
related activities such as roads, stockpiles, and spoil piles in
1978--yet the company planted 270 acres of land that year. On
a cumulative basis, 55 percent of the lands disturbed at the mine
have been planted.[62] Kent Crofts said Energy Fuels planned to
seed about 350 acres of land with grasses and forbs in the fall of
1979.[63]

The seed mixture being used is still composed of many introduced plant species, but the diversity of its composition is good,
making it a fairly good mixture. The plants seeded by Energy
Fuels represent a range of grasses, forbs and shrubs. In the
fall of 1979 Energy Fuels planned to seed 12 grass species at
8.32 pounds per acre, nine (ten at Mines 1 and 2) forb species
at 3.5 (3.51 at Mines 1 and 2) pounds per acre, and three shrub
species (five at Mine 3) at 0.53 pounds (1.03 at Mine 3) per acre.
As a result, 24-26 plant species were to have been seeded on
lands revegetated in 1979. This is substantially more than the
nine grass and forb species seeded in 1976. Since that year, the
number of plants being seeded has increased; this should help to
establish a more diverse plant community. About half the grasses
being used are introduced species; almost all are cool-season varieties. The forbs and shrubs are a mixture of introduced and
native species and represent a mix of cool- and warm-season
plants.

Energy Fuels contends that although the plants are not native to the United States, "all of the exotic plants [being seeded
were] currently found growing on the site according to a plant
inventory of the site....Therefore, no new species are being introduced onto the site....Rangelands are currently planted almost exclusively to exotic species due to their higher production
levels."[64] However, the use of introduced plants can affect the
diversity of plant cover and the length of time required for native plants to invade the area (see Guideline 19). In June 1979
Kent Crofts said Energy Fuels believed that these introduced
species would become established faster than native species, that
native species will invade reclaimed lands naturally, and that
grazing animals eat exotic plants in other parts of the world and
would do so in Colorado.[65] However, in August 1978 an inspector for the Mined Land Reclamation Board wrote that the company
"is possibly relying too heavily on introduced species,"[66] and in
July 1979 another Board representative told INFORM the company
should seed "more native species."[67] This would provide plants

TABLE A

1979 SEED MIXTURE
FOR LANDS BEING RECLAIMED
AT ENERGY FUELS MINES 1, 2 AND 3

Plant Species	Introduced/ Native	Cool/ Warm	Pounds Applied Per Acre
Grasses			
Slender wheatgrass	N	C	2.0
Intermediate wheatgrass	I	C	1.5
Pubescent wheatgrass	I	C	1.5
Smooth brome	I	C	2.0
Desert wheatgrass	I	C	0.25
Streambank wheatgrass	N	C	0.25
Basin wildrye	N	C	0.25
Orchardgrass	I	C-W	0.125
Timothy	I	C	0.05
Kentucky bluegrass	I	C	0.02
Hard fescue	N	C	0.125
Western wheatgrass	N	C	0.25
			8.32 Pure live seed
Forbs			
Alfalfa	I	C-W	0.5
Cicer milkvetch	I	W	0.5
Emerald crownvetch	I	W	0.5
Small burnet	I	C	0.5
Sweetanise	N	C	0.5
Rocky Mountain penstemmen	N	C	0.25
Lewis flax	I	C	0.25
Western yarrow	N	C	0.25
Annual sunflower	N	W	0.25
*Arrowleaf balsamroot	N	C	0.01
			3.51 Pure live seed
Shrubs			
Big sagebrush	N	C	0.025
Antelope bitterbrush	N	C	0.25
Lanceleaf bitterbrush	N	C	0.25
†Saskatoon serviceberry	N	C	0.25
†Black chokecherry	N	C	0.25
			1.025 Pure live seed

*at Mines 1 and 2 only
†at Mine 3 only

Source: Energy Fuels Corporation

better adapted to the climatic extremes of the western United
States and require less maintenance, producing a more self-sustaining plant community. Energy Fuels increased the number of
native species in the 1979 seed mixture.

The company does not vary the seed mix or amount according to either the slope of the land or the quality of its soil. Research on the use of native and introduced plant species is being
done at the mines in cooperation with the U.S. Soil Conservation
Service (SCS). Energy Fuels and SCS have established a study
area where 36 varieties of native and introduced plant species
are being monitored for their establishment and growth characteristics. INFORM was not able to determine how the company
has incorporated its research into the choice of plants being
seeded.

Preparations for seeding and actual seeding practices are
good. Energy Fuels employs a drill seeder for almost all seeding, and seeding follows the contour of the slopes. Broadcast
seeding is done only in areas where the company wishes to establish a temporary cover--such as stockpiles, sediment pond
dikes, and areas where roads have been built.[68] In the winter of
1976 the company tried broadcasting seed from airplanes on the
theory that, as the snow thawed, the seed would become embed-
ded into the ground and provide a good base from which plants
could grow in the spring. The 516 acres seeded in this way had
good seed distribution but only poor to fair vegetation.[69] That
year (in October) Energy Fuels began seeding grasses and forbs
in the fall, the commonly accepted seeding period, as seeds sown
then are in place, ready to germinate early in the spring.

Amendments and Maintenance

Energy Fuels, unlike many coal mine operators in the West,
has used contour furrowing for erosion control for years. In the
past, mulching was not required, and the company felt that its
manipulation of the ground surface was sufficient to achieve the
same ends as mulching. But, as required by the Colorado MLRB,
Energy Fuels began using mulch on reclaimed lands in 1978. It
is following an apparent western trend to seed annual grains as
mulch, instead of applying either hay or straw to the surface and
then attempting to anchor it to the soil. The company creates
mulch by seeding an annual grain, then mowing it before it ma-
tures and before the standard seed mixture is sown. Energy
Fuels drill-seeded 72 pounds of winter rye seed per acre of land
in 1978.[70] Company officials disagree with the Colorado MLRB's
mulch requirement. They contend that mulch may make it more
difficult to establish vegetative cover because seeds in the mulch
may compete with the plants a company is trying to establish;
they believe seeding an annual grain and mowing it is 10 times

cheaper than applying a straw mulch.[71] (They say they can seed
and mow annual grains for $30 an acre.[72] Also, the use of annual
grains will help to prevent some of the problems that may occur
if hay or straw mulches are used. Because the company mows
the grains, it needs to do this only before seeds are produced to
prevent further competition with establishing grasses.

Although Energy Fuels says it has documented the superior-
ity of creating furrows along the contours of graded lands as
more effective than mulching, a spokesman for the Colorado
Mined Land Reclamation Board said the company's evidence was
not directly applicable to the Routt County mine site.[73]

Energy Fuels has fertilized reclaimed lands in the past but
no longer does so.[74] The company says it stopped using fertilizer because "several hundred soil tests had been performed and
none revealed any deficiencies in plant nutrients."[75] No irrigation is performed beyond the directing of natural runoff into the
areas of tree transplants. Energy Fuels also sprays herbicides
on reclaimed lands to control noxious weeds but applies the herbicide only to the plant to be killed, which is a good practice.

Trees and Shrubs

Energy Fuels has had considerable success in transplanting
trees and shrubs. The transplants are from nursery stock and
from trees and shrubs salvaged from lands that are to be disturbed. It also seeds shrubs with the grasses and forbs of the
basic seed mixture used on the site. Trees and shrubs are
planted together rather than as single units surrounded by
grass, creating a small forest covering 15-20 square feet. Energy
Fuels has created small channels about 2 feet wide to direct
water running off slopes to the base of the trees. This irrigation
system has proven quite effective. The transplants are not only
surviving, but thriving. New trees are sprouting from their
roots, and shrubs and other plants are growing beneath the
trees. The trees also provide a refuge for wildlife. Kent Crofts
reports that Energy Fuels has planted and transplanted 58 different varieties of tree and shrub, of which some 60 percent are
native. He says that in the last three years the company has
been using 80-85 percent native trees and shrubs.[76] Although
most of the direct transplants have the soil surrounding their
roots, nursery stock generally comes with bare roots and must
be handled more carefully. Although some of the transplanted
trees seen during a tour of the site in the summer of 1979 had
not yet become established, Energy Fuels' success in transplanting trees makes it likely that they will survive and flourish.

Clumps of transplanted trees are surviving well. Path in foreground helps direct natural runoff to the base of the stand, not only aiding tree and shrub growth, but reducing scouring by the water and subsequent erosion and sedimentation.

VEGETATION

In general, lands that have been seeded with an annual grain to provide mulch are well protected against erosion. About 240 acres were seeded with winter rye in 1978 for this purpose.[77] As the grain is mowed and grasses are planted, these grasses also begin to provide protection from erosion and to stabilize the soil. Twenty-six acres of land seeded in the fall of 1977 with the grass, forb and shrub mixture, where topsoil had been immediately replaced, showed diverse growth of both grasses and shrubs. In addition, in mid-1979, more varieties of native plants were growing in the areas where topsoil had been immediately replaced than in areas where it had been stored before laying-- possibly because seeds contained in the soil had lost their fertility in storage.

The Mined Land Reclamation Board predicts that the dominance of introduced species on reclaimed lands may cause problems.[78] For these reasons Energy Fuels may find it difficult to win release of the land from its reclamation bond, as requirements for release include diversity of plants. The company has won release of some lands from its bond at the mines. In July 1978, 25 acres of land at Mine 3 and 10 acres at Mine 2 were released from bond under a 1973 Colorado reclamation law. In March 1979 an inspector for the Mined Land Reclamation Board told INFORM these lands would not qualify for release from bond under the new state law or the 1978 federal standards.[79] Energy Fuels officials said the productivity of these lands was good (see below) but the diversity of plant species was not.[80] The company planned to seek release of 100-150 acres of lands from bond under the 1973 Colorado law, late in 1979 or in the spring of 1980.[81] Mining activities had been completed on these lands before May 1978; thus the lands are not held to the requirements of the federal Surface Mining Control and Reclamation Act.

Four vegetative communities dominate the Energy Fuels mine site. Mountain shrub communities cover 30-40 percent of the area; sagebrush communities cover 30 percent, and aspen or wheat stubble communities cover 15 percent.[82] Aspen and mountain shrub do not produce commercially valuable wood, but trees and shrubs in these communities yield 1,500-3,000 pounds of dry forage per acre, and 4-4.5 acres of this land can support one animal unit (a cow and calf, or five sheep) per month.[83] Other land can produce 1,200-1,800 pounds of dry forage per acre and seven to eight acres can support one animal unit per month.[84] Energy Fuels has stated that some lands on the mine site produced 3,500-4,000 pounds of dry forage per acre before mining began.[85] It says wheat production on undisturbed lands at the mine site is 23-28 bushels per acre.[86]

In June 1979 Kent Crofts told INFORM he expected the company's reclamation program to produce stable, revegetated areas
within three years of seeding. He said 25 acres of land at Mine
3, whose productivity levels have been tested, are producing
forage up to three times their pre-mining levels.[87] INFORM was
unable to verify this.

Productivity data, collected by the company from 1975 through
1977 and in 1979, support these claims. The 25 acres released
from bond at Mine 3 produced 425 pounds of forage per acre before mining began in 1974. After seeding for three to four years
they produce 1,263 pounds of forage. In 1979 ten acres that had
been released from bond at Mine 2 produced 2,262 pounds of forage five years after seeding. This land produced 917 pounds of
forage before mining began.[88] However, this forage consists
generally of non-native plant species. It may not provide the
proper type of nutrients for grazing animals, and it may not be
able to resist long-term climatic variation.

Several studies of the productivity of reclaimed land at the
site are under way. These include a joint Energy Fuels-Soil
Conservation Service test area, where the ability of 36 native
and introduced plants to establish themselves at the site is being
tested. The Agricultural Research Service, a branch of the U.S.
Department of Agriculture, is conducting experiments on nitrogen cycling (between soils and plants) in mine spoils, evaluating
plant species, testing fertilizer and carrying out studies of the
water system in the mine area.

Energy Fuels also plans, in cooperation with the United
States Bureau of Land Management, to begin cattle grazing experiments on lands released from bond.

WILDLIFE

Wildlife concerns have high priority at the Energy Fuels
mines. Tree and shrub establishment is good. Yet is is unclear
whether the dominant introduced grass species on older reclaimed
lands will be as valuable as native plants for wildlife. Transplanted trees and shrubs will provide sufficient food and refuge
for wild animals. Thus Energy Fuels does not plan to create
other habitats on the mine site.

The company plans to avoid mining lands south of Mine 2
that are dancing grounds for sharptail grouse.[89] According to
the environmental statement prepared for coal development in
northwest Colorado, mining in the vicinity may displace these
animals.[90] As the dancing grounds are areas where the birds
mate, disturbance could upset the delicate natural balance that
keeps this species in the area.

RECLAMATION COSTS

TABLE B*

RECLAMATION COSTS AT ENERGY FUELS MINES 1, 2 and 3

Operation	Cost Per Acre	Approximate Percentage of Cost†
Earthmoving		78%-81%
Pit reclamation grading	$ 997-$3,200	
Topsoiling (replacement)	565	
Revegetation		2%-5%
Seeding	62	
Mulch: seeding and mowing annual grain	30	
(20% added as an administrative charge)	331- 771	20%
Subtotal:	$1,985-$4,628	

*These costs and percentages are not strictly comparable to those of the same operation or activity at other mines. See the Cost of Reclamation chapter for a detailed explanation.

†Because Energy Fuels did not disclose a total reclamation cost, approximate percentages have been calculated based on a subtotal of defined costs.

Source: Energy Fuels Corporation

Seneca II/Peabody/CO

Mine: Seneca II Mine

Location: Hayden (population 1,750), Colorado, in Routt County (population 12,200).

Operating company: Peabody Coal Company

Parent company: Peabody Holding Company*

Parent company revenues: *

Parent company net income: *

First year of production: 1969

Coal production in 1978: 1,372,251 tons

Acres for which a mining permit has been issued: 3,155 as of December 31, 1978

State regulatory authority: Colorado Department of Natural Resources, Mined Land Reclamation Board

Reclamation bond posted: $1,489,548 as of December 31, 1978

Reclamation cost: reclamation cost per acre, $5,065; reclamation cost per ton of coal, $0.70

Expansion plans at mine: Production to reach 1.8 million tons per year by 1981

Additional acres for which the company will seek a permit: 3,100

Parent company U.S. coal reserves: Owns or leases 8.9 billion tons of coal reserves

Development and expansion plans for surface coal mines in western United States: At present, Peabody Coal operates five surface mines in the western United States: Big Sky Mine (2,064,886 tons in 1978) in Montana, Nucla Mine (102,393 tons in 1978) and Seneca II Mine in Colorado, Black Mesa (4,800,000 tons in 1978) and Kayenta (6,771,768 tons in 1978) Mines in Arizona. Peabody plans to produce 5 million tons of coal per year at the Big Sky Mine by 1985. The company is also developing three new mines in Wyoming which will produce a total of 21 million tons per year by 1986.

---

*Peabody Holding Company is a consortium of six companies including: Newmont Mining Corporation (27.5 percent ownership), Williams Companies (27.5 percent), Bechtel Corporation (15 percent), Boeing Company (15 percent), Fluor Corporation (10 percent), and Equitable Life Assurance Society (5 percent).

Sources: Peabody Coal Company; 1977 and 1979 Keystone Coal Industry Manuals; "New Coal Mine Development Survey 1978-1987," Coal Age, February 1979; Rand McNally Road Atlas, 54th Edition; U.S. Department of the Interior, OSM, Region V On-Site Inspection: Seneca II Mine, August 17, 1978.

SITE DESCRIPTION

The Seneca II Mine is located approximately 7 miles southeast of Hayden, Colorado in Routt County, a mountainous region of northwestern Colorado.

In 1964 Peabody opened the first Seneca Mine just north of the Yampa River near Hayden. Between 1968 and 1969, mining operations shifted to land south of the river, the area now known as the Seneca II Mine. The original Seneca Mine is closed, because all coal reachable by surface mining has been removed.

Operations at Seneca II take place on some of the steepest slopes mined in the West. At an elevation of 7,000 feet above sea level, the incline of slopes in the area ranges from 8.5-24 degrees[1]; that of most slopes being mined ranges from 14.5-18 degrees.[2]

Eighty percent of the vegetation in the area of the mine consists of mountain brush communities, including shrubs, gambel oak, and grasses.[3] Trees such as aspens and conifers cover 5 percent of the mine site, while sagebrush and grass communities cover the remaining 15 percent.[4]

Precipitation in this area is high for the West. Average annual precipitation ranges from 16 - 19 inches.[5] Forty percent falls during the growing season.[6] Peabody reports that the

growing season at the mine is approximately 169 days, extending
from late April to early October, a much longer period than the
102-day growing season at Energy Fuels's mines nearby. The
climate in this mountainous region can change radically from one
location to the next, due to variations in slope aspect and alti-
tude.[7] Annual precipitation in the area has ranged from 12-24
inches over the last 30 years.[8]

Before mining began, the Seneca II mine site was primarily
grazing land for sheep and a wildlife habitat. An increasing
amount of land in the area is now also being used to graze cattle.
Although the flat lowland areas surrounding the mine are used
for farming, the mine site itself encompassed no croplands before
mining began.[9]

Peabody plans to return the site to its original use as grazing land.[10]

OVERVIEW

Peabody is developing a fair to good program for reclaiming
lands disturbed by mining at Seneca II. However, serious problems exist with the vegetation established by Peabody on the
mine site, due in part to poor reclamation practices in the past.
Tom Ehmett, an inspector with the federal Office of Surface Mining in Denver, told INFORM in March 1979 that "in terms of what
is on the ground [now], [Seneca II] is the worst mine in Colorado."[11] The new seeding program should improve vegetation on
lands now being seeded, but the effect of these changes will not
be known for several years.

Since fall 1977, Peabody has done a good job in upgrading
the revegetation program at Seneca II. The seed mixture now
applied to reclaimed lands contains more grass species and fewer
legumes by weight than that of previous years. This should
prevent legumes from dominating reclaimed lands and inhibiting
the growth of other plants. Seeding is done using drill seeders
and broadcast seeders following discing machines, which till the
soil before the seed is dropped, a vast improvement over the
former practice of seeding lands from the air. Seed is more
evenly distributed and is placed in the soil, not on it.

Although erosion on most parts of the mine is about the same
as that on surrounding undisturbed lands, in 1979 Peabody
started to upgrade water control on the site. The company began placing furrows in the earth along the contour of graded
slopes, a practice which should help reduce the rate at which
water flows down the mine's steep slopes, thereby controlling
erosion. Most erosion that may cause potential problems is limited to cuts in the sides of hills where roads have been created
and to areas where water flowing down steep slopes is collected

and directed through culverts to pass under these roads. The
company has not experienced problems with sedimentation off the
mine's permit area. Mining at Seneca II is disturbing groundwater aquifers on the mine site, but no information is available on
the off-site effects of this disruption.

Earthmoving operations including grading and topsoil removal are basically good. Until 1977 Peabody graded steep
slopes to a topography unlike the site's pre-mining contour, increasing the potential for erosion and making revegetation very
difficult. Since the fall of 1978 when Colorado's Mined Land Reclamation Board, the state regulator, fined Peabody $18,000 for
failure to save topsoil in an area at the mine, the company carefully removes and replaces the materials. However, very little
topsoil is immediately replaced on graded lands, increasing the
possibility that this nutritive material may be lost through erosion or decrease in quality while it is stockpiled.

Other than planting trees and shrubs on reclaimed lands,
Peabody does only a fair job of creating wildlife habitats on the
site. Many shrub plantings and all tree plantings have been un-
successful, unlike the tree transplants Energy Fuels has done at
its nearby mines. Peabody plants bare-root plant stock, while
Energy Fuels transplants shrubs and trees keeping the soil surrounding the roots intact.

Peabody has collected little data on the productivity of either
reclaimed or undisturbed lands on the site. Productivity figures
on lands seeded with alfalfa show that forage production is extremely high for the area. However, the alfalfa is at its productive peak and should begin to decline within the next few years.
As mentioned earlier, the plant is not as suitable for the site's
proposed post-mining use, cattle grazing, as are the plants Peabody now seeds.

Regulators have had a major impact on reclamation operations
at Seneca II. The suggestions and requirements of Colorado's
Mined Land Reclamation Board (MLRB), the state agency which
oversees mining and reclamation, have required Peabody to improve grading, topsoil removal, and seeding programs at the
mine. The MLRB inspects Seneca II quarterly. The U.S. Office
of Surface Mining (OSM) has also emphasized these same areas
and issued fines for violations in 1979, as did the MLRB. OSM
inspected the Seneca II Mine three times between June 1978,
when the Denver regional office began inspecting surface coal
mines in the six states in INFORM's study, and mid-October
1979. Since 1977, Peabody has generally responded diligently to
both state and federal rulings.

EARTHMOVING

Earthmoving operations at Seneca II are now good, compared to fair to poor practices in the past. Due in part to regulatory pressures, Peabody has improved its earthmoving operations in the last 2 years. Grading of spoil piles follows mining closely, and the contour of graded lands now more closely resembles that of the surrounding undisturbed topography. In the past, grading left slopes which were too steep to revegetate or to protect adequately from erosion. The company has experienced problems with toxic substances moving into surface soils after grading was completed. Although these problems are confined to small areas, Peabody has not developed a plan to deal with them. Because of the site's steepness, seeding and other reclamation equipment cannot operate on the contour of all graded slopes returned to approximate original contour, increasing the potential for erosion of newly seeded lands.

Grading

As of June 1979, 978 acres of land had been disturbed at Seneca II. Of these 978 acres, 630 acres (or 64 percent) have been graded. According to Peabody, the remaining 308 acres cannot be graded, as they encompass open mining pits, roads, sediment ponds, topsoil stockpiles, spoil ridges, and ramps leading to the mine pits.[12] In 1977 Peabody mined 103 acres of land and graded 108 acres. In 1978 the company mined 193 acres of land and graded 53 acres.[13]

Peabody's current grading procedures are good. Grading follows mining very closely. A December 1977 inspection conducted by the Mined Land Reclamation Board found spoil piles were graded to within one and one-half ridges of the active mining pits.[14] During INFORM's tour of the mine site in June 1979, grading was within three spoil ridges of active mine pits.

Graded lands, however, do not always approximate the contour of the pre-mining topography. The Mined Land Reclamation Board suggested in December 1977 that Peabody grade land it was reclaiming to a more rolling topography. The MLRB also cited a letter from Colorado's Department of Wildlife recommending the same procedure.[15] In July 1978 an MLRB inspector wrote that the "biggest problems at this mine are a result of the steep slopes [and the company should be] regrading back to approximate original contours (NOT to uniform slopes across the entire permit area)...."[16] If Peabody fails to create a more rolling topography (rolling up and down the slope), the steep slopes of the reclaimed land will increase the flow of water and thus increase the chances of erosion. Rolling topography alters the pattern in which water flows from the mine site, and prevents it from cascading down in one wide sheet. Rolling topography also provides a wider range of habitat for animals and allows snow to accumulate, yielding more water in the spring for establishing plant life. The company has changed recent grading to create these conditions.

In 1979 Peabody officials said that they wish to leave some permanent water bodies on the mine site as part of their final reclamation plan.[17] The company will submit a new mining and reclamation plan requesting such permission to the Mined Land Reclamation Board and OSM in late 1979 or early 1980. One such body of water may be the 554 acre-foot sediment pond located at the bottom of the hill where mining is taking place. A representative of the Colorado Mined Land Reclamation Board doubts that the MLRB will allow the company to leave a body of water that large on the site,[18] as it may not have been built to meet design criteria and state and federal requirements. If Peabody is permitted to leave permanent water impoundments on the site, these bodies of water would alter the pre-mining topography but may enhance the value of the land.

Topsoil

Removing topsoil on the site is difficult, but Peabody is trying to salvage as much of the material as possible. From 4 inches to 4 feet of topsoil cover the mine area. However, a layer of shale lies beneath the topsoil. This shale layer is irregular, near the surface in some areas and deep beneath it in others. As a result, scrapers which pick up topsoil sometimes collect shale as well. According to a state inspector, it is better to remove all topsoil and gather some shale with it than to lose topsoil by trying to avoid shale. Although not usually toxic, shale is not a fertile medium in which plants can grow.[19]

Peabody reports that 10-15 percent of all the topsoil removed at the mine is immediately replaced on graded lands,[20] which is more than many of the mines in INFORM's sample achieve.

The mine site contains no subsoil, making it unnecessary to segregate topsoil and subsoil.

Removing topsoil at the mine has not always been done completely. In 1979 a representative of the Mined Land Reclamation Board stated that removal operations at Seneca II had been "sloppy" in the past. He feared that Peabody may run out of stockpiled topsoil before it can cover all graded lands.[21] In October 1978 the Mined Land Reclamation Board fined Peabody $18,000 for failing to remove topsoil and filling other mine pits with overburden and topsoil removed during mining.[22] When questioned about this fine, an inspector with the MLRB said, "Had they [Peabody] been fined for the value of the topsoil, the fine could have been $150,000 to $250,000."[23] He also criticized

Peabody's entire topsoil operation, stating that the company "had a real topsoiling problem."[24] When interviewed again in July 1979, this same inspector said that Peabody's topsoil removal operations were "getting better." He added that the company was collecting topsoil well in advance of mining.[25]

When saved the company replaces topsoil evenly on reclaimed lands. Peabody first began saving and replacing topsoil on graded lands in 1976. Company officials told INFORM that the company spreads from 8 - 12 inches of topsoil on all graded areas at Seneca II.[26] Before 1976 Peabody reported that in areas where topsoil was insufficient, overburden would be suitable for revegetation.[27] A state inspector familiar with the Seneca II operations was incredulous: "How it [the proposal to substitute overburden for topsoil] got approved is beyond me.... They think they can revegetate spoil, and that's just plain garbage. There was not in my opinion, sufficient data...to support the proposal."[28]

Peabody has also improved procedures for stockpiling topsoil at the mine. The company has enlarged the size of these stockpiles, improving the chances of keeping topsoil from being lost as small stockpiles scattered over the mine area tend to be more difficult to remove when needed. In addition, less of the topsoil's surface area will be exposed to air and water, decreasing the potential for erosion.[29] During a visit to the site in June 1979, INFORM noted that stockpiles to which topsoil was being added showed vegetation along their sides and few signs of erosion. Peabody has also improved the location of these stockpiles. Older piles placed near haul roads risked contamination from dust and road scrapings pushed aside when roads are plowed in winter. These materials can prevent vegetation from growing on the stockpiles. Contaminants would also be picked up along with the topsoil when it is respread over graded lands. Newer stockpiles are not as close to haul roads and are placed in more protected areas.

Toxic Material

Although Peabody is not mining the Wolf Creek coal seam at Seneca II, the seam is exposed during mining operations, and the overburden atop the seam is acidic. The Mined Land Reclamation Board's inspections of September 1977 and July 1978 both indicated that acidic materials were forming "hot spots" in the ground.[30] These hot spots are areas where acidic material leaches to the surface and can create conditions preventing vegetative growth. In July 1979 Alten F. Grandt, reclamation director of Peabody's Western Group, told INFORM that the company has never taken a core sample of the soils in the areas where this problem exists. However, he said that the mine's reclamation supervisor believes that the shale around the seam and not the coal seam itself is the source of the problem.[31] Although this problem may only affect areas 4 feet in diameter over a few acres of land, as of July 1979, the company has not altered its handling of overburden to control the contamination acidic soils produce.[32]

WATER QUALITY

In the last 2 years, Peabody's program for ensuring water quality both on and off the mine site has been good. However, the company had problems previously with erosion of lands at the site where surface water flow is concentrated, such as culverts through which passes water under haul roads. Peabody has contained sediment on the mine site, much like the neighboring Energy Fuels mines. Steep slopes and summer rainfall, higher than that which many mines in the West receive in a full year, combine to create the potential for harmful erosion on disturbed lands. But Peabody has successfully controlled erosion in most areas on the mine. Both the U.S. Geological Survey and the company operate surface and subsurface water monitoring stations in the area, and INFORM found no evidence that mining disturbs surface waters. Peabody is mining through aquifers, but the quality of their water is considered poor. No alluvial valley floors have been identified on the mine site.[33]

Surface Water and Groundwater

The mine's water control network consists of two sediment ponds, and drainage channels which run alongside main haul roads on the site. Though Seneca II contains fewer sediment ponds than most mines in this part of the country, the main pond is extremely large with a capacity of 544 acre-feet of water.[34] The main pond traps virtually all the water which runs off the mine site and is large enough to hold this runoff. It has adequately controlled sedimentation. Under the current National Pollutant Discharge Elimination System (NPDES) permit for the mine, Peabody can discharge water from two points, one at each of the mine's two sediment ponds.[35] Water is discharged into natural drainages which eventually carry the water to the Yampa River. Streams near the mine site include the Grassy Creek and Little Grassy Creek, tributaries of the Yampa River. While Peabody considers these creeks intermittent along most of their length, perennial seepage of water from underground springs keeps portions of the creeks flowing throughout the year.[36] Both creeks receive the water discharged from the mine site. However, water must be contained and sediment allowed to settle before being released from the mine site. The mine's water con-

trol network must be capable of handling rainfall up to 1.7 inches
in 24 hours, the amount the U.S. Environmental Protection
Agency defines as the maximum daily rainfall or its equivalent
likely to occur within a 10-year period.[37] The system is adequate, but according to the Mined Land Reclamation Board, melting
snow, channelled into the main sediment pond, fills it near capacity in the spring.[38] Peabody uses water in the pond to spray
on roads at the mine in order to control dust. The rest is held
in the pond and discharged from the site when water reaches the
level of the discharge channel. The other sediment pond collects
runoff in a small area and is unimportant for most of the year as
it handles very little runoff.

In 1979 at least two or three water channels at Seneca II
were used to direct water through culverts under haul roads.
Since these roads run along the contours of the hills and water
channels flow downhill, water must pass through these culverts
to travel under the roads. In November 1978 an OSM inspection
found that some of these culverts might not be sufficient to allow
the runoff from a 10-year/24-hour precipitation event to flow
through the system without causing erosion and sedimentation.[39]
The inspection report also said that Peabody had made no provisions to slow the speed of flowing water in order to minimize the
potential for erosion around these culverts. OSM found that no
devices for controlling the flow of water "were installed at the
culvert [installed under the haul road] intake to avoid plugging,
and no form of energy dissipators were evident at the culvert
discharge. Excessive erosion...would be expected to occur during periods of surface water runoff."[40] In December 1978 Roy
Karo, the mine's reclamation supervisor, said the company was
drawing up plans to deal with these problems as a result of this
finding.[41] In July 1979 an inspector from Colorado's Mined
Land Reclamation Board told INFORM that Peabody had improved
placement of culverts under haul roads.[42] During a tour of the
mine site in June 1979, INFORM saw very little erosion of the
water control system but noted very little vegetation along either
these water channels or the main sediment pond. Although the
system was carefully maintained and sufficient to handle summer
rains, it is unclear what effect this lack of vegetation will have
on erosion of the system during seasons when water flow is
rapid.

Water discharges from Seneca II have been within NPDES
permit limits for the last 2 years. Peabody reported that it did
not discharge water from the mine site from April 1977 through
March 1978. From April 1978 through July 1979 all water discharges were within permissible limits. No water was discharged
from the mine site between July 1, 1978 and January 28, 1979.[43]
Thus, although the water control system may be stretched to its
limits, no off-site disturbance has occurred.

The impact of Peabody's mining through aquifers at Seneca II will probably not be significant, though it is still not certain. The company has reported that aquifers exist within, above and below the principal coal seam it is mining.[44] In mid-1979 company officials said that water from these aquifers seeps into the mining pits where the coal seam and earth are exposed.[45] However, Peabody reports that the water in these aquifers is unsuitable for domestic use.[46] The company adds that the soils above these groundwater systems are extremely porous, allowing water to infiltrate readily from the surface,[47] so that the aquifers are recharged, thus replacing water lost during mining. However, as of late 1979, Peabody had not completed its first analysis of the area.[48]

Peabody and the U.S. Geological Survey monitor surface water and groundwater on and off the Seneca II site. The U.S. Geological Survey operates two stations monitoring surface water and one station sampling groundwater. Peabody employs one surface water sampling station and has sunk ten wells--four in 1978--to evaluate groundwater.[49] According to Alten F. Grandt, director of reclamation for Peabody's Western Group, a hydrologist monitors these wells regularly.[50] Peabody is studying the amount of water available on the mine site, the control of erosion and sedimentation in the area, and conditions during peak flow of surface water at the mine.[51] Yet, when INFORM questioned both Grandt and mine superintendent Don Zulian about the effects of mining operations on groundwater in the area, neither commented on the subject.[52]

Alluvial Valley Floors

The Seneca II site contains no alluvial valleys.[53]

Erosion

Severe erosion may occur at Seneca II. A July 1978 inspection by the Mined Land Reclamation Board noted that the steep slopes in the area made erosion one of the "biggest problems at [the] mine."[54] Alten Grandt disagrees.[55] In June 1979 he said that the roads built along the contours of hills on the mine site act as terraces to slow the flow of water and decrease erosion.[56] INFORM's visit to the site in June 1979 supported the view that roads would slow water flowing downhill. However, most erosion on the site has occurred along these roads. Where roads are cut into the hillside, no vegetation grows, and much of the soil along these roads is severely eroded. In July 1978 the Mined Land Reclamation Board stated that "some...roadcuts may need intensive revegetation efforts" to prevent erosion.[57] As noted earlier, some erosion has also occurred in areas where water passes through culverts under these roads. Vegetation on lands which have not been disturbed by mining largely controls erosion on these lands.

The greatest erosion at Seneca II occurs in areas where the land has been disturbed and reclamation has not begun, mostly roads and culverts. Graded and topsoiled slopes are carefully seeded and maintained. Once vegetation has been established, the potential for further erosion is reduced. INFORM saw very little erosion in reclaimed areas where vegetation had begun to grow. Lands without vegetation contain small gullies 2-4 inches deep running downhill at a distance of 2 feet from one another. These gullies could increase in size if unattended. During the visit, Peabody officials told INFORM that the mine's staff would create small furrows along the contour of slopes once erosion does occur.[58] Yet, a good reclamation program would create these furrows before erosion began, much the way Energy Fuels does at its mines.

In September 1979 Alten Grandt said that the company had just begun furrowing along the contour of graded slopes to control erosion.[59] Unlike Energy Fuels, Peabody did no furrowing before 1979. In December 1977 an inspector with the Mined Land Reclamation Board "strongly recommend[ed] future seeding be done...only *AFTER* [MLRB's emphasis] some erosion control or water diversion ditches have been constructed on long, steep reclaimed slopes."[60] Again, in July 1978, the MLRB suggested that "some of the steeper slopes should be cross-ditched along the contour to mitigate the erosion of soil."[61] Responding slowly to these suggestions, Peabody now places furrows along the contour of lands being reclaimed,[62] but, according to Alten Grandt, the company also plans to transform roads on the mine site to act as terraces once mining is completed. Peabody will alter the roads to slope inwards toward the hill to collect water running down the hillside. This measure could effectively control erosion in steeply sloped areas, as long as the roads themselves did not erode because they lacked stabilizing vegetation.

SEEDING AND AMENDMENTS

Pressure from regulators and a more conscientious attitude toward revegetation have led to many recent improvements in Peabody's fair to good seeding and amendment program at Seneca II. Past seeding relied heavily on legumes, such as alfalfa, which dominate lands seeded in the early 1970s. Very few other plants, such as grasses, have been able to grow, although these plants are more suitable for grazing, the site's proposed post-mining use. The new seeding program instituted in the fall of 1977 relies much more heavily on native grass species. This mixture will produce a much more diverse selection of grasses on the mined land, providing a wide variety of nutrients to animals as well as a less homogeneous, and therefore hardier, ground cover. In 1979, Peabody began creating furrows along the contour of slopes to be seeded. This procedure should help control erosion and improve the chances for successful revegetation. Energy Fuels has successfully used these practices at its mines since 1976. Under OSM pressure, Peabody has also begun to add mulch to revegetated lands, another improvement over past performance. Although the results of these changes will not be evident for several years, they should greatly improve Peabody's chances for successful and environmentally sound reclamation at Seneca II.

Seeding Procedures

Peabody has yet to seed as much land as it grades at the Seneca II Mine, but the gap between these two figures is shrinking. In 1977 Peabody graded 108 acres of land at the site and seeded only 19 acres. The next year, the company graded 53 acres but seeded 125 acres with grasses, while planting trees and shrubs on twelve of these acres.[63] The difference between graded and seeded lands is decreasing due to improved reclamation practices and compliance with the requirements of the Mined Land Reclamation Board and OSM.

Most seeding is done in the fall,[64] the generally accepted time for seeding throughout most of the western United States (except the Southwest). Seeding at this time allows seeds to begin germinating and growing in the spring--before the ground is hard enough to support seeding machinery--thus increasing the chances of successful growth. In the spring of 1978 and 1979, the company also seeded certain areas of the site. Neither James Jones, Peabody Coal's director of environmental quality, nor Alten Grandt knew the reason for this change in seeding practice.[65]

Under a 1977 directive from the Mined Land Reclamation Board, Peabody improved the seed mixture it uses on reclaimed lands. The company seeds 10 species of grasses and 3 species of legumes on lands at the mine, a change from the mixture of 6 grass species and 3 legume species it used in 1976.[66] Decreasing the number of legumes seeded per acre from 10 pounds to 4 pounds and increasing the number of grasses seeded should decrease the number of leguminous plants--which do not provide the adequate grazing grasses do--in the site's vegetative cover. Leguminous plants like alfalfa also tend to prevent grasses from becoming established in large quantities.

Peabody seeds native and introduced grass species at the site. Most of these grasses are cool-season plants, which could mean that vegetative cover on reclaimed lands will be sparse in the summer, allowing invasion of seeded areas by undesirable plant species. Preventing undesirable plants from invading an area enhances the opportunity for desired vegetation to grow in the fall. Weeds are especially hardy and may compete with desired vegetation, decreasing the chances of successful reclamation.

Peabody does not experiment with different seed mixtures at the mine site. The company reports that the seed mixture used there has been "developed from experience since 1965."[67] Peabody officials state that mine staff seeds topsoil stockpiles with the regular seed mixture if the stockpile is to remain in place for more than 1 or 2 years. If the stockpile is to be removed sooner, it is seeded with grains such as winter wheat, wheat or oats.[68] These grains provide quick temporary cover, while grasses provide longer lasting protection from erosion.

TABLE A

1979 SEED MIXTURE
FOR LANDS BEING RECLAIMED AT SENECA II

Plant Species	Introduced/ Native	Cool/ Warm	Pounds Applied Per Acre
Grasses			
Slender wheatgrass	N	C	2
Western wheatgrass	N	C	2
Pubescent wheatgrass	I	C	1
Intermediate wheatgrass	I	C	1
Whitmar wheatgrass	N	C	1
Hard fescue	N	C	1
Bromegrass (Smooth brome)	I	C	3
Orchardgrass	I	C-W	1
Canada bluegrass	N	C	1
Russian wildrye	I	C	1
			14 Pure live seed
Legumes			
Alfalfa	I	C-W	2
Sweetclover	I	W	1
Cicer milkvetch	I	W	1
			4 Pure live seed

Source: Peabody Coal

In June 1979 Peabody told INFORM it usually uses 14 pounds
of grass seed per acre and 4 pounds of legume seed per acre at
the Seneca II Mine, an amount of seed comparable to that used
by other companies in this study.[69] However, when using a
broadcasting seeder on steeper slopes (about 50 percent of the
site), as opposed to a drill seeder, reclamation staff increases
the amount of grass seeds to 18 pounds per acre because more
seeds are blown away, and it is more difficult to place the seed
evenly.[70] As of December 31, 1978 Peabody reported that it
had seeded 25,420 pounds of seed on 490 acres of the Seneca II
site, an average of just under 52 pounds of seed per acre.[71] In
September 1979 Alten Grandt explained that the company reseeds regraded lands frequently to improve vegetative cover.[72]
Peabody does not vary the amount or type of seed planted on
varying slopes or soils at the mine site. Because of the variation
in slopes, it might be better to seed faster-growing plants on
steeper lands to prevent erosion and then to reseed with the
normal plant mixture once vegetation has been established.

Since 1977 Peabody has greatly improved seeding methods
at the Seneca II Mine. In the early 1970s, mine staff broadcasted
almost all seed from the air. In 1979 Alten Grandt said Peabody
used a drill seeder for about 50 percent of seeding. Grandt
stated that a converted fertilizer spreader, which broadcasts the
seed over the ground after a discing machine has cut grooves
into the soil, carries out most of the rest of the seeding. This
method does not put seeds into the soil and, as mentioned earlier, when broadcast seeding is employed, the company adds 4
pounds of grass per acre to compensate for this loss.[73] Peabody
says that between 75 and 85 percent of the seeding is done along
the contour of the slopes, a good practice in these steep lands,
because tracks left by the machinery help reduce erosion and
trap water for use by plants.[74] A spokesman for the Mined Land
Reclamation Board agreed.[75] Peabody also manually seeds steep
slopes, such as the sides of stockpiles, where machinery cannot
be operated. Seeds broadcast are not generally covered by soil
after their application. The company reports that reclamation
staff does very little harrowing (covering of seed with soil) after
seeding.[76]

Amendments and Maintenance

As reported earlier, Peabody now furrows land at the site
after topsoil has been applied. The company states that when
mine staff operates reclamation machinery along the contour, it
creates furrows which can collect moisture and provide a good
seedbed in which plants can grow, as water and roots are better
able to penetrate the loosened soil.[77] According to Alten Grandt,

in 1979 Peabody purchased a larger tractor to replace a much smaller one that was incapable of performing the jobs required of it.[78]

Peabody employs few amendments at the site. The company does not irrigate, fertilize, or use herbicide on reclaimed lands.[79] Under federal mandate, Peabody began using mulch in the fall of 1978.[80] Mulching machinery moves up and down slopes rather than along the contour.[81] Mulch is blown onto the ground and then crimped into the soil when seeding machinery passes over it.[82] Since only one tractor is used in reclamation at the site, and mulching is done up and down the slope, while most of the seeding is done along the contour, the mulch is not crimped into the soil of reclaimed lands immediately after it is placed on them, increasing the chance it will be blown or washed away. Peabody uses hay mulch, consisting of dried grasses, on the site. As a result the company has experienced problems with undesirable species of grass growing on newly topsoiled lands. Peabody plans to use a straw mulch, which consists of dried annual grain, to avoid this problem, but as of mid-1979, had not yet done so.[83]

Trees and Shrubs

Peabody Coal has planted both trees and shrubs, including both Ponderosa pine and conifers, at Seneca II, to establish the type of vegetation that existed in the area before mining began. However, in the fall of 1979, Alten Grandt reported that almost all conifers had died after planting.[84]

Peabody plants almost all shrubs from small, bare-root sprigs, not full transplants, the method employed at the Energy Fuels mines where that company has had much better success with its transplants. In April and May 1979, Peabody planted 2,500 bare-root shrubs. However, as the summer progressed, the plants died due to lack of moisture.[85] In July 1979 an inspector with the Mined Land Reclamation Board told INFORM that "bare root planting of shrubs should work in theory, but in practice around Colorado, it has yet to show success."[86] In a summer 1978 inspection, another state inspector recommended "transplanting trees and shrubs [to] better achieve final reclamation goal[s]--wildlife and domestic grazing."[87] In the spring of the second year after seeding, reclamation staff at the mine plants the bare-root shrubs by hand.[88]

Although Peabody has had limited success in planting trees and shrubs at the site, the company has conducted studies on the subject. In a September 1974 memorandum, Russell T. Moore of Ecology Consultants told Peabody that Russian olive and old-man wormwood sage showed the most vigorous growth that year. Other successful species were caragana, willow and green ash. The memorandum also stated that after 3 years of growth, some

native shrubs had invaded the area.[89] Peabody currently includes old-man wormwood sage in the shrubs it is planting. In
1975 William A. Berg, a scientist from Colorado State University
who experimented with shrubs on the Seneca II site, said that in
order to increase the chances for survival of shrubs planted on
reclaimed land, Peabody would have to continue managing the
area. He added that browsing by wildlife, competition with other
plants for moisture, the invasion of weeds, and erosion in areas
without grassy ground cover would doom most plant species.[90]
In the late 1960s, Peabody established a test plot at the mine to
evaluate the planting of shrubs. The company is also fencing
many areas to stop cattle from grazing in them. However, Peabody has not had as much success establishing shrubs and trees
in large areas as it has had in small test areas.

VEGETATION

To date, reclamation activities at Seneca II have failed to
produce good grazing vegetation. Peabody officials say that
they expect that 10 years after mining began reclaimed lands will
have achieved only 50 percent of pre-mining productivity, far
short of OSM's mandated 90 percent.[91] Peabody first replaced
salvaged topsoil on graded lands in 1976 as required by state
law. This topsoil was seeded in 1977. Lands seeded before 1977
are covered with alfalfa, a crop which can bloat cattle if eaten at
the wrong time in the plant's life cycle. Although these lands
have enough cover to prevent erosion, the diversity and thus
the quality of the cover suffers because it consists solely of alfalfa. The company has had some success in establishing a few
shrubs in one area of the mine, but more are needed to re-establish the vegetative community that existed before mining began.
Areas seeded since 1977 showed very little growth of grasses in
mid-1979. Company officials say that they plan to reseed these
areas in a year or two, a practice they believe necessary over
wide areas of the mine, because initial seeding has yet to establish a cover adequate to prevent erosion.[92]
Peabody has established a few grass species in small quantities on approximately 450 acres of land seeded in the early and
middle 1970s. Although alfalfa dominates the vegetation on these
lands, some crested, western and intermediate wheatgrasses
have appeared. Smooth brome and yellow sweetclover also have
appeared.[93] Western and intermediate wheatgrasses are native
grass species. Their presence shows that such plants can be
established but their growth may be hindered by excessive
amounts of alfalfa.[94] Shrubs such as Russian olive, caragana
and golden willow will also grow on these reclaimed lands.
Chokecherry and serviceberry may also become part of their
vegetation.

Lands seeded in the fall of 1977 contain a good deal of Russian thistle, an undesirable species which cannot serve as forage, and uses much water and soil nutrients. However, this species provides some shelter for grasses such as Canadian bluegrass, bromegrass, orchardgrass, western and slender wheatgrass, and sideoats grama. Once the thistle withers, these grasses may grow faster and expand to cover more surface. Vegetative cover on this land has failed to control erosion, and in mid-1979, Peabody said that it planned to reseed this area in the fall of 1979.[95]

Vegetative growth on a 10-acre area where topsoil was immediately spread over reclaimed lands has shown that direct use of the material is desirable. In the fall of 1978, Peabody seeded this land and covered it with straw mulch. Although Russian thistle dominates vegetation on this land, as it does other recently seeded areas on the site, grass species which were not seeded have also appeared there. Alten Grandt theorizes that this grass sprang from live seeds carried in the topsoil which was scraped and immediately placed on reclaimed lands.[96]

Since Peabody reduced the number of legumes it seeds on reclaimed lands in the last 2 years, far fewer legumes have appeared on newly seeded lands. Although they provide the soil with nitrogen, legumes tend to prevent other plants from becoming established in areas where they grow. Because alfalfa dominates older reclaimed lands, company officials predict it will be 10 years before native plant species account for more than 50 percent of vegetative cover on these lands. Yet, since newer lands are seeded with fewer legumes such as alfalfa, this should allow native grasses to increase in number and provide a better ground cover, eventually yielding higher quality grazing.

Alfalfa covers lands which have enough vegetation to resist erosion and produce forage approximating pre-mining levels. This crop may not be compatible with cattle grazing because it can cause the animals to bloat and die. Alten Grandt told INFORM that this was not a problem. He said that the loss of a few cattle from bloat would be more than compensated by the increased weight gain of other animals.[97] Wallace McRae, a rancher in Montana and a former board member of the Northern Plains Resource Council, an organization representing western ranchers and environmentalists, disagreed: "You wouldn't kill a few cattle, you'd kill a lot. The person who said [that the increase in productivity would compensate for the death of a few cattle] doesn't know anything about the economics of ranching. Alfalfa is good when it is dry and fed to animals as hay, but when it is at its best [high in protein and vitamin A] during the growing season it is also the most dangerous for cattle."[98]

Grandt also said that Peabody may be able to produce more plant forage on reclaimed lands than the levels produced before

mining began.[99] Prior to mining, productivity on the site ranged
from 6 acres per animal unit month (AUM) in forb vegetation
areas to 3 acres of land per AUM in aspen tree vegetation areas.
areas.[100] (An AUM is equivalent to the amount of land necessary to support one cow and her calf, or five sheep for one
month.)

Data collected on the productivity of lands reclaimed in the
early 1970s showed that productivity has increased over pre-
mining levels. Company consultants William A. Berg and Russell
T. Moore found that forage produced on reclaimed land was twice
that of adjacent unmined rangeland.[101] However, most of this
vegetation is alfalfa. Moore's report, written in 1974, said:

> Non-bloating legumes, such as cicer milkvetch,
> could likely be used in place of the alfalfa to pro-
> vide rapid growth of a nitrogen-fixing species
> ....Grass stands from the initial seedings will
> likely increase in productivity for several years
> as the nitrogen level of the soil is gradually
> built up, but they will not likely match the peak
> productivity of alfalfa.

It concluded that "The reclamation procedures [reported on]...
have not been successful in developing a productive browse component [edible plant] as part of the revegetation program."[102]

Peabody reports reclaimed lands at the site will have a carry-
ing capacity of 1-1.5 acres per AUM. E. F. Sedgley, a Resource
Conservationist with the federal Soil Conservation Service, told
INFORM that soils in the area of the Seneca II Mine could produce from 800-2,500 pounds of dry forage per acre annually
before mining began.[103] In mid-1979 Alten Grandt said that
Moore's latest studies conducted in 1978 showed forage production at the site to be from 3,550-7,100 pounds per acre.[104]
These extraordinarily high figures for alfalfa growth support
Moore's earlier contention that "Alfalfa stands will probably
reach their peak density and productivity between 5 and 10
years after planting and then decline to a relatively stable low
level until displaced by the invasion of native species."[105] This
stable, low level will most likely fall in the accepted range of
forage production for the site of 800-2,500 pounds per acre.

Other than the data on alfalfa, Peabody has collected little
information on productivity at the mine site. However, in 1979,
the company hired Maria Associates, a Wyoming-based consulting
firm, to establish 250 fenced plots, 1 meter in diameter, on unmined portions of the site to determine pre-mining productivity.[106] Peabody reports that it plans to begin collecting data on
species type, forage production, plant cover and density. But
the company does not plan to conduct trial grazing on reclaimed
lands in the immediate future.[107]

WILDLIFE

Peabody is doing only a fair job at Seneca II to make reclaimed lands attractive to wildlife. The company is not constructing wildlife shelters or habitats, nor does it plan to do so.
Peabody officials state that unmined areas at the site contain
enough rock outcroppings and piles of brush to serve as shelter
for animals.[108]   Larry Damrau, an OSM inspector, agreed.[109]
The company adds that topsoil stockpiles seeded with grain attract small wildlife and that this constitutes enhancement of wildlife refuge land uses. However, topsoil stockpiles are temporary
structures and will not sustain animals once they are removed.
Tree and shrub planting should provide wildlife both food
and shelter once a stable vegetative community can be created.
But browsing wildlife may prevent the shrubs from ever taking
hold, and to date Peabody has had little success in planting
trees. The company has fenced the entire mine site since 1976
to prevent browsing by sheep and cattle but has not yet found a
way to control wildlife damage to plantings.[110]

RECLAMATION COSTS

TABLE B*

RECLAMATION COSTS AT SENECA II

Operation	Cost Per Acre	Percentage of Total Cost
Earthmoving		96%–100%
Removal, storage, and stabilization of topsoil	$1,500	
Grading	2,200	
Topsoil replacement	1,500	
Revegetation		0%–4%
Seeding	25	
Tree Planting	65	
Mulching†	140	
Total Cost:	$5,430	

*These costs and percentages are not strictly comparable to those of the same operation or activity at other mines. See the Cost of Reclamation chapter for a detailed explanation.

†The company reports this cost for 15 acres mulched out of a total of 125 acres seeded in 1978.

Source: Peabody Coal Company

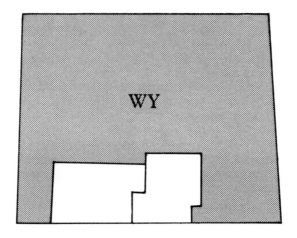

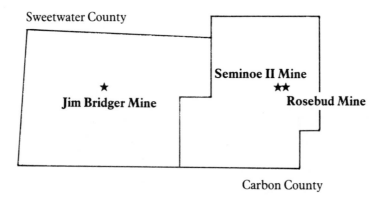

Sweetwater County

Seminoe II Mine

★ ★★

Jim Bridger Mine

Rosebud Mine

Carbon County

Southern Wyoming

This area is one of the drier and more difficult to reclaim regions in the northern parts of the West. The region receives less than 10 inches of precipitation annually. Like the Southwest, probably less successful reclamation is to be seen here than in Montana, Colorado or North Dakota. Reclaimers battle to beat the weeds and establish grasses and other desirable plants on the land. The Wyoming Department of Environmental Quality (DEQ) enforces a fair regulatory program. Aside from the DEQ, there is little activity by environmentalists; notable exceptions include the Wyoming Outdoor Council and the Sierra Club. In addition, Lander is the home of the High Country News, one of the premier environmental newspapers in the West.

Jim Bridger/NERCO/WY

Mine: Jim Bridger Mine

Location: Superior (population 885), Wyoming, in Sweetwater County (population 44,284)

Operating company: Bridger Coal Company

Parent company: Bridger Coal Company is a joint venture between Idaho Energy Resources (1/3 ownership) and Pacific Minerals (2/3 ownership). Pacific Minerals is a wholly owned subsidiary of Northern Energy Resources Company (NERCO), which in turn is wholly owned by Pacific Power & Light. Idaho Energy Resources is a wholly owned subsidiary of Idaho Power Company.

Parent company revenues: 1977, $156.4 million* and $404.6 million†; 1978, $195.3 million* and $506.1 million†

Parent company net income: 1977, $24.8 million* and $88.6 million†; 1978, $33.5 million* and $105.8 million†

First year of production: 1973

Coal production in 1978: 5,200,000 tons

Acres for which a mining permit has been issued: 23,471 (7,000 - 9,000 of which will be mined)

State regulatory authority: Wyoming Department of Environmental Quality

Reclamation bond posted: $6,950,000

Reclamation cost: reclamation cost per acre, $7,936; reclamation cost per ton of coal, NA

Expansion plans at mine: Annual production of 7,300,000 tons by 1980

Additional acres for which the company will seek a permit: none

Parent company U.S. coal reserves: NERCO owns or leases 1.4 billion tons of coal reserves in the West and 1.7 billion tons of coal in the United States.

Development and expansion plans for surface coal mines in western United States: In 1980, NERCO plans to open the Spring Creek Mine in Montana to produce 10 million tons per year by 1982. By 1984, the company plans to open the Cherokee Mine in Wyoming to produce 6 million tons annually by 1985; and in 1983, NERCO plans to open the Antelope Mine in Wyoming to produce 10 million tons per year by 1985.

NA = Not available

*Idaho Power Company
†Pacific Power & Light Company

Sources: NERCO; Pacific Power & Light Form 10-K, *1978;* Idaho Power Form 10-K, *1978;* 1977 Keystone Coal Industry Manual; *"New Coal Mine Development Survey 1978-1987,"* Coal Age, *February 1979;* Rand McNally Road Atlas, *54th Edition*

SITE DESCRIPTION

The Jim Bridger Mine lies at the western edge of the Red Desert in Sweetwater County, southwestern Wyoming. It is 35 miles northeast of Rock Springs. Bridger Coal Company says the mine, opened in 1973, contains about 40 years' worth of commercial coal resources.

The terrain consists of gently sloping, flat-topped bluffs with moderate to steep ridges, and broad, gently sloping valleys.[1] Slopes range from 3-15 degrees in the mine area, but a few 45-degree slopes occur.[2]

Local water drainages include Deadman Draw, which runs through the site; Nine-Mile and Ten-Mile draws, and Bitter Creek. But the area is generally arid, with annual precipitation averaging only 8.7 inches. About half of this precipitation falls as rain, with about 35 percent falling in April, May and October.[3] Total precipitation from 1970 through 1973 was above average, but in 1974, 1976 and 1977 precipitation was below average, placing stress not only on natural vegetation in the area but also on the company's revegetation efforts.[4] The ground remains

frost-free for about 100 days a year; the growing season is from
late May through August.[5]

Before mining began, the land served as rangeland for
sheep, cattle and horses, as well as a habitat for wildlife.[6] The
company intends to return the site to these uses.[7] The mine site
contains no farmland.[8]

OVERVIEW

In 1978 the Jim Bridger Mine produced 5.2 million tons of
coal--more than all but five other surface mines in the country.
However, as of May 1979 its record in reclaiming lands disturbed
was only fair to poor. Only 13 percent of the 1,850 acres disturbed by mining and related activities had been seeded. Poor
planning and the layout of the mine are the principal reasons
behind this serious delay. The problem could worsen if, as
planned, production increases by 40 percent by 1980 and the
eventual length of the active pit is increased to 16 miles. In mid-
1979 Tom Schreeg, the mine's reclamation specialist, told INFORM,
"Bridger Coal is putting the land back together," but he also
said the company was "raping the hell out of the land,"[9] as the
area is so hard to reclaim. If these conditions persist, the Jim
Bridger Mine may prove to be one of the biggest and most difficult surface coal mines in the country to reclaim. To date, revegetation and reclamation have been minimal. Although the
company plans to upgrade its mulching practices and to institute
use of its irrigation system, continued management of the land
could be required for many years after mining and basic reclamation practices cease.

The water quality control system at the mine--it had none
before 1979--is minimal. In this area of low precipitation, surface water runoff occurs only a few times each year. It was only
under pressure from the federal Office of Surface Mining (OSM)
that Bridger Coal finally began to construct a water-control system, which includes sediment ponds to control surface water
runoff. Although the company monitors groundwater, only
sparse data have been collected, and they show that mining has
not harmed water systems in the area.

Earthmoving operations at the Bridger Mine, although carried out well, are poorly planned. This diminishes the quality of
reclamation that can take place at the mine.

Although the grading of spoil ridges closely follows mining,
most mined land cannot be graded because of the location of
roads and ramps entering active mining pits. Bridger Coal removes and stockpiles scarce topsoil and subsoil. Despite this fact
and the fact that, in 1977, Wyoming's Department of Environmental Quality found that prompt replacement had reduced undesir-

able vegetation on land reclaimed at the mine in the past,[10] the
company does not replace topsoil on graded lands immediately.
Moreover, to accommodate expansion, topsoil stockpiles have
been moved and replaced on graded soils at inopportune times,
increasing the potential for erosion and loss of soil by destruction of any vegetative cover that may have been protecting the
soil.

The revegetation program at the mine has failed. Neither
state nor federal regulators have found otherwise. In August
1978 an inspector for Wyoming's Department of Environmental
Quality stated:

> I became quite disturbed following [a tour of the
> site] as I reconsidered the lack of success of the
> revegetation program at the Jim Bridger Mine...I
> am thoroughly convinced that the geographic lo-
> cation of the Jim Bridger Mine poses very great
> challenges to achieving reclamation in the time
> limits of the State Laws....[11]

Vegetation growing on reclaimed lands is not dense; nor does
it consist of desirable plant species. Weeds invade seeded areas,
and grasses are usually eaten by rabbits. Although the mine's
seeding and amendment program is being revised, the company
has prepared very little land for seeding. Bridger Coal is now
beginning to use an enormous irrigation system to provide water
to all reclaimed lands, but the system will be effective only as
the amount of land to be irrigated is increased. At present,
Bridger Coal seeds very few plants at the mine, and even fewer
are actually growing. Irrigation should help to establish plants
on newly graded and topsoiled lands, but the question remains
whether it will need to be continued indefinitely if these plants
cannot survive under normal conditions.

Dick Kail, director of reclamation at the mine, sees improved
planning as the key to reclamation success at the mine: when
mining is complete in 40 years, he says, there is "still going to
be a lot of work to do."[12] Without the planning Kail seeks, the
16 miles of mining pit at the Jim Bridger Mine may leave a monu-
ment to the ravages of surface mining on the western edge of
Wyoming's Red Desert.

The impact of regulatory enforcement by the Wyoming state
regulators is harder to measure than that of OSM. The Wyoming
Department of Environmental Quality (DEQ) inspected the Jim
Bridger Mine seven times between February 1977 and July 1978.
Although problems with topsoil stockpiling, lack of vegetative
establishment and invasion of weeds were noted, none of DEQ's
reports indicated that any enforcement actions had been taken.
The inspections appear to be basically a means of reporting on

the state of operations at the mine at a given time. The DEQ
refused to allow INFORM to interview the inspector most knowledgeable about Bridger. It also declined INFORM's request for
copies of more recent inspections. Other states, notably Montana, Colorado and North Dakota, were more responsive to these
requests, as was the Office of Surface Mining.

OSM inspected the mine twice between March 1979 and June
1979. On its first inspection, in March 1979, the agency cited
the company for three violations concerning potential water degradation from erosion and sedimentation and lack of a sedimentation pond system.[13] No fines were levied, but Bridger Coal has
built sediment ponds and repaired the immediate erosion problems
cited in the violations.

Because much of the coal being mined at the Jim Bridger
Mine is federal coal, the U.S. Geological Survey visited the mine
site at least 15 times between July 1976 and November 1978.
However, these visits were similar to those made by the DEQ,
and no orders or recommendations resulted.

EARTHMOVING

Grading

Hundreds of acres of land have been disturbed at the Jim
Bridger Mine, but few have been graded and seeded. From May
1978 to May 1979, 462 acres of land were disturbed by mining
and related activities, and only 82 acres were graded. Of the
1,850 acres affected by May 1979, only 246 acres (13 percent)
had been graded, covered with topsoil and seeded.[14] Though in
part due to the mine's having opened only recently, the fact remains that less land (on a percentage basis) has been graded
and seeded at Bridger than at most other mines in INFORM's
sample.

When done, grading is good, approximating the contour of
surrounding lands and closely following the active mine pit and
mining operations. In June 1979 the company was grading spoil
piles within two spoil ridges of the open mine pit, and in some
areas spoil piles had been graded almost to the edge of the working pit. Graded lands are easily traversable by seeding and
other reclamation equipment. However, OSM reports that spoil
materials from the initial "box cut" or first mine pit cannot be
graded to the contour of the surrounding area because these
piles are so large. The result is a change in the topography.[15]
Gary Deveraux, director of mined land reclamation for Northern
Energy Resources Company (NERCO), one of Jim Bridger's parent companies, says this spoil will eventually be graded to approximate the original contour of the land.[16]

As noted above, the mine's layout is a key factor in the company's failure to keep grading current with mining. In November 1979 the mine consisted of four mine pits, which by mid-1981
will be connected to form one pit 16 miles long, dissected only by
Deadman Draw. As this pit grows longer so will each of the
spoil ridges, and careful monitoring of the disturbed land will be
required.[17] Access ramps enter the pits from their non-active
side. Thus, as mining progresses and the mine pit advances,
the access roads become longer. As a result, grading cannot be
completed; the roads would be graded as well as the spoil piles.
Only small areas of land between roads are graded and reclaimed.
In March 1979 OSM suggested that roads enter the mine pit from
the highwall or active side and be eliminated as mining progresses.[18] Dick Kail told INFORM in June 1979 that the mine's reclamation staff agrees with OSM's recommendation. But the company
has not yet carried it out.[19]

Another grading problem will occur in the middle of the mine
area. According to the mine plan, about 60 acres of land there
will be left unreclaimed for 10-15 years in order to allow mining
machinery to travel from the northern section of the mine to the
southern section.[20] Again, the mine's design accounts for this,
and alternative plans might have prevented this situation, which
delays timely reclamation.

Topsoil

Topsoil removal operations at the mine are good, but stockpiling and replacement practices are only fair to poor. Bridger
Coal segregates topsoil and subsoil collected from disturbed lands
and eventually replaces them on graded lands to enhance plant
growth. However, like some other companies, it strips topsoil
and subsoil one year before mining begins, usually in the summer, when the soil is easiest to remove.[21] This means most of
the soil must be stockpiled for long periods of time, risking the
loss of much of its nutritive quality. Because topsoil in many
areas of the mine is thin, reaching a depth of only 3 inches,[22]
Bridger Coal removes the topsoil and enough subsoil to replace
8-12 inches over the entire mine area.[23] In addition, according
to Tom Schreeg, reclamation specialist at the mine, 5-7 percent
of the land contains sagebrush draws where topsoil and subsoils
reach depths of 28 feet. Schreeg says the company salvages all
these soils.[24]

The reason Bridger stockpiles all topsoil and subsoil disturbed at the site, says Kail, is the size of the mining operation:
immediate replacement is impractical given the company's lack of
grading equipment and its procedures for grading small areas,
he explains.[25] Gary Deveraux says the company can immediately
replace about 30 percent of the topsoil and subsoil, but that

operational difficulties prevent it from doing more. For example,
Deveraux says, the topsoil would have to be transported long
distances (because it is removed so far ahead of mining) to reach
graded lands that were ready for its application, and the expense
would be too great.[26] However, the company has incurred additional expenses in stockpiling, too, as vegetation has been poor
and must be replaced, and stockpiles--as discussed below--must
sometimes be moved because of poor planning. The company has
not attempted to improve its topsoil handling operations, and,
according to Kail, no topsoil had been immediately replaced by
June 1979.[27]

The U.S. Geological Survey inspected the mine and found
that "stored topsoil banks appear to have been relatively stabilized by grasses seeded on them." But it also reported "excessive
wind erosion of stored topsoil banks."[28] In March 1979 OSM said
hand seeding of steep stockpiles was not producing an effective
plant cover on the sides of the stockpiles.[29] In June 1979 IN-
FORM found two stockpiles too steep to be seeded effectively to
produce a plant cover and at least six stockpiles--in place for
more than $2\frac{1}{2}$ years--lacking vegetative cover and beginning to
slide in places. In March 1979 OSM said some stockpiles "were
too steep to seed with a range drill"[30] and suggested changes
such as altering the method by which the topsoil was unloaded in
order to create flatter slopes. Bridger now creates flatter stockpiles, which can be seeded on the contour.[31]

Coal production dictates reclamation progress to a higher
degree at the Bridger Mine than at many other mines. Mining
operations have made it necessary to move stockpiles of topsoil,
incurring the risk of loss of this valuable material. An April
1978 inspection of the mine by Wyoming's Department of Environmental Quality found that between January and April, four topsoil stockpiles had been relocated and another nine would have
to be moved later in the year.[32] Kail explains that the economics
of mining, as well as planning at the mine, have required the
movement of topsoil stockpiles. For example, he says, economics
may dictate that mining be conducted only to depths of 100 feet.
In this case, Bridger Coal will stockpile topsoil on areas where
coal exists only at the 200-foot level. But if the economics of
mining (related to the price of coal) change by the time mining
reaches this area, these stockpiles must be relocated.[33] On occasion, Kail said, the advance of mining operations has forced him
--against his judgment--to apply topsoil at the wrong time of
year or to lands not ready for the soil.[34]

The U.S. Geological Survey found another instance where
poor planning hindered reclamation. A March 1978 inspection
report said, "Reclaimed areas were being disturbed [topsoil is
being stripped] to make room for spoil piles."[35] Constant move-

ment of topsoil prevents material from being vegetated and stabilized; storage increases the risk of loss or contamination.

A plan to incorporate immediate replacement of topsoil might
solve some of these problems. Such a plan could involve changes
in mine layout that would allow haul roads to enter mining pits
from the highwall (or active) side, or the use of conveyor systems to move subsoil and topsoil.

Toxic Materials

The mine has experienced no problems with toxic materials.
The company uses fly ash from the Jim Bridger Steam-Electric
Plant--which burns most of the coal produced at the mine to
generate electricity--to fill pits after coal is mined. This procedure has caused no harm, but OSM warns that no ash should
be disposed of (buried) in roads and haul ramps which act as
paths for surface water drainage.[36] This ash could contaminate
water, as it is rich in minerals that might hinder plant growth
and pollute water systems.

WATER QUALITY

Surface Water and Groundwater

Until September 1979 the mine had no sediment ponds. In
its mine plan as updated a year earlier, Bridger Coal said such
ponds were unnecessary to comply with standards for permissible
sediment levels (TSS) in waters flowing off the mine site.[37] The
company said runoff could be controlled by the use of straw
bales and sod to slow down and filter water.[38] But in March
1978 a U.S. Geological Survey inspection found that "in several
instances, silt has been removed in runoff water and redeposited
in natural drainages."[39] Once in the natural drainage, this silt
is more easily picked up when waters flow through the drainage,
and could affect lands and waters farther downstream. An OSM
inspection in March 1979 found that drainage from the mine had
flowed into Deadman Draw, carrying enough sediment to increase
the potential for killing fish and destroying aquatic habitats in
both Bitter Creek and Green River.[40] As no sedimentation ponds
existed at the time, water discharges have not been monitored on
the site, and the Environmental Protection Agency (EPA) has no
record of any having been monitored before 1979.[41] By February 1980 EPA had yet to issue a permit for the mine to discharge
waters from the site.[42]

Bridger Coal had planned to protect Deadman Draw. In its
1978 mine plan modification, the company said no operations
would be allowed within 100 feet of the drainage unless it was diverted.[43] However, in 1979 mining machinery crossed the drain-

age; this disturbed only about 100 square feet of land within the planned buffer zone,[44] but the impact of this erosion could have extended to any lands reachable by sediment carried in the water. OSM found three violations: sedimentation caused by the failure to pass water runoff through sediment ponds; disturbance of the buffer zone surrounding Deadman Draw; and failure to install culverts in natural drainage areas where haul roads and access roads were built to permit machinery to cross the mine site.[45] No fines were levied, and the company was ordered to clean up the situation, which it did.

By mid-1979 Bridger Coal had built 20 sediment ponds on the mine site and OSM was to have certified this system by the end of that summer.[46] Though not formally certified, the company's water-control plans are now adequate to meet OSM requirements.[47] The company also said it plans to retain, and not discharge, all water flowing into these ponds.[48] By February 1980 the company still had not been issued an EPA permit for discharged water.[49] By mid-1979 Bridger Coal was to have completed a hydrological study of both surface and subsurface water systems in the area. This was to serve as a baseline of information against which to measure any changes in water quantity or quality that may result from surface mining and reclamation operations. No other data exist on water systems in this area.[50]

The Jim Bridger site is a water recharge area for subsurface water systems that feed the Fort Union Water Formation.[51] Coal seams in the area are at least intermittent aquifers.[52] To monitor the effect of mining and other operations upon groundwater, the company has sunk 34 wells both on and off the mine site; these are monitored by a hydrologist under contract to the company. In June 1979 Bridger Coal said it plans to assume this task in the future.[53] Kail told INFORM the mine staff lacks trained hydrogeologists, as the company's hydrologists are based in NERCO's Portland, Oregon offices. Kail emphasizes that whenever questions arise about the location of roads along water drainages, or monitoring for the correct types of hydrologic data, he must contact Portland, more than 650 miles away. Many of the issues raised cannot be answered by telephone, he says, and this can create delays.[54]

Alluvial Valley Floors

No alluvial valley floors have been identified in studies done on the Jim Bridger site.[55] However, in late 1979 Gary Deveraux said the company was studying the Deadman Draw area to determine whether it is in fact an alluvial valley floor according to OSM criteria.[56]

Erosion

Because of the low rainfall, very little water erosion is evident at the Jim Bridger mine. However, wind erosion is common
on stockpiles, spoil piles, and lands lacking vegetative cover.
Erosion can occur in areas such as the disturbance zone along
Deadman Draw if the land is not stabilized before melting snows
in the spring bring water flowing through the area.

SEEDING AND AMENDMENTS

Bridger reclamation director Dick Kail told INFORM: "In the
past, there was not enough equipment for reclamation, and not
much was done. The company has taken a new look at reclamation and now backs me up. Our approach is that of a whole new
program."[57]

Seeding Procedures

Because of the mine's design and the earthmoving problems
cited earlier, very little seeding has been accomplished at Bridger. Between May 1977 and May 1978, only 58 of the 490 acres
disturbed in that period had been seeded,[58] and only 13 percent
of all the lands disturbed by mining and related activities had
been seeded, less than all but one other mine in the INFORM
sample. From May 1978 to May 1979, no seeding took place at the
mine because all lands within one spoil ridge of the active mine
pits had already been seeded, though much land is used for haul
roads and topsoil storage[59] (see Earthmoving). The company
seeded and reseeded about 200 acres in the fall of 1979.[60] The
company seeds in the fall to allow the seeds to germinate early in
the spring, long before the land would be dry enough to allow
machinery to work on it. This is a common practice in the northern portions of the western United States.[61] Bridger Coal also
reseeds land a year or two after initial seedings because of the
generally poor establishment following the first applications.[62]
The seed mix is only fair. It is made up of fewer grass species than all but four other mines in the INFORM sample, and
experimentation at the mine is directed more toward the effects
of irrigation than the types of plants to be used. The seed mix
consists of three native grasses, one native shrub and one introduced grass. This mixture is not sufficient to produce a diverse
mixture of plants on reclaimed lands, and it may rely too much on
the ability of plants in the region to invade reclaimed lands. The
company also seeds three species of shrubs in selected areas of
the mine. All of the grasses used are cool-season varieties and
will not provide forage in warmer summer months.

TABLE A

1979 SEED MIXTURE
FOR LANDS BEING RECLAIMED AT JIM BRIDGER

Plant Species	Introduced/ Native	Cool/ Warm	Pounds Applied Per Acre
Grasses			
Thickspike wheatgrass	N	C	3.0
Western wheatgrass	N	C	3.0
Streambank wheatgrass	N	C	3.0
Pubescent wheatgrass	I	C	3.0
Shrubs			
Fourwing saltbush	N	W	0.5
			12.5 Pure live seed
*Gardner saltbush	N	W	0.5
*Antelope bitterbrush	N	C	0.5
*Rubber rabbitbrush	N	C	0.5

*Used in selected areas only, not used on all reclaimed lands

Source: NERCO

Seeding operations take advantage of natural precipitation and serve to control erosion by following the contour and creating furrows on topsoiled lands. In 1978 Bridger Coal purchased machinery that allows it to create these furrows along the contours where topsoil has been replaced, to help prevent erosion and increase the lands' capacity to retain water and enhance plant establishment.[63] A drill seeder is used almost exclusively at the mine, except on the steep sides of stockpiles where seed is broadcast by hand. In 1978 the company said that it planned to use a "hillside combine," which it is converting to plant seed and spread mulch along the contour of reclaimed slopes.[64] The machine was used in 1979.

Amendments and Maintenance

Experience has shaped the mulching program at Jim Bridger. Until 1978 Bridger Coal used a mulch made of hay applied at a rate of two tons per acre and crimped into the soil. However, the combination of extremely dry climate and brittle hay caused much of the mulch to dry out and blow away,[65] leaving little protection from erosion.[66] In 1979 the company switched to a standing mulch of seeded grains, which the mine's reclamation director believes will work better.[67] The technique of using a seeded grain as mulch is gaining in popularity in the West because it is cheaper than spreading and crimping and appears to be more effective in stabilizing the soil.

Bridger Coal does not use fertilizer. However, it was to begin varied applications of different fertilizers in the spring of 1980 as part of a pilot program to determine future fertilizer use on a broader scale.[68] The company is also experimenting with various systems to ensure that adequate water is available to seeded areas. Argonne National Laboratory, a federal research facility, has experimented with the use of snow fences to trap snow and provide more water for plants in the spring, but these experiments are incomplete.[69]

Construction of an extensive irrigation system was under way in June 1979. Kail says Bridger Coal would supply enough water to provide the equivalent of a good year's rainfall on all seeded lands, about 5-6 inches of water each summer, starting in 1980. Experience will ultimately determine how much water is supplied and when.

In 1977 Argonne National Laboratory began experiments involving supplemental irrigation on specially seeded test plots at the site. Broadcasting seed mixtures at very heavy levels (24-28 pounds per acre) and providing 4-6 inches of water by irrigation, the laboratory found that plants germinated and native species, as well as introduced species, responded to the treatment.[70]

Trees and Shrubs

Bridger Coal does not have a good program for establishing trees and shrubs on reclaimed lands. Although few trees exist on undisturbed areas, shrubs are common. Trees and shrubs are not transplanted as part of the regular reclamation program at the mine, but shrubs are seeded as described earlier. Bridger Coal has transplanted shrubs only in the Deadman Draw buffer zone, where the company was required by the state DEQ and OSM to return the area immediately to its previous, native form.[71] Gary Deveraux reports that Bridger Coal has had some success in growing native fourwing saltbush and shadscale shrubs. This should be continued. The University of Wyoming and Utah State University have been investigating shrub seeding and shrub transplanting at the mine.[72] However, Kail told INFORM he has found no reports on this work and no findings have been determined.[73]

VEGETATION

The company must share the blame with climate and terrain for the failure of its revegetation program.

Climate and soils combine to create lands in the Bridger Mine area that are not very productive for grazing. According to Kail, it takes 18-24 acres of land to produce enough forage for one Animal Unit (a cow and calf, or five sheep) for one month. (By contrast, only three acres of land are needed in the north-eastern parts of Wyoming [see Belle Ayr profile].) The only areas that are more productive are the 5-7 percent of mine lands near small creeks and water drainage areas--and these are only slightly better.[74] In fact, Kail may be optimistic. Howard L. Millsap, district conservationist for the U.S. Soil Conservation Service, told INFORM:

> Most of the range land [in the area of the Bridger Mine] would probably be in "fair" condition....
> The rangeland is not very productive. A lot of this area requires *50 to 100 acres* to provide forage for one animal unit month [emphasis added].[75]

In mid-1979 company officials said that they planned to complete their first study of the type and quality of vegetation on undisturbed lands in early 1980.[76] In late 1979 Bridger had no information on the productivity of reclaimed lands at the site.

INFORM found that most seeded lands at the mine show very little vegetation. Weeds, including halogeton, Russian thistle, various mustards, and kochia, dominate.[77] Of the plants seeded by the company, only the saltbush has been successful.[78] Even this vegetation is usually concentrated in areas where mulch has been crimped into the soil or where drill seeder wheels have created furrows.[79] In 1977 the Wyoming DEQ found that undesirable invaders did not appear as frequently on lands where topsoil had been immediately replaced and mulch applied. It theorized that stockpiled topsoil supported heavy infestations of weeds and weed seeds, which germinated when the soil was respread.[80] Yet--as noted above--as of late 1979 Bridger Coal was not replacing topsoil immediately.

Improved reclamation practices--including the furrowing of topsoiled lands, and use of drill seeders--have been more successful on recently seeded lands. These areas show the greatest growth of grass species--but also of rabbits and other animals, which may undo this success as they begin to eat these grasses. The reclamation program still has serious problems with weeds. Kail suggests that the grasses now being seeded will eventually overwhelm the weeds. But, he adds, "all [this is] conjecture; we don't have a policy to deal with it."[81] DEQ and U.S. Geolog-

ical Survey inspectors confirmed these problems.[82] The company
has no quick solution to the rabbit problem. Officials say fences
are not the answer because a rabbit-proof fence would be pro-
hibitively expensive and would not deal with animals already in-
side the fence. Over the long term, Bridger feels, coyotes and
hawks, the rabbits' natural enemies, will control the rabbit population. However, mining has driven most of these wild predators from the mine. Only one coyote has been seen regularly in
the Jim Bridger area since mining began.[83]

Erosion by water is not a major problem at this site, but
wind erosion is, and the general lack of vegetation on reclaimed
lands increases the risk of both. As the wind carries some soils
off reclaimed lands, it may create small ridges that could accelerate erosion by flowing water. The loss of the soil itself is
critical as so little exists in the area to begin with. A DEQ inspection found that soil in reclaimed areas tended to form a
crust, both reducing the potential for water to infiltrate these
soils during rains, and increasing the amount of water flowing
over the land--thus threatening serious erosion.[84]

WILDLIFE

Bridger Coal has made one excellent and innovative effort to
enhance wildlife habitats on the site. In a few areas near the
mine's first pit, the company left intact rock outcroppings and
small rocky cliffs. Bridger has also graded spoil around these
rock outcroppings to blend with the surrounding topography.
Shrubs being seeded provide both good and poor forage for
wildlife, and the company left substantial vegetation on and
around the rocks. Gary Deveraux said the company will build
shelters for wild animals by piling rocks together, a common
practice at western surface mines.[85]

RECLAMATION COSTS

TABLE B*

RECLAMATION COSTS AT JIM BRIDGER

Operation	Cost Per Acre	Percentage of Total Cost
Earthmoving		95%
Topsoil removal and replacement	$2,646	
Grading and contouring	4,883	
Revegetation		5%
Seeding	72	
Mulching and maintenance	335	
Total Cost:	$7,936	

*These costs and percentages are not strictly comparable to those of the same operation or activity at other mines. See the Cost of Reclamation chapter for a detailed explanation.

Source: Bridger Coal Company

Rosebud/Peter Kiewit/MT

Mine: Rosebud Mine

Location: Hanna (population 3,450), Wyoming, in Carbon County (population 22,545)

Operating company: Rosebud Coal Sales Company

Parent company: Peter Kiewit Sons' Company

Parent company revenues: 1977, NA; 1978, NA

Parent company net income: 1977, NA; 1978, NA

First year of production: 1973 (by Peter Kiewit)

Coal production in 1978: 2,868,048 tons

Acres for which a mining permit has been issued: 3,193

State regulatory authority: Wyoming Department of Environmental Quality

Reclamation bond posted: $8,700,000

Reclamation cost: reclamation cost per acre, NA; *reclamation cost per ton of coal,* $2.60-$2.80

Expansion plans at mine: Unknown

Additional acres for which the company will seek a permit: Plans to lease more land, but doesn't know how much

Parent company U.S. coal reserves: NA

Development and expansion plans for surface coal mines in western United States: Peter Kiewit is developing three mines in Wyoming as a joint venture with the Rocky Mountain Energy

Company. The three mines will produce 12.1 million tons of coal
per year by 1981.

NA = Not available

Sources: Peter Kiewit Sons' Company; 1979 Keystone Coal Industry Manual; *"New Coal Mine Development Survey 1978-1987,"*
Coal Age, *February 1979;* Rand McNally Road Atlas, *54th Edition; Peter Kiewit Sons' Company,* Annual Reclamation Report,
July 1, 1977-June 30, 1978

SITE DESCRIPTION

The Rosebud Mine lies in a very dry part of south-central
Wyoming. It is about two miles north of the small town of Hanna,
and is adjacent to Arch Mineral Corporation's Seminoe II Mine.
Hanna is also surrounded by three other large surface mines,
and two or three more mines are being developed in the area.
Rosebud is situated among high, rolling plains. The area is
in the Hanna Basin, with mountainous ridges along its borders.
To the south of the basin rise the Medicine Bow Mountains.
The average precipitation reported by Peter Kiewit is from
8.75-12 inches annually.[1] (Next door, Arch Mineral reported
10-12 inches.) Kiewit also reports that the range in precipitation for the area has been as low as 4.3 inches and as high as
19.3 inches per year.[2] About 25 percent of the precipitation
falls as rain in July, August and September, 38 percent as snow
from October through March, and between April and June either
rain or snow make up the remaining 37 percent of moisture.[3] The
frost-free period ranges from 66 - 114 days, from about mid-
June into September. The average is 94 days.[4]
Like Seminoe II and the rest of the area, land is primarily
used for grazing livestock, and also as wildlife habitat. Peter
Kiewit intends to return the mine site to a predominant use as
sheep grazing land.[5] Three basic plant communities make up the
mine area. They include big sagebrush, low sagebrush-salt-
bush, and greasewood.[6] According to George F. Brown, acting
conservation manager for the U.S. Geological Survey in Denver,
no prime agricultural lands exist on the Rosebud Mine site.[7]

OVERVIEW

Peter Kiewit's Rosebud Mine produced just under 2.9 million
tons of coal in 1978. Only 16 other mines in the West produced
more. But the arid climate and the steeply pitching coal seams
in the area keep reclamation from being either current or very

successful. In the face of these natural obstacles, Rosebud's
entire reclamation program is fair to good, although vegetative
success has been only fair to poor--for which the company must
share the blame with nature.

Earthmoving practices are only fair, but under federal and
state pressure some aspects are being improved. Slopes graded
since 1978 have gentler contours, where more operation of reclamation machinery can occur. Grading lags far behind disturbance; this is a function of 1) production requirements, as many
coal pits are left open for the mixing of coal, and 2) steeply
pitching seams that do not allow grading to follow these mine pits
closely. It is not clear what could be done to alleviate the problem, but at Seminoe II, Arch Minerals has kept more current
than Peter Kiewit. Only 18 percent of the land mined through
June 1978 had been graded at Rosebud. Most topsoil is stockpiled. Fires occur from time to time in pits where coal has been
left exposed for long periods--common when pits remain open in
dry, hot climates--and more needs to be known about the possible toxic or sodic elements to be found in the interburden between coal seams and some of the overburden found on the site--
even though the company has been mining here for over 15 years.

Rosebud, like the neighboring Seminoe II Mine, has no trouble with surface water control on the site as little water is to be
found in this climate. Rosebud's sedimentation system is adequate to contain runoff, and erosion control is good. The company is not certain to be free of groundwater problems, however.
Only in 1979 did Kiewit begin to gather data on background levels and quality of the site's groundwater resources. Against this
data future monitoring data will be compared. Whether an alluvial
valley floor exists on this site is being debated by Peter Kiewit,
the Office of Surface Mining (OSM), and the U.S. Soil Conservation Service. For the moment, one area the company suspects
might be one is being left untouched.

The seeding and amendment programs at Rosebud are medio-
cre. Seeding lags far behind land disturbance. The company
properly seeds along the contour of the land and creates furrows
to aid in controlling and providing water for growing plants. But
its seed mixture is dominated by cool-season plants, and at times
--as witnessed by INFORM--as a result of careless practice machinery has run out of seed and areas have been left bare instead of being planted immediately. Experiments are providing
some information to the reclamation staff on the value of using
mulches and snow fences to retain moisture on the site, but the
company has yet to measure the results of this work. Shrub
transplants predominantly involve bare-root stock, and these
have not shown a high degree of success. More containerized

plantings would be beneficial, as they have been found to enhance transplanting in arid climates.

Vegetative success, owing largely to the harsh, dry climate,
has been limited at Rosebud. Weeds dominate, and there is little
vegetation to show for all the land that has been disturbed at
the mine. Of 3,832 acres of land mined at the site--200 acres of
which were disturbed before Peter Kiewit bought the mine--only
726 acres of land have been reclaimed.[8] Little information exists
on the productivity of reclaimed lands. Although weeds may
provide shelter for grasses, no areas can genuinely be said to
be showing good growth dominated by non-weedy species. It
remains to be seen whether land reclamation can ever succeed in
areas like this one to create a viable plant community that does
not fall prey to undesirable plants.

Before mining, the Rosebud site was home for mule deer and
pronghorn antelope. But little is being done to insure an adequate post-mining wildlife environment at Rosebud. The company
says it will create rock habitats for small animals, but it has been
content merely to leave large boulders--which roll from surrounding spoil piles onto graded lands--as micro-habitats. Greater
success with shrub growth will be necessary to provide shelter
and forage for wildlife.

Regulatory activities at Rosebud are limited. The Wyoming
Department of Environmental Quality (DEQ) inspects quarterly--
sometimes more often--and the U.S. Geological Survey has made
some cursory inspections. Though inspections by the state are
more descriptive than prescriptive, this has begun to change as
the Wyoming state program comes under the scrutiny of the federal Office of Surface Mining. OSM visited the mine once in August 1978 and again in June and December 1979.[9] Grading has
improved, and the company has kept within the bounds of state
requirements or reached agreements with regulators when problems have arisen.

EARTHMOVING

Grading

In a report to the Wyoming DEQ in June 1978, Peter Kiewit
said that of 3,493 acres that had been disturbed at the mine
since the early 1960s, only 632 acres had been graded.[10] Only
200 of the 3,493 acres disturbed were mined prior to the mine's
purchase by Peter Kiewit. Nine open mining pits exist on the
site, and only four were producing coal in June 1979. Dave
Evans, reclamation supervisor at the mine, explains that many of
these pits must remain open because the company mixes different
quality coals before shipping coal to customers.[11] Thus grading
at the mine is often years behind coal production.[12] In some

areas many spoil ridges remain where grading has not been started. Between the mixing of coals and earthmoving problems related to the steep coal seams, grading is slowed. Yet much less land (on a percentage basis) has been graded here than at the neighboring Seminoe II Mine, where coal seams also are steeply pitched.

Early grading at the mine produced terrain that was steeper than either the surrounding lands or the pre-mining contour. However, with pressure from regulators, newer grading is blending well with undisturbed lands, according to OSM.[13] Sam Scott, reclamation director for Peter Kiewit, concedes that earlier-graded slopes could be seeded only by machinery working up and down the slopes. Today, seeding and other machinery traverses graded slopes.[14]

Topsoil

Topsoil has been saved at the mine since 1973. Lands seeded in 1974 were the first to receive topsoil replacement. Topsoil depths range from 3 inches to 3 feet over the mine area. It is removed in two separate lifts and stockpiled separately as topsoil and subsoil.[15] From 6-8 inches of topsoil and subsoil are spread over graded lands.[16] Scott indicates that, although the company has replaced very little topsoil directly (instead of stockpiling), it should have followed this practice more, as it requires having to move the material only once. Direct replacement also allows more organic materials (biomass) to be moved along with the soil, and fewer windblown seeds of weeds to collect in the soil.[17]

Stockpiles are generally free of erosion, but they lack protective vegetation. Many older stockpiles--in place for more than 2½ years--had little if any vegetation when INFORM visited the mine in June 1979. This leaves them more susceptible to wind and water erosion. Evans says the company has overridden Wyoming DEQ's wishes and determined not to seed stockpiles until all materials have been placed in them. According to Evans, this included piles where additional materials might not be added for a full year. "The DEQ didn't like it, but they went along with it in the end,"[18] he says. This delay seems unnecessary. Even in the steep mountainous region of Colorado, Peabody Coal is able to add topsoil to existing stockpiles that are already amply vegetated along their sides. (See Seneca II profile.)

Toxic Material

INFORM found no evidence that toxic substances were hampering reclamation at Rosebud. It is important to note, however, that like Arch Mineral, Kiewit is mining multiple coal seams, and

the interburden material between these seams may contain potentially harmful components.

Also, as determined at Seminoe II, salts in the overburden
itself could affect vegetative growth if concentrated within plant
root zones. This potential problem should be monitored at Rosebud.

Coal fires have occurred in one of the open pits at Rosebud.
U.S. Geological Survey inspection reports from as early as September 1976 indicate that coal fires in mining pit No. 8 have been
a recurring problem.[19] The fire discovered that month extended
for 50 feet in the pit.[20] In March 1977 another fire was found to
extend more than 350 feet,[21] and the debris from this fire was
still smoldering in August 1977.[22] Again, fire was found in the
pit in October 1977[23] and April 1978.[24] The mining company
pushed the burning materials away from the exposed coal and
buried them, but fire again broke out in October 1978.[25] These
fires break out because the coal is left exposed to the air for extended periods of time, wasting the resource and polluting air in
an area already dirtied by the many coal mines operating in the
vicinity. If pits were not left open as long as they are, this
problem would be reduced; however, Peter Kiewit continues to
battle the fires as they break out.

WATER QUALITY

Surface Water and Groundwater

Peter Kiewit has had a permit from the Environmental Protection Agency (EPA) since 1977 to discharge water from sediment
ponds on the mine site. The EPA permit allows the mine to release water from six discharge points into tributaries--Big Ditch,
Hanna Draw and Pine Draw[26]--that flow into the Medicine Bow
River. Under permit requirements, Peter Kiewit's sedimentation
and water control system must be able to control and contain water from a 24-hour precipitation event with a recurrence interval
of 10 years. This 10-year, 24-hour event is deemed to be equal
to a rainfall of 1.6 inches.[27]

The sediment ponds have been able to control water discharge quality very well, for the most part. Although poorly
vegetated, the discharge points are well constructed to resist
erosion. The mine seldom actually discharges water, and most of
it remains in the sediment ponds. One discharge in the spring
of 1978 was slightly above the allowable limits for suspended solids, but the amount was minimal.[28] No water was discharged
from the mine from April 1978 to April 1979, and water discharged from April to July in 1979 was within effluent limits.[29]

Aquifers running beneath the Rosebud Mine occasionally discharge into active mining pits. Especially in areas to the south

where the coal seams are cracked and faulted, and where aquifers are recharged from surface water soaking into the ground, more water can be found in open cuts in the earth.[30] Peter Kiewit says it is monitoring 15 wells for changes in subsurface water quantity and quality. Four are located off the mine site; three more are planned.[31] In December 1979, however, Sam Scott told INFORM that Peter Kiewit had just begun analysis of groundwater resources information for the Hanna area,[32] and that it had no data on the effects of mining on groundwater at Rosebud. The number of large surface mines in the area suggests that water monitoring should become a joint effort. However, if problems arise, it may be difficult to ascertain which mining operation in the Hanna area actually is causing the problem, or if several or all are contributing. At Seminoe II, groundwater monitoring has yet to be done, and it appears that concern for this issue has been minimal in the past.

Alluvial Valley Floors

Although preliminary information from the U.S. Soil Conservation Service indicates the area contains no alluvial valley floors,[33] the company and OSM are concerned that Pine Draw may be one. The company does not plan to mine into the draw-- but it will be mining near it. Company reclamation officials say they do not know the potential effects of these mining operations on alluvial valley floors and groundwater systems.[34] Like the situation at the Seminoe II Mine, this warrants further investigation.

Erosion

Although the Wyoming DEQ said in 1978 that erosion is a problem at Rosebud, INFORM found little evidence to support that contention.[35] The use of planted grains for mulch on steep slopes and contour furrowing in a few areas have controlled erosion well. Spoil piles erode a bit, and occasionally this sediment flows into water drainages. But most sediment from eroding spoils remains trapped within the spoil ridges. Closer grading, with less-exposed spoil ridges, would help.

SEEDING AND AMENDMENTS

Seeding Procedures

Seeding of disturbed lands at Rosebud has not been kept current with land disturbance. By the end of December 1979, the company had seeded only 726 acres (20 percent) of the 3,832 acres of land disturbed by mining.[36] From July 1977 to June

1978, 303 acres were disturbed, but only 71 acres were seeded.[37] Much of the unseeded land has also not been graded, owing to the steeply pitching coal seams in the area and the practice of leaving pits open so that coals can be blended before being shipped (see Grading).

Seeding is generally conducted in the fall, a common time for most reclamation operations in the West (excluding the Southwest). When grains are planted for standing mulch, they are seeded in the spring, also a standard practice.[38] The company is not experimenting with the timing of seeding; nor would this appear necessary in this region and climate.

TABLE A

1979 SEED MIXTURE
FOR LANDS BEING RECLAIMED AT ROSEBUD

Plant Species	Introduced/ Native	Cool/ Warm	Pounds Applied Per Acre
Seed Group E (for loamy soils)			
Grasses			
Bluebunch wheatgrass	N	C	1.5
Green needlegrass	N	C	1.5
Indian ricegrass	N	C	3.0
Streambank wheatgrass	N	C	2.25
Western wheatgrass	N	C	3.75
Thickspike wheatgrass	N	C	1.5
Forbs and Shrubs			
Fourwing saltbush	N	W	1.5
			15.00 Pure live seed
Seed Group F (for sandy soils)			
Grasses			
Bluebunch wheatgrass	N	C	1.5
Indian ricegrass	N	C	3.0
Thickspike wheatgrass	N	C	4.5
Streambank wheatgrass	N	C	4.5
Forbs and Shrubs			
Fourwing saltbush	N	W	3.0
			16.50 Pure live seed

Source: Rosebud Coal Sales Company

The seed mixture contains only native plant species. At most mines in Wyoming, DEQ has required that companies seed predominantly native plants. Peter Kiewit's seed mixture, however, consists almost entirely of cool-season varieties. This means that without invasion of other plant species, plant growth and forage productivity will be low in the warmer months of the growing season. Establishing warm-season plants is concededly difficult, as they are usually outcompeted by cool-season plants that germinate and grow earlier in the year. Nonetheless, a diverse mosaic of plants native to the region must include some warm-season species growth. Fourwing saltbush--a shrub--is the only warm-season plant being seeded.

Peter Kiewit uses two different seed mixtures, depending upon the types of soils encountered. If the soils are loamy, seed group E (see table) is used, whereas in sandy soils, western wheatgrass and green needlegrass are removed from the mixture and the proportions of other plants being seeded are raised. The company has been experimenting with native species growth, effects of soil manipulation on vegetative growth, and other aspects of reclamation work, but, according to Scott, the company needs more data. Furthermore, initial analysis of the data collected so far, Scott says, is incomplete.[39]

Amendments and Maintenance

Although seeding at Rosebud was formerly conducted up and down graded slopes, after pressure from regulators to grade to flatter slopes (see Grading), it now runs along the contour, providing small tracks on all seeded lands. These serve to retain water and provide extra moisture for plants, while also slowing the flow of water downslope and minimizing its potential for erosion.[40] The company also creates contour furrows along the slopes being reclaimed, to retain water and prevent erosion.[41]

Except for limited areas of extremely steep slopes where broadcast seeding is required, Rosebud uses drill seeders. Following a discer, which breaks up the topsoil and prepares the seed bed, the drill seeders place seeds in rows and cover them with soil. The company has sometimes been careless in its seeding. On INFORM's visit to the site in June 1979, areas seeded in 1976 and in the spring of 1979 were observed to be lacking seed in certain places. Company officials attributed this error to the machines' having run out of seed, and to inattention by the machine operators. The company said it would be more careful.[42]

Two different types of mulch are used at the Rosebud site. The company seeds grains--such as barley or wintergraze (a sterile wheat hybrid)--as a live, standing mulch, and spreads hay mulch made from native grasses.[43] The choice of these two mulching techniques has been developed through experimentation.

One limited practice was the technique of hydromulching, where
water and mulch were mixed and sprayed onto the ground. However, the company reports that wind caused uneven distribution
of the material, and it discontinued the practice.[44]

Reclaimed lands are fertilized two years after seeding.
Amounts used are determined by soil tests,[45] but the company
says it generally spreads the equivalent of 40 pounds per acre
of nitrogen and 20 pounds per acre of phosphorus, applied aerially.[46]

Irrigation is not practiced at Rosebud. The company told
INFORM that in its single experiment with irrigation, surface
soils hardened, inhibiting plant growth and infiltration of natural
precipitation. Irrigation also caused salts in the overburden to
migrate to the surface, to the detriment of vegetative establishment.[47]

Peter Kiewit has also concluded that weed control with herbicides is not a good practice because weeds help trap snow, providing additional moisture in the spring, and provide cool shelter
for grasses growing amidst the weeds. The company also assumes
that grasses will have established sufficiently in three years to
outcompete weeds on seeded lands[48] (see Vegetation).

Some experiments have been conducted using fences to trap
blowing snow in the winter, providing additional moisture in the
spring. However, the company feels that even though snow retention and resulting water infiltration are enhanced, water runoff is also accelerated, and the risk of increased erosion is
greater.[49] Sam Scott told INFORM that the company will probably
use fences only on flat areas after initial vegetation has established, to prevent erosion.[50]

Trees and Shrubs

Trees are rare and shrubs are common in the Rosebud area.
Peter Kiewit has been experimentally planting shrubs since 1976.
At least 19 different shrub species have been tried.[51] The company has also established a shrub-growing facility in the base-
ment of the administration building on the site. However, lack of
control over the building's climate has thwarted this venture.[52]
All the shrubs transplanted so far have been purchased.[53] Approximately 8,000-10,000 shrubs have been planted from bare-
root stock, and only 3,000 have been containerized. The company
says containerized plants--which generally survive transplants
much better than bare root sprigs--are too expensive.[54]

VEGETATION

Two distinct trends emerge concerning vegetative establishment at Rosebud. The first is that lands where topsoil was not replaced are heavily dominated by invading weeds. Although some grasses appear in clumps, lands that were seeded in 1973 and 1974 show little hope for substantial grass growth. Significantly, lands where topsoil has been replaced have much better vegetation.

The second trend, somewhat more tentative, is that after three to four years, lands seeded with grasses but dominated by invading weeds in their first years are now showing greater growth of desired plant species.[55] This would bear out the company's predictions. The best reclaimed lands observed by INFORM in June 1979 were those seeded in 1976. There, more grasses are growing and weeds and undesirable species are giving way. Lands seeded in the fall of 1977 or 1978 show more invading weeds, with small clumps of desirable grasses growing among them. It remains to be seen whether this trend will continue. As might be expected, weeds are most dominant on lands where planting has missed entirely, as described above (see Seeding Procedures).

Pre-mining productivity was about 25 acres per Animal Unit Month[56] (the amount of land needed to graze one cow and calf, or five sheep, for one month). According to the U.S. Soil Conservation Service, forage production ranged from 500-1,200 pounds of air-dry forage per acre.[57]

Peter Kiewit has established six reference areas to measure more accurately the land's productivity.[58] The company told INFORM it would begin collecting data on the productivity of reclaimed lands on the site in July 1979. But it said that information would be for internal use only.[59] These data include the density and frequency of plants established in reclamation, and the productivity of annual and perennial grasses, forbs and shrubs.[60] The company's only productivity data so far pertains to a 1977 sampling of its best reclaimed land, which had undergone three years of growth since seeding. Productivity was 682 pounds of oven-dried forage, which is difficult to compare with Soil Conservation Service figures expressed as air-dried forage.[61] Until Peter Kiewit decides to make the information public, we can know little about the success of revegetation operations at the mine. The company plans no grazing trials on reclaimed areas soon, it says, because it would have to hire a sheep herder.[62] Perhaps it could investigate controlled grazing of its lands by the stock of local ranchers instead.

WILDLIFE

Plants being seeded and transplanted at Rosebud are suitable for a wildlife habitat. The company also plans to build rockpile shelters for wildlife in the future, but told INFORM on its June 1979 visit that boulders, which randomly roll from spoil piles onto graded areas, are sufficient for the moment to provide micro-habitats for animals in the area and would remain.[63] This is a weak argument for leaving boulders on the land instead of removing them and either grading them into mine pits or placing them together to create a shelter with better spaces for habitat. Many other mines in the West do this.

The company plans to leave sediment ponds as permanent water impoundments, which would provide habitat for waterfowl and a water source for domestic and wild animals. Sam Scott reports that local wildlife managers and ranchers want these impoundments to be left for this purpose.[64] It is up to the regulatory authorities to determine whether the company should be allowed to leave the ponds, or must reclaim them. It has yet to be proven that these impoundments will in fact contain water all year. Low precipitation and the dry climate might cause these impoundments, without recharge from groundwater sources, to dry up in those seasons when they are most needed.

RECLAMATION COSTS

TABLE B*

RECLAMATION COSTS AT ROSEBUD

Operation	Cost Per Acre	Approximate Percentage of Cost†
Earthmoving		93%
Topsoil replacement▽	$1,100	
Grading	819	
Revegetation		7%
Seeding and seedbed preparation	112	
Fertilizing	30	
Subtotal:	$2,061	

*These costs and percentages are not strictly comparable to those of the same operation or activity at other mines. See the Cost of Reclamation chapter for a detailed explanation.

†Because Rosebud did not disclose a total reclamation cost, approximate percentages have been calculated based on a subtotal of defined costs.

▽Topsoil removal is considered a production cost, not a reclamation cost.

Source: Peter Kiewit Sons Company

Seminoe II/Arch Mineral/WY

Mine: Seminoe II Mine

Location: Hanna (population 2,460), Wyoming, in Carbon County (population 22,545)

Operating company: Arch Mineral Corporation

Parent company: Ashland Oil, Inc. (48.9% owner)

Parent company revenues: 1977, $5.104 billion; 1978, $5.675 billion

Parent company net income: 1977, $164 million; 1978, $245 million

First year of production: 1973

Coal production in 1978: 2,800,000 tons

Acres for which a mining permit has been issued: 6,286

State regulatory authority: Wyoming Department of Environmental Quality

Reclamation bond posted: $7,247,000

Reclamation cost: reclamation cost per acre, NA; reclamation cost per ton of coal, NA

Expansion plans at mine: Production to increase to 4 million tons per year by 1986

Additional acres for which the company will seek a permit: 1,704 over the life of the mine

Parent company U.S. coal reserves: Owns or leases 150 million tons of coal reserves

*Development and expansion plans for surface coal mines in west-
ern United States:* Aside from operating the Seminoe I Mine and
Medicine Bow Mine in Wyoming, Arch Mineral is developing two
surface mines in New Mexico to produce 600,000 tons of coal annually by 1987.

NA = Not available

Sources: Arch Mineral Corporation; Ashland Oil Form 10-K,
*September 30, 1979; "New Coal Mine Development Survey 1978-
1987,"* Coal Age, *February 1979; Written communication from
Sheridan A. Glen, Director of Environmental Affairs, Arch Mineral Corporation, to INFORM, November 15, 1979*

SITE DESCRIPTION

The Seminoe II Mine is in south-central Wyoming, two miles
north of Hanna. Four other surface mines operate in the area,
and two or three more are being planned. One nearby mine is
Peter Kiewit's Rosebud Mine, also profiled in this study.

Elevation varies from 6,500 - 6,900 feet above sea level.
Slopes vary from flat to gently rolling, and a few slopes being
mined at Seminoe II exceed 20 degrees.[1] High rolling plains are
broken by sandstone ridges and rock outcroppings.[2]

Average annual precipitation for the area is 10.5-12 inches.
Most moisture is recorded in April and May.[3] For 30 years,
yearly precipitation has ranged from 8.2-15.8 inches.[4]

Land in the area has been predominantly used for grazing of
sheep and other livestock.[5] Sixty-one percent of the land is
made up of big-sagebrush communities. Arch Mineral plans to
return the land to its premining uses. Neither the company nor
the U.S. Geological Survey has encountered prime farmlands in
the area.

OVERVIEW

One of several mines in the Hanna area, Seminoe II produced
2.8 million tons of coal in 1978, placing it among the West's top
20 mining operations. Although Arch Mineral's revegetation plan
at Seminoe II is good, practice has been fair to poor. From 1973,
when the corporation began producing coal from Seminoe II, to
June 1979, only 24 percent of the land disturbed by mining had
been seeded.[6]

The combination of a dry climate and minimal reclamation
success to date does not place Arch Mineral's practices in a good
light. However, two unique problems--two steeply dipping coal

seams and the region's arid environment--burden any company
trying to reclaim plant and animal communities disturbed by surface mining. Plans for upgrading programs to revegetate disturbed lands, upgrading water control on the site, and salvaging and replacing topsoil all should aid reclamation. Results to
date do not show that native plant communities can be established, or that land and water resources can be protected completely when disturbed by surface mining in this area. According
to the Wyoming Department of Environmental Quality:

> Looking at the Seminoe II Mine from a historical
> perspective, reclamation effort has improved con-
> siderably....Significant deficiencies still exist in
> the reclamation effort at [the mine], however.[7]

To establish native plant communities on reclaimed lands,
Arch Mineral relies heavily on native species in its good seeding
program. With the use of mulch and fertilizer and the replacement of topsoil, this should help create better vegetation on reclaimed lands. Shrub seeding and transplanting will enhance the
quality of vegetation on the mine site. Yet the small areas of land
whereon revegetation has taken place so far allow no more than a
guess at the results of this program.

Arch Mineral is only now building sediment pond and water
control systems at the mine under pressure from the Wyoming
Department of Environmental Quality (DEQ) and the federal Office of Surface Mining (OSM). Although sedimentation and water
erosion have apparently not been major problems, erosion of
portions of a haul road and sides of topsoil stockpiles has required the building of this water control system. In general,
surface water runoff is quickly soaked up by the dry soil. According to Greg Bierei, Arch Mineral reclamation director for
its western division, an aquifer appears to be disturbed by mining in the South Simplot area of the mine, and Arch Mineral must
investigate the matter.[8]

The large area of ungraded land is a major reason for Arch
Mineral's failure to stay current with its seeding. By June 1979
only 457 of the 1,247 acres of land disturbed by mining had been
graded. An additional 268 acres of land had been disturbed by
haul roads, stockpiles and other facilities at the mine.[9] This is
partly due to specific earthmoving problems in the North Simplot
area--but inadequate planning must share the blame. According
to OSM, better planning would aid reclamation efforts considerably.[10] OSM inspector Tom Ehmett told INFORM that the "key to
problems at the (Seminoe II) mine is planning."[11] Grading has
improved somewhat: only 31 percent of the land disturbed by
mining had been graded in June 1978,[12] and about 37 percent

had been graded by June 1979.[13] Newly graded slopes are
closer to approximate original contours than they once were.

OSM visited Seminoe II four times between August 1978 and
September 1979.[14] As a result, OSM has required water control
on the site and salvaging of topsoil and minimization of its contamination. The Wyoming DEQ inspects Seminoe II quarterly,
with other inspections from time to time. Although reclamation
problems are noted, the inspections appear to be basically a
means of reporting on the state of operations at the mine at any
given time. DEQ declined to allow INFORM to interview the inspector most knowledgeable about the Seminoe II site. DEQ also
denied our request for copies of more recent inspections. Other
states, notably Montana, Colorado and North Dakota, were more
open to these requests, as was the Office of Surface Mining. Because much of the coal being mined at Seminoe II is federally
owned, the U.S. Geological Survey inspected the mine 12 times
between September 1976 and September 1978. These inspections
were cursory; they did not substantively address reclamation issues.

EARTHMOVING

Grading

Land is being disturbed at Seminoe II faster than it is being
reclaimed. In 1977 82 acres were mined; 136 were graded. In
1978 161 acres were mined, only 7 graded.[15] By June 1979
1,247 acres had been affected by mining operations, but only 457
had been graded, leaving 790 acres to be reclaimed. Arch Mineral had said in 1976 it would seek to reclaim as much land in a
given time period as it disturbed.[16] It has failed to do so. A
1978 inspection by the Wyoming DEQ found that "reclamation is
not concurrent with mining operations at this mine. Mining operations have been accelerated so that disturbance is now far
exceeding reclamation....There is a lot of 'catch-up' reclamation
to be done in the Monolith area with spoils dating back 3 to 4
years yet to be graded."[17] Only in the North Simplot area is
Arch Mineral encountering mining conditions that would prevent
it from grading as it mines. At this site the coal seam plunges
steeply into the earth. As successive cuts are made to expose
it, the amount of removed overburden increases. Only after the
second of the two seams being mined is removed may grading be
done. However, more earth and rocks must be moved with each
cut, and this creates a delay from the time the first coal seam is
removed to the time when the area can be final-graded. Thus,
more than three spoil ridges exist in some of these areas. (A
similar problem occurs at the Rosebud Mine, next door, where
the same dipping coal seams are mined.) A member of Arch Min-

eral's reclamation staff said the company was investigating the
use of conveyor belt systems to speed movement of materials on
mines in the Midwest. However, no decision has been made on
their potential for use in the West.[18]

Grading generally approximates the original and surrounding
contours of the land. Older grading has been steeper than pre-
mining topography,[19] but this practice appears to be changing.
One change to the contour and topography may occur if Arch
Mineral is allowed to leave final highwalls without grading, to
provide a habitat for small mammals and nesting areas for raptors
(local birds of prey).[20] This proposal is now under consideration
by the OSM and DEQ.

Topsoil

Topsoil in the Seminoe II area generally ranges from 6-24
inches in depth. In some areas as much as four feet of topsoil
can be found,[21] and Arch Mineral is removing all of this material.[22] About 18 inches of topsoil is to be replaced evenly on
graded lands at the mine site.[23]

All but approximately 5 percent of removed topsoil is stockpiled; the 5 percent is hauled directly back onto graded lands,
according to Greg Bierei.[24] But an OSM inspection in August
1978 found that no topsoil had yet been replaced on graded
lands; it was all in stockpiles.[25] The company is improving its
care of these stockpiles to prevent their contamination or erosion.
Older stockpiles have been small, widely dispersed and poorly
vegetated. Many in the North Simplot area are contaminated by
dust, water runoff and road scrapings owing to their proximity
to haul roads.[26] Why did the company place these stockpiles so
close to haul roads? The Bureau of Land Management (BLM)
owns the surface rights to the land where stockpiling had been
proposed. Arch Mineral contends that by refusing to allow it to
stockpile on BLM lands, BLM forced the company to locate its
stockpiles next to major haul roads.[27] However, OSM contends
that the company should use other, privately owned surface lands
near these sites for topsoil storage.[28] Sheridan Glen, director
of environmental affairs for Arch Mineral, reports that by September 26, 1979 all stockpiles had been moved and consolidated.[29]

Some stockpiles have eroded. OSM has suggested that using
a seed mix to vegetate stockpiles doesn't work. It suggests that
the seeding of a grain would better protect against erosion.[30]
Many of the mines in INFORM's sample are now seeding grains as
mulch and as protective cover for stockpiles. Some of the two-
year-old stockpiles next to the haul road to the North Simplot
area are vegetated solely by Russian thistle, a weed that, when
dry, turns into tumbleweed. Annual grains are now used to seed
stockpiles at the mine. These stockpiles are larger, exposing

less surface area to wind and water and consolidating materials
to prevent their loss. The company has received a violation notice from OSM because of erosion of some of its stockpiles.[31] The
company reports this notice of violation was rescinded after an
informal conference with OSM and DEQ.[32] However, unless eroding materials flow into water drainage areas, this problem will be
limited to loss of materials needed for revegetation and some on-
site sedimentation. The potential for off-site sedimentation is
slim if materials do not enter water drainage areas, and few of
these cross the mine area.

Toxic Material

Salts in soils at Seminoe II may be harmful to the growth of
plants on lands being reclaimed. According to the DEQ, materials
harmful to plants have been found on the surface of graded lands
at the mine.[33] Greg Bieri told INFORM these problem soils were
being mixed with other soils at the site and would actually lower
the concentration in any given area, thus relieving any substantial detriment to plant growth.[34] Arch Mineral told INFORM it
does not handle problem soils specially; where found, such soils
would be buried before topsoil was applied to graded lands.[35]
Yet, in June 1978 DEQ found that:

> Spoiling techniques at this mine are such that
> zones of carbonaceous and otherwise undesirable
> overburden materials do appear on the regraded
> spoil surface. The operator has shown very little
> concern for this problem. Overburden analyses
> have shown strata of low pH [a measure of acidity
> where the lower the number, the more acid the
> material] and high salts, but mining and spoiling
> techniques do not take this into account.[36]

When queried on DEQ's finding, the company told INFORM it was
aware of the salt concentrations on the soil, but felt that no
acid-forming materials existed.[37] The company will have to monitor the effects of these materials on vegetative growth, and it
may have to institute special burial of this material in future
mining operations.

WATER QUALITY

Surface Water and Groundwater

Surface water drains into small tributaries that feed Big
Ditch, the Seminoe Reservoir and the Medicine Bow River.[38]
However, not much surface water flows over the mine for most of
the year. Until 1979 the Seminoe II Mine site had no sedimenta-

tion ponds. After a notice of violation[39] (later rescinded) from OSM and another from the Wyoming DEQ, Arch Mineral began to build four sediment ponds and one water treatment pond for groundwater pumped from mine pits.[40] These ponds have been designed to hold the water runoff that would occur in a 24-hour storm with a 100-year chance of occurrence, more than the required ten-year occurrence level. The company wants to contain all water on the site and use it to suppress dust on haul roads, and eventually to serve as watering areas for livestock and wildlife at the request of the landowner.[41] This will require approval by OSM and/or DEQ, because leaving water impoundments is considered an alteration of land use and topography. A permit to discharge water under the NPDES program was issued to Seminoe II in April 1979. No discharges have yet been reported from the mine[42]; thus INFORM cannot determine the success of the company's effort to control surface water with these newly built sediment ponds.

Arch Mineral insists no groundwater is present around Seminoe II, but this may not be true. The company reports it has found no water (except for surface runoff) in its mining pits.[43] It also says the water table in the area is not high enough to be tapped by mining operations; nor is it mining any aquifers that are tapped for livestock watering.[44] However, on a visit to the mine in June 1979, INFORM was told by mine staff that mining operations in the "five-block" area of the South Simplot portion of the mine were disturbing a subsurface aquifer.[45] Arch Mineral should begin a groundwater monitoring program to evaluate the effects of its operations on this aquifer.

Alluvial Valley Floors

Arch Mineral and the U.S. Geological Survey contend that no alluvial valley floors exist on the Seminoe II mine site.[46] However, the Bureau of Land Management has said alluvial valley floors may exist in the area.[47] Sheridan Glen, director of environmental affairs for Arch Mineral, told INFORM the Seminoe II Mine was mentioned as having some of the properties necessary for the positive determination of alluvial valley floors. Examples of these characteristics include the occurrence of silver sagebrush, possible intermittent streams and deposits of alluvium. However, none of the areas currently affected or to be affected by any of Arch Mineral's three coal mines in the Hanna Basin has the criteria necessary to be labeled alluvial valley floors. Glen also told INFORM that OSM and DEQ had both determined that no .alluvial valley floors exist.[48]

Erosion

Some erosion has occurred along water drainage paths on the mine site, but very little is apparent on reclaimed slopes.[49] Because the land is so dry, rain or melting snow often soaks into the soil.[50] Even lands graded to a 4:1 slope show little or no erosion. Most erosion that does occur is found on spoil piles awaiting grading. However, this erosion and the resulting sedimentation are contained on-site between the spoil ridges. Other erosion is found on stockpiles where vegetative cover is minimal at best (see Topsoil). Yet, except for erosion along the edges of water drainageways, no sediment is carried into water courses moving off the site.

SEEDING AND AMENDMENTS

Seeding Procedures

Despite the promising seeding plan, very little land has been seeded at Seminoe II. By June 1979, of 1,247 acres of land that had been disturbed by mining, only 299 acres (24 percent) had been seeded.[51]

When Arch Mineral seeds, it does so in the fall. This practice is common in most of the West (except the Southwest), as companies prefer to have their seeds in the ground ready to germinate in the spring, long before the ground would be dry enough to cope with seeding machinery.

The seed mixture now being used consists of seven grasses and forbs seeded at 16 pounds per acre, with an additional 2 pounds of seed selected from one of seven other plant species. Almost every one of the plants seeded in a given year is native. However, very few warm-season plants are being seeded, and the result may be that productive vegetative cover all but disappears in warmer periods. Moreover, one pound of pure live seeds made up of three shrub species is broadcast on each acre of land being revegetated. The amount of the mixture applied to different slopes or soils does not vary. No seeding experiments were conducted in 1979 to assess the need for variation.

In such harsh climates experimentation would be the key to helping determine proper reclamation seeding mixtures and other reclamation techniques. Depending upon the similarities among Seminoe II and the other two Arch Mineral mines in the area, experience at the two other sites may provide much of the needed information. However, initial vegetative establishment does not indicate that successful revegetation is guaranteed (see Vegetation).

TABLE A

1979 SEED MIXTURE
FOR LANDS BEING RECLAIMED AT SEMINOE II

Plant Species	Introduced/ Native	Cool/ Warm	Pounds Applied Per Acre
Grasses			
Western wheatgrass	N	C	3.5
Thickspike wheatgrass	N	C	3.5
Intermediate wheatgrass	I	C	3.0
Slender wheatgrass	N	C	2.0
Indian ricegrass	N	C	1.0
*Experimental species	*	*	2.0
(grass or legume)			15.0 Pure live seed
*Smooth brome	I	C	
Green needlegrass	N	C	
Russian wildrye	I	C	
Bluebunch wheat-			
grass	N	C	
Blue grama	N	W	
Pubescent wheat-			
grass	I	C	
Yellow sweetclover	I	W	
Shrubs and Forbs			
Winterfat	N	C	0.5
Rubber rabbitbrush	N	C	0.25
Big sagebrush	N	C	0.25
Fourwing saltbush	N	W	1.5
Shadscale	N	W	1.0
			3.5 Pure live seed

Source: Arch Mineral Corporation

Amendments and Maintenance

Most seeding is done on the contour; this provides small
ridges that retain water and sediment that may flow down the
slopes. Contour furrowing has been done in the past on selected
areas; according to DEQ it has provided excellent terrain for
catching snow and keeping it in these furrows.[52] This, in turn,
provides a wetter micro-climate, enhancing plant growth.

Contour furrowing is practiced on approximately 50 percent
of the lands being seeded, based on the steepness of the slopes
and whether a slope's exposure is to the south. Southern slopes

lose water faster to evaporation; thus the use of furrows to retain more water is beneficial to plant establishment and growth.[53]

All Seminoe II seeding is done with a rangeland drill, a good
practice. The equipment belongs to a local rancher who is employed by Arch Mineral for the seeding season. If broadcast
seeding is ever required, Arch Mineral doubles the amount of
seed per acre.[54] This is usually done only on stockpiles.
Shrubs are also seeded by broadcast methods.[55]

Many types of mulch have been tried at Seminoe II. Use of
hay mulch began in the spring of 1978. Now the company would
like to begin planting a standing grain mulch to be mowed into
the ground before fall seeding.[56] It attempted to use a hydro-
mulch in summer 1979. The water, wood fiber mulch and fertilizer were mixed and sprayed over graded lands. Some 1,500-
2,000 pounds of wood fiber mulch were applied per acre, along
with 20-30 pounds of nitrogen fertilizer.[57] However, according
to Greg Bierei, Director of Reclamation for Arch Mineral's Western
Division, this proved to be labor- and time-intensive, and the
benefits did not justify the costs.[58]

The company has a good mulching policy. In the fall, when
seeding is done, straw or hay mulch is applied to the ground
and tacked in. If the land is not ready for seeding by fall, the
company plants annual grains there in the spring to provide a
standing, live mulch, then mows this grain and seeds grasses in
the fall. The company told INFORM the stubble mulch of annual
grains appeared to be the best treatment on lands being revegetated.[59]

In 1979 Seminoe II reclamation included the application of
fertilizer on all seeded lands. About 20 pounds of nitrogen fertilizer per acre were applied on 23 acres of land at the site. No
other fertilizer components were used.[60] No test plot experiments
are being conducted using fertilizer on the Seminoe II site. The
company plans no irrigation at the mine, as it lacks sufficient
water to do so.[61]

Trees and Shrubs

Arch Mineral is making a good effort to establish shrubs on
lands being reclaimed. It is not attempting to plant trees, which
are uncommon to the area.

Shrubs are being planted and transplanted at the mine. On
each acre of land seeded with the standard seed mixture, an
additional one pound of seeds made up of three shrub species is
being planted, using a hand-broadcasting seeder as the company
has had difficulty seeding with a drill seeder.[62] Shrub seedlings
and tubelings are also being planted at the mine. They include
antelope bitterbrush, big sagebrush, currant, horsebrush, rose,
rubber rabbitbrush, serviceberry, spring hopsage and winter-

fat.[63] In addition, Arch Mineral maintains a shrub-tubeling
test-plot on the mine.

VEGETATION

Vegetation on reclaimed lands at Seminoe II is poor. Native
species grow on some of these lands, and erosion is largely being prevented. However, most of these areas are being invaded
by weeds. The DEQ found in 1977 that revegetation efforts have
not all been successful and that reseeding would be necessary.[64]
Arch Mineral is reluctant to say that reclaimed lands will make
better grazing lands than those that existed before mining.[65]
Premining productivity of the land ranges from 261 to 1,877
pounds of forage per acre.[66] Over the whole mine area, the
average productivity of the land is 916 pounds of forage on each
acre of land.[67] Most of the company's information on pre-mining
productivity of the land comes from the Soil Conservation Service's data for the area, according to Bierei.[68] Although the
company did some clippings of vegetation on undisturbed lands
to test productivity levels, Bierei says, the effort was minimal.[69]
In terms of animal grazing capacity, Arch Mineral reports that
the land is capable of feeding one animal unit (one cow and calf
or five sheep) for one month on every 9-14 acres of land. This
contrasts sharply with the neighboring Rosebud Mine, where
Peter Kiewit reports that the land can support only one animal
unit per month on 25 acres. (See Rosebud profile.) According
to Bierei, the difference may be due to steeper topography at
Rosebud, as well as the lack of good data.[70]
The most common plants found on reclaimed lands are not accounted for by seeding. Russian thistle, halogeton and kochia
are the most common invaders of seeded lands. In early years
after seeding, the seeded plants establish themselves only in
small amounts. Lands seeded in 1977 were found to be covered
predominantly with fourwing saltbush and native grass species
one year later.[71] Crested wheatgrass dominated land seeded in
1976.[72] However, on land seeded in 1975 a remarkable change
has occurred. Inspection by DEQ in January 1977 found that
several wheatgrass species and fourwing saltbush were appearing with high cover values. DEQ also reported that the presence
of Russian thistle and halogeton was decreasing.[73] But another
inspection of the same area in 1978 found that native plant
growth was poor and that Russian thistle and kochia abounded.[74]
When INFORM visited the site in June 1979 lands seeded in
1975 had been invaded by greasewood (a highly unpalatable
plant); however, Indian ricegrass, fourwing saltbush, crested,
western and thickspike wheatgrasses, and some spots of yellow
sweetclover were growing well. Greg Bierei concedes that four-

wing saltbush and shadscale are establishing slowly. He recognizes that Russian thistle abounds on reclaimed lands, but points
out that grasses can be found protected under and in the shade
of these plants. The thistle finally dies after one year and becomes tumbleweed, blowing away and allowing the grasses growing underneath to flourish.[75] Although one can find small clumps
of grass under the thistle, it remains to be seen whether further
growth of these grasses will occur, but growth at the mine to
date has been poor. Bierei also feels it will take at least four to
five years to establish a good plant cover; then growth will accelerate.[76] Attempts at growing shrubs from seeds have not
been evaluated, but test plots have been fenced off to study
shrub tubeling growth.

Arch Mineral has developed little information on the productivity of reclaimed lands. It reports that a vegetation survey
was conducted in the summer of 1977 by Mine Reclamation Consultants, Inc., of Laramie, Wyoming.[77] This information was to
help ascertain reclamation success. The company would not release any of the report's findings to INFORM. Benton Kelly,
permitting director for the mine, says some vegetation has been
clipped on reclaimed sites to ascertain their productivity, but
the work is irregular. He adds that no one has studied nutrient
value, species composition or grazing trials at the site, and none
is planned.[78] Cover on most of the reclaimed slopes ranges from
8-23 percent and varies.[79]

Erosion of the seeded lands is not a problem at this mine,
primarily because of the soil and climate conditions. As mentioned above, water readily infiltrates the soil and does not run
down the slope. Regardless of whether the slope's vegetation is
made up of desirable or undesirable plants, it can protect the
ground from erosion in light rains. However, INFORM found
little evidence that desirable vegetation was growing in sufficient
quantities to prevent erosion on reclaimed lands.

In the Environmental Statement for Southcentral Wyoming
Coal resources, the Bureau of Land Management said the impact
of mining in the Hanna region could be such that "vegetative
composition similar to present-day composition is expected to be
attained in the area approximately 40 - 50 years after initial
reclamation efforts."[80] Revegetation at the Seminoe II Mine has
not yet shown this assessment to be false; however, failure to
establish a stable cover of native vegetation may prove even 40
or 50 years to be optimistic.

WILDLIFE

Arch Mineral has made good efforts to enhance reclaimed
areas for wildlife, planting shrubs and building rockpiles for

small animals.[81] INFORM found these rockpiles well built--and probably suitable for wildlife.

Some shrubs being planted at the mine are good forage for wildlife; others are poor. Winterfat and antelope bitterbrush provide good forage. According to the Cooperative Extension Service of Montana State University,[82] rubber rabbitbrush is thought to be poor, big sagebrush fair to poor. This means some shrubs may not be grazed as heavily, thus providing better ground cover and habitat for small animals.

RECLAMATION COSTS

TABLE B*

RECLAMATION COSTS AT SEMINOE II

Operation	Cost Per Acre	Approximate Percentage of Cost†
Earthmoving		95%-97%
Grading	$2,661-$3,960	
Topsoil stripping	600	
Topsoil replacement	600	
Revegetation		3%-5%
Seeding	200	
Subtotal:	$4,061-$7,360	

*These costs and percentages are not strictly comparable to those of the same operation or activity at other mines. See the Cost of Reclamation chapter for a detailed explanation.

†Because Arch Mineral did not disclose a total reclamation cost, approximate percentages have been calculated based on a subtotal of defined costs.

Source: Arch Mineral Corporation

The only variation in costs for reclamation at this mine results from the amount of overburden that must be graded. Grading costs vary among the three mining areas, ranging from $2,661-$5,960 per acre.[83]

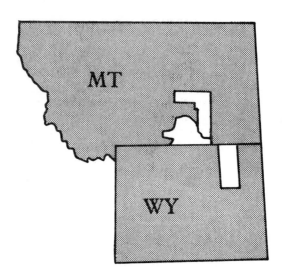

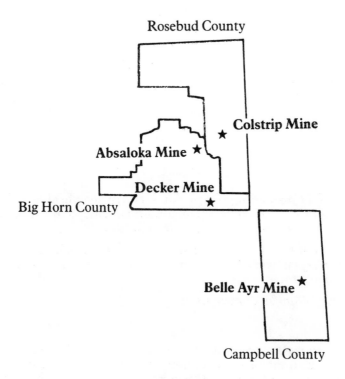

Rosebud County

Colstrip Mine ★

Absaloka Mine ★

Decker Mine ★

Big Horn County

Belle Ayr Mine ★

Campbell County

The Northern Plains

With an average annual precipitation rate of about 14 inches, this region is not among the easiest to reclaim, with its low amounts of topsoil, regular dry years, and the potential impacts of disturbing groundwater systems. However, conditions are better than in areas like southern Wyoming or the Southwest, and regulatory oversight has been stringent for many years. The Montana Department of State Lands has maintained a high level of concern for proper land reclamation at surface coal mines. In northeastern Wyoming, regulation comes under the auspices of the Wyoming Department of Environmental Quality, whose regulatory program, while not as good as Montana's, nonetheless is fair. Watching over the area from an environmental and land-use perspective are the Friends of the Earth and rancher-public interest coalitions like the Northern Plains Resource Council and Powder River Basin Resource Council.

Absaloka/Westmoreland/MT

Mine: Absaloka Mine

Location: Hysham (population 373), Montana, in Big Horn County (population 10,688)

Operating company: Morrison-Knudsen Company

Parent company: Westmoreland Resources Incorporated (owned by Westmoreland Coal Company 60%, Morrison-Knudsen Company 24%, and Penn Virginia Corporation 16%

Parent company revenues:* 1977, $352,970,000; 1978, $296,613,000

Parent company net income:* 1977, $17,864,000; 1978, $1,379,000

First year of production: 1974

Coal production in 1978: 4,529,053 tons

Acres for which a mining permit has been issued: 1,270 as of June 1979

State regulatory authority: Montana Department of State Lands

Reclamation bond posted: $6,175,000

Reclamation cost: reclamation cost per acre, NA; reclamation cost per ton of coal, NA

Expansion plans at mine: To expand production at the mine to 15 million tons per year by 1984

Additional acres for which the company will seek a permit: In August 1975 a 30-year mining plan was approved by the U.S. Geological Survey for mining on 2,100 acres in the east fork of the Sarpy drainage. Will have to receive a mining permit if this area is to be mined. A total of 15,000 acres included in the mine's leasehold could be mined.

Parent company U.S. coal reserves: Westmoreland Coal Company owns or leases 849 million tons of coal reserves.

Development and expansion plans for surface coal mines in western United States: None

---

NA = Not available

*for Westmoreland Coal Company

Sources: Westmoreland Resources Incorporated; Westmoreland Coal Company Form 10-K, 1978; 1977 Keystone Coal Industry Manual; "New Coal Mine Development Survey 1978-1987," Coal Age, *February 1979;* Rand McNally Road Atlas, *54th Edition*

SITE DESCRIPTION

The Absaloka Mine, operated by the Morrison-Knudsen Company (the contractor) and Westmoreland Resources, is located in the northeast corner of Big Horn County, Montana, about 26 miles east of Hardin, a town of 2,733 residents.[1] The terrain is one of rolling hills with occasional sandstone outcroppings and other rocks. Fragile and distinct sandstone formations, called "mushroom rocks" or "medicine rocks," also dot the land just east of the mine boundaries. Rising to the south and east are the Little Wolf Mountains. Flowing near the mine are the Middle Fork and the East Fork of Sarpy Creek, a tributary of the Yellowstone River.

Precipitation averages 16 inches annually[2] and has ranged from 10-21 inches in the last 30 years.[3] About half falls between March and June[4]; the wettest months are May and June.[5] From July through September the site is very dry.[6] The ground is unfrozen from about mid-May through mid-September.[7]

Populated by ranchers and farmers, land around the Absaloka Mine is generally used for cattle grazing and consists of grasslands with ponderosa pine growing on approximately 35 percent of the area.[8] Some agricultural lands are used for the production of grains and hay near the site, and some grazing lands have been planted with introduced grasses. The land around Absaloka produces many more trees than the other Montana mines studied by INFORM.[9] More than 80 percent of the land in the Sarpy Creek area is native rangeland.[10]

OVERVIEW

Although more coal was produced at Absaloka in 1978 than at all but six other mines in the West, and although much of its reclamation work is good, a major controversy threatens the entire operation's very existence.

That controversy centers on groundwater: the potential impact of mining on the underground aquifer systems in the area. Areas of lush vegetation have resulted from the surfacing of this groundwater in springs. Two large alluvial valley floors lie along two forks of the major water drainage, Sarpy Creek. Friends of the Earth (a national environmental group), regulators, the Crow Indian Tribe, local residents and members of Congress have focused their attention on Absaloka and the debate between the company and those who feel the mine should be shut down (or kept from expanding) before groundwater and potential alluvial valley floors are destroyed. The evidence to date does not fully support either side over the other. The debate continues.

Overall, the reclamation program at Absaloka--which began operations in 1974--indicates that considerable thought has gone into producing some rather innovative practices--including state-of-the-art grading and special soil salvage for transplanting ponderosa pines. Apart from the controversy over water issues, small problems persist with topsoil recovery and replacement, and erosion of sloping lands is preventing much of the reclaimed land from stabilizing.

Earthmoving practices at Absaloka are generally good, although small situations consistently arise where better performance is necessary. Grading has improved in recent years, creating slopes that are more easily traversable by machinery. Jack Schmidt, former inspector for the Montana Department of State Lands, told INFORM the new grading and the resulting drainage patterns at Absaloka were "state of the art" practices, and a "great improvement on Westmoreland's part."[11] For a new mine, a relatively large area has been graded. By January 1979, 49 percent of the lands disturbed by mining had been graded.

Topsoil removal practices vary in quality. A good practice at Absaloka is the immediate backhauling of about 50 percent of all the topsoil being stripped. But the company has often failed to take all the existing topsoil, a lapse for which it has been cited by regulatory authorities. Handling of toxic or sodic materials on the site is innovative in that the company uses the material for construction of pads from which the dragline operates. This practice is monitored, however, as it could result in harmful materials leaching or eroding into the surface waters or groundwater.[12] Most earthmoving procedures are being improved, and the company's segregation of special soils for tree-transplant

areas and immediate re-use of topsoil are positive steps toward a
good reclamation program.

Control of surface water quality has been good, and Westmoreland has not exceeded discharge limits for water leaving the
site. But, again, the big question at Absaloka centers on
groundwater and alluvial valley floors. The company's extensive
groundwater monitoring system has not until recently been kept
up to date. Preliminary findings show few adverse effects from
mining, but the potential is great for groundwater supplies to be
interrupted and for nearby alluvial valley floors to lose the shallow, subsurface aquifers critical to the land's high productive
potential for grazing and pastureland. Erosion control is only
fair to poor. On-site erosion has been serious--yet not as bad
as Colstrip's Pit 6, according to Dennis Hemmer, an inspector
with the Montana DSL (see Colstrip profile)--and the company is
finally implementing long overdue measures to combat it. However,
until erosion is completely controlled, vegetation will never completely succeed, and off-site sedimentation may begin to occur.

The seeding and amendment program at Absaloka may now be
termed fair to good. By the end of 1978, however, only 25 percent of disturbed lands had been seeded. The seed mixture
consists of a good sampling of native grasses, divided between
cool- and warm-season species. Application of straw mulch, as
well as the seeding of grains for a standing cover crop, should
help to control erosion better, and the fertilizer used to get
plants growing will probably not create a plant community dependent upon man for its continued growth and stability. Shrub
and tree transplants are fair to good. The company still relies
heavily on bare-root seedlings of shrubs--even though their
survival rate has been poor. Ponderosa pine transplants--using
the innovative technique of salvaging rocky soils for this pur-
pose--have been better, and the company is taking great pains
to make these transplants take hold and proliferate.

The results of revegetation are promising. If erosion can be
better controlled, and if more shrubs and forbs can be established, a good stand of vegetation will result--not only from the
seed mixture but from plants that volunteer from topsoil that is
immediately backhauled. It is too early to determine whether
productivity or diversity will match pre-mining levels, but initial
results are good. Wildlife such as deer, game fowl and rodents
will benefit from the trees, shrubs and grasses being grown for
forage and habitat.

Regulators have kept Westmoreland on its toes at Absaloka.
Inspectors of the Montana Department of State Lands (DSL) visit
the mine about twice a month and are quick to point out problems
as well as sound practices. Between August 1974 and August
1979, 12 fines totaling more than $18,500 were levied--a common

amount for Montana surface mines, according to Dennis Hemmer.[13]
The company has since tried to comply with state requests for
improvement of some of its operations. The federal Office of
Surface Mining (OSM) visited Absaloka four times between June
1978, when the Region V office began inspecting, and October
1979. OSM issued a notice of violation to Westmoreland for failure to salvage topsoil in some areas, but terminated the notice
after the company resolved the problem. The larger issue for
state and federal regulators will be the question of mining as it
relates to groundwater and alluvial valley floor disturbance both
on and off the site. The resolution of this issue at Absaloka
deserves close attention, as it may set a precedent for areas
where the potential for environmental harm is not fully realized
until commitments to mining are too great to allow termination of
operations.

EARTHMOVING

Grading

Grading and contouring practices at Absaloka are fair to
good as mining progresses uphill. With more experience, these
operations have improved as the mine has expanded. As with
any mine only recently opened, more land has been disturbed
than is being graded and reclaimed in the first years of operation.

Grading of mine spoils follows closely behind the open pit,
usually within $1\frac{1}{2}$-2 spoil ridges.[14] However, much land remains
to be graded. As of January 1979, the ungraded portion made
up 263 of the 516 acres of land disturbed by mining.[15]

As grading and contouring have progressed, Westmoreland
has had to alter the contour of graded slopes slightly from their
pre-mining pattern. In some areas, steep slopes of more than 12
degrees are encountered, along with flatter slopes of 6 degrees.[16]
In recontouring, Westmoreland has followed directives by the
Montana DSL to grade to a flatter slope, then place short, steep
slopes into the contour instead of creating longer, steeper slopes
that would be more prone to erosion and harder to revegetate.[17]
Most of the graded slopes at the mine approach the flatter angles
of surrounding lands, inclining at approximately 7 degrees.[18]
DSL told INFORM that although this somewhat alters the original
surface water drainage pattern, the same general slope configuration, land profile and stability are being maintained.[19]

David Simpson, environmental administrator for Westmoreland
Resources, explained to INFORM the company's efforts to improve grading practices. Where grading used to be done by
sight, the company is now paying much closer attention to drain-

age patterns and is utilizing both civil engineering and geological principles in designing post-mining slopes.[20] Roads used to haul coal from the mine pits are being designed as the predominant path for surface flow over regraded areas. They are being graded to drain water away from active mining pits as the mine progresses uphill. It is important to note that the planning of grading and contouring is much easier in areas like this one, where the coal seam is relatively flat under the surface, and where overburden depths to not change drastically as they do in areas such as the Hanna Basin in Wyoming. (See Rosebud and Seminoe II profiles.)

One grading problem has arisen with spoils from the initial mine cuts at Absaloka. These "boxcut spoils" were poorly graded at first, and the company has had to regrade these areas twice.[21] This has created some steep slopes and erosion (see Erosion) into an area within the permit boundaries that the company has been told not to disturb. The so-called "critical fragile" area was protected by the state of Montana from mining because of its unique rock bluffs and concentrations of ponderosa pines.[22] However, DSL has recognized the problems this restriction has caused for grading up to the boundaries in the area, and it is watching to ensure that long-term difficulties do not arise.[23]

Slopes on some older, more steeply graded lands did not allow seeding or other reclamation machinery to be operated on the contours. This led to increasing erosion in some parts of the mine.[24] However, about 95 percent of the newly graded areas are traversable along the contour.[25]

Topsoil

Both topsoil and subsoil are being stripped and segregated at Absaloka. Although this is a good practice, performance has not been completely satisfactory. The company has frequently been found lax in its topsoil removal. Montana DSL inspection reports from late 1977 and 1978 cite numerous instances where topsoil was insufficiently salvaged, or was lost owing to the proximity of the mine pit to areas where removal had not yet been undertaken.[26] David Simpson told INFORM these were "minor problems, involving small amounts of topsoil."[27] However, the company has not yet solved them. An inspection by the Office of Surface Mining in July 1979 found that topsoil had not been properly salvaged in at least three different areas; it promptly issued a notice of violation to the company.[28] A followup inspection in October found that the company had rectified the deficiencies cited, and the violation notice was terminated.[29] Although these failures to salvage topsoil properly are relatively minor, they can add up. Westmoreland's approach to this issue warrants continued scrutiny.

Another innovative practice is the removal of topsoil in winter. Many mine operators will not remove topsoil when the ground is frozen, but Simpson says he prefers this season because plant materials in the soil are dormant and there is less dust.[30] To remove the topsoil when the ground is frozen, the company first rips the soil with a tractor or other machine, then follows with standard scrapers. This is more expensive than summer salvage, when ripping is unnecessary.

A considerable amount of topsoil has been directly hauled back for immediate replacement on graded lands. In 1978 the DSL reported that only 35 acres of land had received this treatment.[31] The company now reports that about 50 percent of the removed topsoil is immediately replaced and that even more could be handled in this fashion in the future.[32] Westmoreland also attempts to salvage soils with special characteristics for use in areas slated for tree and shrub transplanting. Some soils are used specifically for replacement in water drainage areas to support the vegetation the company will try to establish.[33] This practice is unique and, although long-term success is not yet assured, the company's efforts are well conceived and commendable.

In general, topsoil replacement has been satisfactory. But OSM found that in at least one instance topsoil was replaced in an area before grading was finished. An inspection in July 1979 found that Westmoreland had replaced topsoil in a low-lying depression and used it for fill, instead of first filling the area with spoil and grading it before replacing the topsoil.[34] This incident, although minor, again tends to illustrate an occasionally careless attitude by the company toward salvaged topsoils. These valuable soils belong on the surface of graded land, to provide a good seedbed and growing medium, and should not be used for filling depressions.

The six topsoil storage areas at the mine cover from 2-19 acres each.[35] INFORM found that stockpiling of salvaged soils at Absaloka is generally done properly, and that topsoil stockpiles are well maintained. Inspections by the Montana DSL in 1977 and 1978 found small instances where road scrapings had been pushed onto stockpiles,[36] grading of overburden had allowed some materials to end up on stockpiles,[37] or erosion was creating small gullies on the sides of a stockpile.[38] Most of these occurrences appear to have been minor.[39]

Toxic Material

Materials unsuitable for plant growth are sometimes encountered at Absaloka. Mining removes coal from four coal seams

separated by varying amounts of interburden materials. The two
main seams being mined are separated by 50-100 feet of interburden.[40] In 1977 and 1978, DSL inspectors found that toxic
materials were not being buried (Westmoreland's Simpson told
INFORM he was unaware of this[41]) and that grasses planted on
soils with a high sodium content in a test area were turning yellow (a sign that undesirable minerals are present in the soil).[42]
However, by late 1978 the company was properly covering highly
sodic spoil materials with a $3\frac{1}{2}$-foot layer of subsoil and 6 inches
of topsoil.[43] The company told INFORM the former practice at
the site was to place interburden materials near the top of refilled areas and cover them with more spoil if necessary. However, it now salvages potentially toxic or sodic material and
either buries it deeper in the old mine pits, or uses it to build
the flat areas upon which draglines rest while digging.[44] This
allows these materials to be buried very deep and should protect
surface soils and vegetation from any harmful effects. Dennis
Hemmer told INFORM these practices are adequately preventing
any problems.[45] Groundwater monitoring will show whether water quality in the area is affected; if so, alternate methods will
have to be employed.

WATER QUALITY

Surface Water and Groundwater

Mining at Absaloka now takes place between the Middle Fork
and the East Fork of Sarpy Creek, a tributary of the Yellowstone
River. Present plans call for mining to proceed toward the East
Fork. Extensive aquifer systems provide the groundwater used
by local residents (more than 300) for domestic and stock watering purposes. Five significant aquifers lie in and around the
coal seams being mined.[46] These groundwater systems also provide water to the surface through springs, of which more than
30 have been identified within the 23 square-mile lease area
Westmoreland plans to mine. Other springs flow just outside the
lease boundaries.[47] Many of these springs occur in "coulees"--
wet, lush areas that drain toward the two forks of Sarpy Creek.
Thus surface water and groundwater are related extensively
throughout the mine area.

Because water flowing from Absaloka eventually finds its way
into either the Middle Fork or East Fork of Sarpy Creek, it is
diverted into sediment ponds on the site and retained to allow
collected sediment to drop out before the water is discharged
from the permit area. Under a National Pollutant Discharge Elimination System permit issued by the Montana Department of Health
and Environmental Sciences/Water Quality Bureau, Westmoreland

may discharge water from seven points. An extensive system of
sediment ponds has been constructed, and four ponds were on
the site as of January 1980.[48] These ponds have a capacity of
about 175 acre-feet of water[49] (an acre-foot equals the amount of
water that would cover one acre of land to a height of one foot).
Although the company is required only to treat water to meet effluent guidelines during any runoff resulting from storms with
an intensity equal to that of a 24-hour precipitation event with a
recurrence interval of 10 years (a 10-year/24-hour storm), the
two large sediment ponds on the site are designed to hold and
treat water that would result from a 100-year/six-hour storm.
This means much of the water held in the ponds will not be discharged at all, and Westmoreland can control the time of discharge.[50] Natural drainage patterns and ditches dug by Westmoreland direct water from the mine area into the sediment ponds.

Very little vegetation can be seen growing on or around the
sediment ponds and drainage ditches at Absaloka. Rocks used
as rip-rap, to protect the ground from flowing water, have been
placed on many of the embankments and at discharge points.[51]
However, discharge ditches on the site rapidly fill with sediment
from extensive erosion in certain areas of the mine (see Erosion)
and must be periodically cleaned.[52] As cleaning requires digging the accumulated sediment from these ditches, vegetation is
not generally found in abundance, and the ditches are themselves
prone to erosion.[53] An OSM inspection in July 1979 found that
many of the diversion ditches had been cleaned of sediment and
that Westmoreland was attempting to maintain the system well.[54]

According to reports by the Environmental Protection Agency
and OSM, discharged water from Absaloka is kept within the limitations of the mine's NPDES permit.[55] However, sediment ponds
have occasionally filled and backed up onto lands, leaving a sediment deposit when the water recedes.[56] These mishaps generally affect only small areas and, according to Dennis Hemmer,
are an allowable feature of the ponds' design.[57] With the larger
ponds being built at the site, the problem should be eliminated.

Given the mine area's complex groundwater systems and its
location between two forks of Sarpy Creek, surface and groundwater monitoring are essential. Absaloka conducts both extensively. OSM inspectors visited the mine in October 1979 and
found that groundwater monitoring was being conducted in 90-
100 wells. The company also has nine surface-water monitoring
stations, and it monitors water discharged from eight springs in
the area. Although OSM found that all monitoring was being
performed properly, it did not discuss the results of these programs.[58] When INFORM visited the site in June 1979, David
Simpson avoided discussion of the program's results, confining
his observations to the extent of monitoring.[59] In January 1980,

however, Simpson told INFORM that Westmoreland's initial findings indicated that neither water quality nor quantity was being
adversely affected by mining. He conceded that past records of
groundwater monitoring were spotty.[60] Inspectors for the Montana DSL told INFORM that, because mining is still confined to
the higher elevations, impacts upon groundwater quantity may
not be seen yet. Later, they say, the consequences of mining
will be better known.[61] Still, the monitoring program is being
implemented; future results, coupled with the data now being
collected, will provide significant information. This is important
in light of findings by three hydrogeologists for the Montana
Bureau of Mines and Geology--that although the coal beds under
Absaloka are not highly productive aquifers (rates of flow are
low), "the small amounts of water in the coal and overburden are
of the most desirable quality of ground waters in the area."[62]

Alluvial Valley Floors

The alluvial valley floor (AVF) provisions of OSM's regulations, requiring that AVFs be protected from mining unless the
integrity of the system can be replicated, raise the most serious
problems for Westmoreland and Absaloka. David Simpson readily
admits that the Middle Fork Sarpy Creek is an AVF under OSM
criteria, and although the company plans only to mine near, but
not through it, the company must monitor the effects of its mining upon that fork.[63] However, the larger concern stems from
plans for mining to proceed toward the East Fork Sarpy Creek.
This is the central controversy at Absaloka.
    As this area is generally considered to be an AVF as well,
the integrity of its combination of soils, surface water and
groundwater must be protected from mining and the effects of
mining.[64] Mining in the direction of the East Fork may destroy
many springs and drastically alter groundwater systems. According to a draft Environmental Impact Statement prepared by
the Montana DSL for Absaloka's mining and reclamation plan:

> Surface water impacts due to mining would be
> limited to those caused by the local loss of the
> springs [in the area]. No direct impact is likely
> on the intermittent surface flow of either East
> Fork or Middle Fork Sarpy Creek. [However,] *a*
> *measurable and significant interruption of ground*
> *water flow would occur to the East Fork basin.*
> *Such a disruption of the hydrologic system could*
> *well constitute a significant impact to any subir-*
> *rigation along East Fork Sarpy Creek near or*
> *downstream from its confluence with the three*

> *tributary coulees, especially during drought*
> *years.* [Emphasis added.][65]

Although the coulees that drain water into the alluvial valley
floor area and support a small but lush vegetative community
would probably not themselves constitute an AVF under OSM
guidelines, they are directly related to the nearby AVFs, providing additional forage and water. Under the 20-year mining
plan now proposed at Absaloka:

> ...impacts on the hydrologic system would be
> significant, because mining would destroy part of
> a recharge area that supports numerous springs
> in three coulees north and northeast of the mine
> area....*Loss of the springs and the associated*
> *drop of the groundwater table would affect vege-*
> *tation, wildlife, and land use in and adjacent to*
> *the coulees....East Fork Sarpy Creek could well*
> *be measurably impacted.* [Emphasis added.][66]

Since publication of the draft EIS, a final statement has been
produced that contradicts these findings completely. The valid-
ity of the new findings has been taken to court and remained
there at the time of this writing.

Westmoreland recently completed construction of a dragline
costing more than $20 million in hopes of being allowed to mine in
this area. But local ranchers and a representative of Friends of
the Earth are fighting to prevent mining from proceeding into
these areas. Federal and state regulators, reporters and members of Congress have been shepherded over the land, and a *New
York Times* article in August 1979 brought national attention to
the issue.[67] Westmoreland faces a battle that pits the company's
interests directly against those of the local citizenry and environmentalists over the long-range detriment of surface coal-min-
ing to alluvial valley floors.

Westmoreland generally disagrees with contentions that
groundwater will be irreparably harmed or lost through mining.
It also believes that vegetation found on lands fed by underwater
sources--such as in coulees--is not unique in its makeup.[68] As
the questions of whether Absaloka will continue mining the East
Fork remains open, debate continues among the company, state
and federal regulators, local residents and environmentalists.

Erosion

Westmoreland has had persistent problems in stabilizing lands
it is reclaiming, and erosion and sedimentation have resulted.
One area that has experienced these difficulties over the past
two years is the "critical fragile" (CF) area (see Earthmoving).

The "critical-fragile area" protected from mining in Absaloka.

This ten-acre area within the mine site was designated by the state of Montana to be protected from all mining activities. However, lands surrounding it have eroded, and sediment has been deposited within the CF area.

In February 1977 the Montana Department of State Lands warned Westmoreland that certain lands surrounding the CF area might erode into the protected boundaries.[69] By May, sedimentation onto the land had occurred[70] and continued to occur into 1978.[71] DSL warned Westmoreland that even more erosion and sedimentation might occur as the year went on (suggesting the company remedy the situation), and in June those predictions were confirmed. Erosion of land just outside the CF area left cuts in the earth 6 feet deep, 6 feet wide and 25 feet long. DSL reported that sediment was carried into the center of the area.[72]

The problem persisted, and following an inspection in late August, DSL reported that work had not yet begun to remedy either this situation or erosion on other lands.[73] An initial OSM inspection in September--although cursory--found erosion still occurring. OSM suggested that the company create ditches along the contour of slopes to slow the flow of water and contain sediment, and that it stabilize the areas permanently with vegetation.[74] DSL told INFORM in April 1979 that erosion and sedimentation into the critical fragile area were continuing,[75] and Westmoreland's environmental administrator, David Simpson, admitted that the company had not left itself enough "elbow room" around the boundaries.[76] Aside from sedimentation into the CF area, INFORM found no sedimentation has been reported off the mine site (see Surface Water and Groundwater).

Erosion has been widely reported on other portions of the mine. Lands that have been graded, topsoiled and seeded have eroded and sometimes been allowed to worsen before the company has begun repairs.[77] One DSL inspection found that machinery operating up and down slopes at the mine, instead of along the contour, was probably to blame for some of the erosion.[78] But recent grading (see Grading) allows machinery to operate along the contour, thus helping prevent further erosion.

Lands around the construction site for the mine's new dragline have suffered repeated erosion, and both DSL and OSM have found Westmoreland negligent in its attention to this situation.[79] The company's general response to all its erosion problems has been that "erosion is a problem in any case where soil is disturbed."[80] However, DSL concluded in June 1978 that "Westmoreland's attitude toward correction of erosion problems once they develop is poor. Two previous inspections...brought significant problems to their attention, yet nothing has been done [to prevent erosion of lands around the critical fragile area]."[81]

Yet DSL took no further action than to suggest the company clean up its operations.

Westmoreland now appears to be making a better effort to control and prevent erosion. On a visit to Absaloka in June 1979, INFORM found several areas where erosion had occurred and repair operations had been completed. The company is instituting the use of sandbags,[82] plastic netting and asphalt to stabilize soils on steeper slopes and in drainage channels.[83] It is also planting annual grains on lands being reclaimed to provide a quick covering crop of vegetation to hold the soil. DSL reports that erosion has been reduced since the company instituted these measures.[84]

Still, because of small slopes that are too steep to allow machinery to operate on the contour, erosion will continue because machinery tracks will help to begin the erosive course of waters flowing downhill. Furrows placed along the contour of these lands (if possible) would help, and closer attention to repair of possible problem areas will also minimize the harm. Hand-seeding might be used to avoid uphill and downhill tracks left by seeding machinery. It is up to Westmoreland to show that it will change its practices to prevent a recurrence of the serious problems that have occurred in the past. Without proper erosion control, establishment of a stable and useful vegetative cover will be virtually impossible. Conversely, establishing a good vegetative cover will help prevent erosion in the long term. Netting and sandbags are temporary control measures, but vegetative growth is the best means of stabilizing soils.

SEEDING AND AMENDMENTS

Seeding Procedures

According to a spokesperson for the Montana DSL, Westmoreland keeps the seeding at Absaloka very current with land disturbance.[85] Because grading is kept current, seeding can also be kept close to the active mining area. OSM reports that lands where topsoil has been replaced never sit for more than 90 days before seeding (or mulching to protect the surface) is begun,[86] as required by Montana regulations. By the end of 1978, however, only 127 of 516 acres of land that had been disturbed at the mine had been planted.[87]

Seeding takes place either in the late fall or early spring.[88] The goal is to have seeds in the ground and ready to take advantage of the longest possible growing season when soils thaw in the spring and melting snows and spring rains provide the needed water for germination and growth. David Simpson told

INFORM he would prefer to seed in the spring, but rains and mud sometimes prevent this.[89]

The mixture of plants includes a number of grasses, with a sprinkling of legumes (nitrogen-fixing introduced plants) and a token amount of the fourwing saltbush shrub.[90] Of the 15 grasses seeded in the mixture, only two are non-native species, and a number of warm-season plants are used. Although cool-season plants still outnumber warm-season species by 2:1, this ratio is better than that found at many of the sites studied by INFORM. Simpson relies on native seeds in the topsoil to provide additional plant growth on lands being reclaimed.[91]

TABLE A

1979 SEED MIXTURE
FOR LANDS BEING RECLAIMED AT ABSALOKA

Plant Species	Introduced/ Native	Cool/ Warm	Pounds Applied Per Acre
Grasses			
Western wheatgrass	N	C	3.9
Green needlegrass	N	C	2.4
Thickspike needlegrass	N	C	1.3
Slender wheatgrass	N	C	0.6
Beardless wheatgrass	N	C	1.3
Indian ricegrass	N	C	0.6
Streambank wheatgrass	N	C	0.6
Sideoats grama	N	W	1.2
Big bluestem	N	W	0.6
Little bluestem	N	W	1.0
Blue grama	N	W	0.7
Sand bluestem	N	W	0.9
Pubescent wheatgrass	I	C	1.2
Kentucky bluegrass	I	C	0.2
Hard fescue	N	C	0.2
Legumes			
Sainfoin	I	W	1.8
Yellow sweetclover	I	W	1.0
Forbs and Shrubs			
Fourwing saltbush	N	W	(token)
			19.5+ Pure live seed
Annual grain (added for cover crop)	–	–	10.0

Source: Westmoreland Resources

While a broad range of seeds is used at Absaloka, the seed
mixture is not varied according to the slope of the land or the
direction in which it faces. Simpson told INFORM in June 1979
that the diverse seed mix would make up for not varying the
amount or type of seed. He believes native seeds will help to
create a more diverse and viable vegetative community.[92] Preliminary study shows that this may indeed be the case (see Vegetation).

Amendments and Maintenance

Because slopes at Absaloka vary from steep to shallow, machinery has not always been able to operate along the contour.
Simpson told INFORM that about 95 percent of the seeding is now
conducted along the contour, and the rest must operate up and
down the slopes.[93] Contour furrowing is not practiced at the
site,[94] but it should be, given the erosion problems cited above
(see Erosion).

Westmoreland employs a broadcast seeder, then pushes the
soil over the seeds after they are sown. This is done primarily
because the company did not like the straight lines of plants that
developed when seeded with a drill seeder.[95] Hand-seeding has
also been infrequently practiced on small areas. The company
does not increase the quantity of seed mixture used when broadcast seeding, but seed is fairly accurately sown, and more seed
should not be necessary to obtain the same amount of germination
during the initial growing season.

Annual grains are used at Absaloka to provide a standing
mulch in the first year after an area has been graded and topsoiled. The annual grain is seeded with the regular mix and allowed to grow for erosion control. Simpson reports that the company was the first in Montana to seed annual grains specifically
for erosion control.[96] Through the use of various grains, including barley, oats and wheat, the company has found that
wheat best serves its purposes.[97] Straw is also applied and
crimped to the soil before the grasses and grains are seeded, to
better control erosion.[98]

Irrigation is not needed at Absaloka. Fertilizer is used in
small quantities. Westmoreland fertilizes once, before seeding.
The company applies 30 pounds of available nitrogen and 30
pounds of available phosphate to each acre, seeking to keep the
rate of application light enough to help promote early growth
without creating a dependence upon fertilizer for plant survival.[99]
The rate at which fertilizer is used depends upon soil analyses
made on the site.[100]

Trees and Shrubs

   The most advanced and ambitious work with trees and shrubs
at Absaloka is the attempt to recreate ponderosa pine stands
similar to those that existed prior to mining. Special soils from
areas where the trees once stood are replaced on graded lands to
provide maximum benefit to transplanted species. Other plants
being transplanted include chokecherry, plum, hawthorne, gold-
en currant, skunkbush and Russian olive.[101]
   The company has used both containerized and tree-spade
transplants of trees and shrubs, as well as planting bare-root
stock. Experiments have been conducted to evaluate the benefits
of one practice over the other. Findings to date are that tree-
spade and containerized transplants survive far better than
bare-root plantings. The U.S. Geological Survey (USGS) found
in May 1978 that bare-root survival was negligible and that tree-
spade transplants were not only surviving, but showing few
signs of stress.[102] By October the USGS inspectors could report
a 75 percent survival rate--considered very good[103] (see Vegetation). A DSL inspection in May 1978 quoted a Westmoreland
official as stating that bare-root transplants were showing only
30 percent survival while 80 percent of containerized transplants
were surviving.[104] In June 1978 a DSL inspector found that
while containerized transplants were surviving steadily, bare-
root stock had dwindled to a 10 percent survival rate.[105] Westmoreland told INFORM in October 1979 that tree-spade transplants
were satisfactory so long as the roots of the tree being moved
were not so large that major portions were lost in the digging-
out process. The company also reported that containerized plantings show approximately 90 percent survival, while bare-root
plantings average 20 percent survival.[106]
   The company has worked to increase the chances for survival of planted and transplanted trees and shrubs. Soils well
suited to tree growth have been segregated and used in areas
where tree transplants were to be conducted, and shrubs have
been pruned and the land around them covered with plastic to
minimize competition during establishment.[107] Should these efforts continue to produce such good results with tree and shrub
establishment, survival and growth, the company may well succeed
in creating a vegetative community that includes plant forms from
grasses through shrubs to trees.

                                                VEGETATION

   The results of revegetation at Absaloka are fair, judging
from the short time that plants have been growing. A basic mixture of plants is germinating and growing, other species are vol-
unteering from the topsoil, and there has been, as noted above,

some success with tree transplants. However, bare spots exist
on some lands and few forbs or shrubs are found. Continued
erosion in the Critical Fragile Area indicates that, as noted,
erosion control by vegetative stabilization has been poor in the
past. Efforts have been increased to control this problem, and
erosion over the whole mine is no worse than at any other in
Montana.[108] In May 1978 the U.S. Geological Survey noted that
revegetation could only be called "moderately fair with localized
spots of low density growth. A lush vegetative cover at this
early stage of development could hardly be expected; however,
Westmoreland recognizes there is room for improvement."[109]

Reports of productivity of the land being mined range from
approximately 2.5 acres per animal unit month (AUM--the amount
of land needed to graze one cow and calf, or five sheep, for one
month) to 3.5 acres per AUM. David Simpson opts for the lower
productivity end of this scale; the Montana DSL and information
contained in a draft Environmental Impact Statement published in
1976 tend toward the higher end.[110] DSL has also stated that
certain lands on the site have especially high productivity values;
Simpson discounts this.[111] Simpson also reports that forage productivity ranges from 1,000-2,000 pounds of forage per acre.[112]
The state and the company should obtain a better picture of the
productivity of native lands in the area and should agree upon
the levels to which Westmoreland should be held in determining
whether reclamation has recreated pre-mining productivity levels.

Simpson explained his theory on reclamation and revegetation
to INFORM in June 1979. He feels that reclamation as practiced
initially strives to stabilize the soil and prevent erosion. Within
three or four years the seeded native plants begin to appear and
to dominate the mulch and some of the introduced species. Native
species whose seeds are already in the topsoil, or have blown in
from undisturbed lands, begin to appear more frequently. Simpson feels productivity will reach a high point of two to three
times the productivity of the native range, then will begin to
drop back to normal levels.[113]

Many of these theories are borne out by initial results of the
sampling of lands seeded in 1975 and 1976.[114] Data sent to IN-
FORM by Simpson show that lands seeded in 1975 and 1976 have
an initial cover dominated by either wheat or sweetclover (which-
ever was used for the initial cover crop), and within two or
three years many of the grasses--especially wheatgrass species--
show greater growth and coverage while the cover crops begin
to be outcompeted. Productivity on these lands begins to drop
within a year or two of seeding. Coverage of the land, thus
protection from erosion, is generally fair, as many areas have
about 50 percent of their land surface shielded by plants.[115]
However, the lands sampled by Westmoreland that are now in

their fourth and fifth growing seasons are different from those lands now being reclaimed. The seed mixture now used is substantially different, relying more on native species and less on highly productive introduced legumes like sweetclover. Further study of these areas, as well as newly revegetated lands, is essential and is being done. Yet, for a company that has been on the site for only a short time, this data collection and its results are generally good. Long-range results will be the final measure of the success or failure of revegetation.

Diversity of plants on lands being reclaimed is improved by the growth of native plants from seeds that are transported with replaced topsoil, or that blow in from undisturbed lands. This is especially true with forb species, which are appearing on the land even though they have not been seeded. A representative of the Montana DSL told INFORM that Absaloka, like many other mines in the state, has trouble getting forbs to grow.[116] DSL is one of the most conscientious of the state regulatory agencies on the matter of plant diversity and the need for growing forb species.

The company has conducted no grazing trials on lands reclaimed at the mine, and it has no immediate plans to do so. Simpson told INFORM such trials might be run in five years or so.[117] DSL representatives said the land did not appear ready for bond release, however,[118] and as OSM requires that the land be used prior to bond release, grazing trials will have to be implemented.

WILDLIFE

Reclaimed land at Absaloka, like most mines in the INFORM study, will be usable by wildlife. Trees and shrubs should provide shelter and forage, and grasses also are palatable. DSL and USGS inspectors report that deer, hawks and rodents have been using the reclaimed lands.[119] The company reports that turkeys and antelope also are adapting to these lands.[120] Simpson is unsure of the importance of shorttailed-grouse strutting grounds found in the area.

The company is building no shelters for wildlife, but tree transplants provide good shelter, and the surrounding areas provide a large habitat for native populations. Simpson says that by 1980 the company will have a firmer grasp of wildlife issues in the area.[121]

RECLAMATION COSTS

Simpson told INFORM in October 1979 that he "cannot provide data on itemized reclamation costs."[122] No independent information was available to INFORM.

Belle Ayr/AMAX/WY

Mine: Belle Ayr Mine

Location: Gillette (population 20,450), Wyoming, in Campbell County (population 26,080)

Operating company: AMAX Coal Company

Parent company: AMAX, Incorporated

Parent company revenues: 1977, $1,170,470,000; 1978, $1,751,120,000

Parent company net income: 1977, $68,990,000; 1978, $160,010,000

First year of production: 1973

Coal production in 1978: 18,065,664 tons

Acres for which a mining permit has been issued: 6,280

State regulatory authority: Wyoming Department of Environmental Quality

Reclamation bond posted: $10,837,000

Reclamation cost: reclamation cost per acre, $4,300; reclamation cost per ton of coal, NA

Expansion plans at mine: Plan to increase coal production to 20 million tons per year by 1981

Additional acres for which the company will seek a permit: Unknown*

Parent company U.S. coal reserves: Owns or leases 5.045 billion tons of coal reserves

Development and expansion plans for surface coal mines in west-
ern United States: AMAX plans to open the East Sarpy Mine in
Montana in 1981 to produce 5 million tons per year by 1985. It
will be expanding production at the Eagle Butte Mine in Wyoming
to 25 million tons per year by 1986.

NA = Not available

**AMAX reports that additional acreage may be permitted, but
that the amount is unknown.*

Sources: AMAX Coal Company; AMAX Form 10-K, *1978;* 1979
Keystone Coal Industry Manual; *"New Coal Mine Development
Survey 1978-1987,"* Coal Age, *February 1979*

SITE DESCRIPTION

The Belle Ayr Mine is about 15 miles south of Gillette, Wyoming, in an area dense with newly developed coal mines. Properties bordering Belle Ayr are owned by Exxon, CONSOL, and
Mobil Oil or their mining subsidiaries. At least 24 mines are in
operation or being planned for the area surrounding Gillette.

The topography in this part of the eastern Powder River Basin is one of low, rolling hills with steepsided gullies, washes
and some broad valleys.[1] Slopes on the mine site range from
nearly flat to approximately 6 degrees, with slopes up to 27 degrees in some of the gullies.[2] The mine site is dissected by Caballo Creek.

Annual precipitation in the area is approximately 16 inches,
having ranged from 11.2-22.9 inches.[3] About 34 percent of the
precipitation falls as snow.[4] The growing season at the mine
lasts from about mid-May through late September and is about
120 days long.[5]

The predominant land use in the area is grazing for livestock, largely on grasslands and sagebrush-dominated areas.
Few trees exist in the area. AMAX reports that slightly more
than 5 percent of the mine area is used for cultivation of wheat
and oats.[6] All the land disturbed by mining since 1974 has been
native rangeland.[7] AMAX plans to return the land to its original
uses.

OVERVIEW

If ever there was a famous surface mine in the West, Belle
Ayr would be it. Since 1977 more coal has been produced from
the mining pit at Belle Ayr than from any other coal mine in the
entire United States, whether a surface mine or deep mining op-

eration. In 1978 more than 18 million tons of coal were produced and shipped from the site.

AMAX conducts a vigorous public relations campaign, which includes air-conditioned bus tours of the mine, situated just outside the boom town of Gillette. Regulators, legislators and environmentalists are constantly shepherded over the haul roads and reclaimed lands at Belle Ayr.

In general, Belle Ayr's reclamation operation measures up to its productivity. It incorporates some of the best procedures known for economical and successful reclamation. The mining operation is totally different from any other in the INFORM sample by virtue of its truck-and-shovel production technique. No draglines are used, making grading, contouring and selective removal and replacement of soils (especially topsoil) much easier. Very little land is disturbed relative to the amount of coal produced from these pits. However, very few areas where coal is mineable from the surface in the West allow the truck-and-shovel method to be employed economically, as it requires a high ratio between the thickness of the coal seam and thickness of overburden. Thus, grading and contouring are good, and approximately 75 percent of the topsoil removed from lands being prepared for mining can be directly hauled back onto graded land, saving both time and money while improving the chances for success in revegetation. Small stockpiles are well maintained.

Good water control is another positive facet of AMAX's mining and reclamation program. Improvements in the holding and diverting of surface waters have helped greatly, after early diversions of Caballo Creek were heavily scoured during rainstorms. Ponds at the site are built with valves to allow controlled discharge of water--uncommon at most surface mines. Natural ox-bows are used to retain water on the site, and AMAX has been within the limits of its National Pollutant Discharge Elimination System (NPDES) permit since mid-1977. A surface water and groundwater monitoring program is in effect and should yield significant data. However, preliminary findings are that groundwater will not be returning to reclaimed areas for two to five years; the long-term effects are unknown. AMAX will probably not have to preserve Caballo Creek, owing to certain characteristics that bring into question whether it is an alluvial valley floor (AVF), as mining and creek diversion began long before the federal AVF regulations were promulgated. Erosion control on the site has been termed the best in Wyoming by an inspector for the state's Department of Environmental Quality.

Although the seeding program at Belle Ayr incorporates a fairly good plant mixture with immediately backhauled topsoil, certain amendments being used are not as good. AMAX still uses a hay mulch, crimped to the surface of the land, and the company has had problems with undesirable seeds infiltrating with the mulch and competing with the desired plants being seeded. The broadcast method is used when speed is deemed necessary; but rushing the seeding of lands being reclaimed is not advisable, as more problems may result than if more time had been spent on careful seeding in the beginning. Seed distribution may suffer, and soil may not be turned over to cover the seed. The company has also had very limited success in establishing trees and shrubs.

Initial results of revegetation efforts at Belle Ayr have tentatively shown that productivity and diversity *may* be matched in the long run, although the mine has not been operating long enough to provide hard data to support this conclusion. Moreover, the use of herbicides--rather than solid revegetation practices--may account for the minimal amount of weeds to be found on reclaimed areas. Grain plantings--though experimental--have not brought productivity up to premining levels, and the real test for AMAX's reclamation program will come in the future as the company gains more time and experience.

Wildlife such as rabbits and rodents will benefit from the construction of rock shelters at Belle Ayr--and also from reconstruction of Caballo Creek, if done properly. Deer, antelope and hawks will also benefit from the availability of forage and prey on these lands. However, more trees and shrubs must be transplanted to provide forage and shelter for large and small animals.

Because most of the operations at Belle Ayr have been sound, and as the company has seldom been cited for violating state or federal reclamation laws, inspection activity at the mine has not had a major impact on operations. The state Department of Environmental Quality inspects the mine at least quarterly. The Office of Surface Mining (OSM) has inspected the mine only three times since 1978, when the Region V office in Denver was established--in July and November 1978 and November 1979.[8] Tom Ehmett, an OSM inspector, told INFORM in March 1979 that the operations at Belle Ayr were fine, and that less enforcement is needed there than at other sites in the West.[9]

EARTHMOVING

Grading

Mining activities at Belle Ayr disturbed more than 200 acres of land in 1977 and again in 1978. In 1977, 241 acres of land were mined and 227 graded. However, disturbance and grading fluctuate; in 1978, 272 acres of land were mined and only 147 acres graded.[10]

Belle Ayr has *no spoil ridges*. Mining is conducted with coal shovels and trucks, which haul overburden from the active pit area and dump it into pits where coal removal has been completed. This procedure allows greater flexibility in handling overburden and leaves large openings after the 70 feet of coal is removed; thus the overburden does not create ridges above the land surface. (See Techniques of Surface Mining and Reclamation.) The company has created two piles of overburden near the area where mining first began. These will be used to fill the hole left from the last mine pit to be dug, around 1995. These piles have slopes of about 11.5 degrees and have been revegetated to protect them over the life of the mine.[11]

Aside from these overburden stockpiles, the grading at Belle Ayr generally approximates the pre-mining contours. However, the entire elevation is lowered slightly, and along the border of the mine property where Exxon's Caballo Mine lease begins, the difference in elevation is such that grading will have to be employed to join the two lease areas. AMAX is doing this work, as well as replacing topsoil and seeding the area, because it presents a strong potential for erosion. Eventually, mining operations at Caballo will redisturb this area, and reclamation will be left to that mine's operator.[12]

One major change in the contour at Belle Ayr has been the temporary diversion of Caballo Creek through the mine area. In order to allow for mining coal lying beneath this intermittent stream, AMAX builds a new temporary path for it as mining intersects the original course. The company uses clayey overburden--an innovative technique--in this diversion in order to prevent water from seeping through the soil instead of flowing along the creek.[13] AMAX has also created permanent water impoundments in the Caballo Creek channel; these are to be used for watering stock, the company says, although the Wyoming Department of Environmental Quality (DEQ) referred to these impoundments in January 1977 as temporary.[14] INFORM was unable to determine what effect the shifting of Caballo Creek's flow pattern would have on the local environment; nor could it judge the effect of lining the diversion with clay on recharge capabilities for groundwater.

All grading is producing slopes that can be traversed by reclamation machinery. On stockpiles, however, and on the steep banks along water diversions, lands cannot be traversed and must be broadcast with seed. However, this is only a small part of the entire mining operation. AMAX will return Caballo Creek to its original path when mining is completed.[15]

Topsoil

Topsoil is stripped from the land in a separate operation at Belle Ayr. Fully 75 percent is immediately replaced on graded

lands--more than at any other mine in our sample. About 25
percent is stockpiled until lands are ready for reapplication.
David Ham, AMAX regulatory affairs counsel, told INFORM that
no topsoil remains in stockpiles for longer than two years, except
for that soil stored in the larger, "life-of-mine" stockpiles, which
will not be used until the mine is eventually closed.[16] Although
the company agrees that immediate replacement of topsoil without
stockpiling should be a goal, it told INFORM that it does so only
when the weather is good.[17] DEQ has termed the topsoil stripping operations at the mine very satisfactory,[18] yet it has questioned the practice of stripping topsoil many months ahead of
mining operations.[19] (AMAX contracts with an outside firm
owned by J. R. Von to strip the topsoil in one operation, either
in the late spring, summer or early fall.[20]) But because stockpiling is so minimal at Belle Ayr, and stockpiles are so well maintained, the company cannot be faulted for this practice.

When topsoil is stripped ahead of the mine pit, the land is
seeded to help prevent erosion by wind and water. The company
uses annual grains to stabilize the exposed soils. Topsoil is replaced uniformly to a depth of 18 inches over the mine site, according to AMAX officials.[21] In June 1979 the company was replacing topsoil from stockpiles that had been created the previous
winter.

The mine has ten topsoil stockpiles, covering less than six
acres of land. According to a survey by the U.S. Fish and
Wildlife Service published in 1978, stockpiles are covered with a
variety of grasses and dominated by alfalfa, affording good protection from erosion.[22]

Toxic Material

AMAX reports that when it encounters shales that may contain toxic or otherwise undesirable materials, it buries them under
8 feet of overburden before replacing topsoil. The flexibility of
the truck-and-shovel mining method, in contrast to the relatively
indiscriminate dragline, allows for much greater accuracy in selectively handling different materials.[23] The U.S. Geological
Survey (USGS) told INFORM that no sodic or toxic materials
were being found on lands being reclaimed at the mine.[24]

Occasional coal fires occur (a natural occurrence due to heat
and exposure to the air) in the mine pits at Belle Ayr. However,
the USGS reports that AMAX is generally quick to extinguish
them.[25]

WATER QUALITY

Surface Water and Groundwater

Control of surface water and runoff at Belle Ayr centers on the Caballo Creek and diversion, which AMAX has constructed since 1972.[26] Sediment ponds have been built, and "ox-bows"-- empty cuts left when the meandering Caballo Creek changes course--are used as water storage areas. Because Caballo Creek is an intermittent stream,[27] much of the water found on the mine comes from precipitation or groundwater pumped from the mine pits into holding ponds or sediment ponds. INFORM found no evidence that off-site disturbances related to poor water control had occurred. Virtually all disturbed areas at the mine drain into either the settling ponds or water storage areas.[28]

The largest project undertaken by AMAX involving surface waters has been the Caballo Creek diversion. Caballo Creek meanders through the mine area. AMAX has diverted it to run straight through the site. Mining operations will move along one side of the lease area, then along the opposite side. As mining progresses down this second side, after 1985 the diversion will be destroyed and AMAX will attempt to return the Creek to its original path bit by bit. Early in the summer of 1978, heavy rains battered the diversion. Portions were scoured down to bedrock.[29] (The company's reaction to requests by USGS for permission to inspect the damage was curious. It refused to allow inspectors to visit the area, and upon leaving the mine the inspectors detoured to view the damage.[30]) Repair operations were begun, and the diversion was well vegetated and rip-rapped with rocks in the summer of 1979 when INFORM toured the site. The vegetative cover provides good protection from erosion as runoff enters or flows through the diversion, and rocks lining the channel slow the flow of water and provide a solid lining for the channel. Other water diversions at the mine were well vegetated.

The 75-foot coal seam being mined at Belle Ayr is an aquifer.[31] When it was monitored in 1974, water was flowing into the active mining pit at a rate of 150,000 gallons per day. Later in the year the flow declined to 100,000 gallons daily.[32] AMAX pumps most of this water from the pit into sediment ponds or water holding areas at the site. Much of the water is then used for dust control on roads at the mine, or allowed to evaporate in the semi-arid climate. The DEQ noted in February 1978 that some water pumped into ponds ahead of the active mining pits might actually be seeping back into those pits[33] (a possible exercise in futility on AMAX's part). Follow-up research is needed. Thus, little water is discharged from the mine, and the company controls discharges with water valves on many of the retention areas. In February 1977 the mine discharged water that did not meet effluent guidelines for total suspended solids (TSS). The company told the Wyoming State Water Quality Division it would install a chemical flocculent system to aid in dropping sediment from retained waters before discharge.[34] However, after experimenting with the system the company decided not to use it, explaining to INFORM that the company making the chemicals had gone out of business.[35] Nonetheless, in 1978 the company met all discharge requirements, and from January through July 1979 it had retained all water on the mine site and remained within the requirements of its NPDES permit.[36]

INFORM was told that little is known about the dewatering of the aquifer and the effects upon groundwater hydrology in the area. AMAX assured us that the water is not used within one mile of the mine (the company's estimate of the range within which water drainage would affect flow in the aquifer).[37] The company has since collected more data and told INFORM:

> It appears at this time that recharge to backfill areas will be slow, specifically, 2 to 5 years before the first indications of reestablishment of groundwater. The high percentage of expandable clays in the backfill materials and the presence of complex, differentially compacted areas throughout the backfill are partially responsible. Low water levels in backfill at the present time are also influenced by the close proximity of the active [mining] pit, which maintains a drawdown effect in the backfill toward the pit. Further, the distance to the recharge area exerts its effect.[38]

It will be some time before firm conclusions can be reached on the effects of mining through this groundwater system. The full impact may not be known until all mining has been completed and all mine pits refilled and graded, as the company observes that open mining pits draw water from the soil.

Groundwater monitoring has been practiced at Belle Ayr since 1976.[39] Surface water is monitored at three points along Caballo Creek: above the mine permit area, in the center of the area and below it.[40] AMAX provided no data on the results of this monitoring.

Alluvial Valley Floors

Mining at Belle Ayr may be disturbing lands bearing the characteristics of an alluvial valley floor as defined by OSM. The U.S. Environmental Protection Agency has said Caballo Creek is an AVF and should be protected from mining.[41] The USGS has also made a preliminary determination that Caballo Creek is an AVF.[42] However, the company told INFORM that "detailed hy-

drological and geological studies conducted...by AMAX indicate
areas are not alluvial valley floors. AMAX does not plan to leave
[these] areas intact."[43] In June 1979 the company told INFORM
that, as yet, no regulatory authorities had made a final decision
on this question, and it expected that, having already begun
operations, they would not be stopped.[44] In any case, protection of Caballo Creek would require that much of the land designated for mining not be disturbed. That would probably mean
shutting down the mine--an unlikely option for AMAX.

Erosion

Erosion control at Belle Ayr is so good that Dennis Morrow,
an inspector for the Wyoming Department of Environmental Quality, calls it the best of any mine in the state.[45] INFORM is unaware of any disturbances off the mine site attributable to erosion.

Some erosion has occurred. The earliest reclaimed area at
the mine is a large rolling hill, known as Hollies Hill. Because of
its long, sometimes steep slopes, AMAX constructed a 25-foot
wide terrace along the contour with the intention that water
flowing down the upper part of the hill would stop at the terrace,
minimizing the potential for erosion. The terrace was designed
to hold and store water until it either soaked into the ground or
evaporated. But early in 1977 AMAX was warned by the Wyoming
DEQ that the terrace might not be sufficient to contain water
from severe storms.[46] On an inspection in June 1977, DEQ repeated this warning and observed that a portion of the terrace
had been washed out and repaired that year. DEQ suggested
that the terrace be sloped so that water would be slowed but not
trapped.[47] AMAX told INFORM in June 1979 that it would eventually remove the terrace when the ground was stabilized well
enough by vegetation and compaction.[48]

SEEDING AND AMENDMENTS

Seeding Procedures

In a truck-and-shovel operation like Belle Ayr's--where
large spoil piles are not created and less grading is required---
seeding should be able to follow mining very closely. Having
only begun to produce coal in 1973, and having expanded production each year, the company has seeded only about 43 percent
of the disturbed land at the mine. Of this land, 16 percent will
be re-disturbed at the end of the mine's life, as production operations return to the starting point, and will have to be reclaimed
again.[49] In 1977 AMAX disturbed 241 acres and seeded 227. In
1978 it disturbed 272 acres and seeded only 118.[50] By December

1979, 541 of the 1,273 acres disturbed by mining and related activities at the mine had been planted. AMAX conducts its seeding
variously from late spring through June--and again in late summer and late fall.[51] Early spring and late fall seeding is timely
and common throughout much of the West--but late-summer seeding appears pointless because plants that establish at this time
may go dormant over the winter and fail to survive. The mine
has not experimented with planting periods.

The basic seed mixture used at Belle Ayr is a fairly good mix
of cool- and warm-season native plants that provide forage in
both cooler and warmer seasons. Eleven plants, including
grasses, forbs, shrubs and legumes are seeded. Part of the
credit for this plant mixture goes to the Wyoming DEQ, which
modified the initial mixture proposed by AMAX.[52]

Sometimes--when specific site characteristics call for it--
AMAX uses an additional 5 pounds of seed from a list of 37 plant
species, although the company complains that these seeds are
hard to find and very expensive. INFORM does not know how
often supplemental seeds are used. A different seed mixture is
used for water diversion ditches and channels.[53]

TABLE A

1979 SEED MIXTURE
FOR LANDS BEING RECLAIMED AT BELLE AYR

Plant Species	Introduced/ Native	Cool/ Warm	Pounds Applied Per Acre
Grasses			
Western wheatgrass	N	C	2
Green needlegrass	N	C	2
Beardless wheatgrass	N	C	2
Blue grama	N	W	1
Canada wildrye	N	C	1
Legumes			
Alfalfa	I	C-W	1
Sweetclover	I	W	1
Forbs and Shrubs			
Fourwing saltbush	N	W	3
Western yarrow	N	C	0.1
Sunflower	I	W	0.5
Winterfat	N	C	1
			14.6 Pure live seed

Source: AMAX

The company is experimenting with different seeding methods and rates of application. These tests will be evaluated on the basis of plant cover, species frequency and density. But hay mulch being used on the test plots, with its potential for seed contamination, may skew some results; better controls should be practiced. Experiments are also being conducted by the U.S. Forest Service, but in June 1979 the company knew nothing about this work.[54]

Amendments and Maintenance

Owing to the generally shallow slopes at Belle Ayr, most seeding is done along the contour except when hydroseeding (see below) is required.[55] The company does not create furrows to retard erosion or trap moisture; it should do so on steeper slopes especially (see Erosion).

AMAX uses both a drill seeding machine and broadcast seed-er. Use of the latter is not advisable, as the company does not cover broadcast seed with soil. David Ham, regulatory affairs counsel for AMAX's western division, reports that between May 1978 and November 1979, 250 acres of land were drill-seeded and 100 acres were broadcast seeded.[56] The company concedes that the practice of broadcast seeding presents problems: when the wind picks up, seeding must stop, as distribution on the ground is poor.[57] AMAX told INFORM that broadcast seeding is done when speed is of the essence.[58] The numbers reflect a situation where the company may be putting itself into rush jobs too often, as 40 percent of the land seeded between May 1978 and November 1979 was seeded in this fashion.[59] Better planning may be nec-essary to remedy this situation. Hydro-seeding is employed where conventional equipment cannot be used,[60] to repair erosion on otherwise vegetated lands and to seed water diversion ditches where slopes are small but steep.

Without experimenting with mulches, the company has used hay mulch and wood chips, sprayed with the hydro-seeder. Hay mulch is tacked to the soil with a crimper, which is dragged be-hind the mulch sprayer. Wood chips are not tacked, as they are too small.[61] Hay mulch often contains the seeds of undesirable plants that might upset revegetation operations. Straw mulch avoids this problem--but very little has been used at the site.[62] Fences have also been used to trap snow on the site, allow melt-ing snow to provide more water to plants in the spring.

Irrigation is not practiced at Belle Ayr--and may not be nec-essary (16-inch annual precipitation)--even though experiments conducted by the company in 1977 showed that irrigation re-sulted in a 70 percent survival rate in test plot applications, while non-irrigated plants had only 30 percent survival.[63] The company does use fertilizers in limited applications when it is

deemed necessary.[64] AMAX applies 20 pounds each of nitrogen
and phosphate at the time of seeding, basing the application on
soil tests.[65] Herbicides have been used annually for weed control
on revegetated lands.[66] This may not allow an accurate assessment of revegetation, as seeded species are not forced to compete with invading species. This practice should be monitored,
or stopped to allow natural competition to show whether increased
land management is necessary.

Trees and Shrubs

     Few trees exist on lands near Belle Ayr, and plantings of
trees and shrubs have not been extensive at the mine--even
though AMAX has had its very own tree nursery on the site
since 1978. Six tree species are being grown from seedlings at
the nursery. The company does not know the exact survival rate
of these plants so far.[67] No tree plantings have been undertaken
since the fall of 1978.[68] Trees and shrubs such as willow, cot-
tonwood, ponderosa pine, penipee, Russian olive, and green ash
have been transplanted using the tree-spade technique or the
much inferior bare-root method.[69] Of approximately 250 bare-
root transplants, very few have survived. Browsing by deer and
periods of low rainfall are believed to have prevented the goal of
20-25 percent survival from being met.[70] Trees planted on land
seeded in 1976 have all died.
     The only areas where tree and shrub transplants have sur-
vived is along the Caballo Creek diversion, where golden willow
is showing the best growth. The company will have to continue
with experimental plantings of trees and shrubs to gain more experience with this practice, and to raise the chances for success.

                                              VEGETATION

     Grazing has been the predominant land use around the Belle
Ayr site. AMAX is attempting to reclaim lands not only to support grazing and provide some wildlife habitat, but also to allow
growing crops should this become a desired land use. More than
70 percent of the land was made up of native rangeland of both
grass and sagebrush before mining began. An additional 20
percent was cultivated range (planted with introduced grass
species), and a small percentage was once used to grow grains
such as oats and wheat.[71]
     Productivity ranges from 2.75-3.5 acres of land needed to
produce adequate forage for one animal unit (equivalent to one
cow and calf or five sheep) for a month.[72] AMAX also reports
that forage productivity averages 560 pounds per acre on the
site,[73] although areas around Caballo Creek (which may be an

alluvial valley floor) are significantly higher, according to the U.S. Soil Conservation Service.[74]

The earliest reclaimed land at the site is an area known as Hollies Hill, where plant growth was in its third season when IN-FORM visited the site in June 1979. Even after only one year of growth, the Wyoming DEQ found in 1977 that plant cover was well established and that permanent species were coming up through the cover crop of winter wheat.[75]   In 1979 the company reported to INFORM that productivity on Hollies Hill was approximately 2,000 pounds per acre in 1978, well above premining productivity.[76] This was to be expected; the movement and break-up of soils in mining releases many nutrients, which allow plants to show high levels of growth in their first few years after germination. Plant diversity on seeded lands has expanded on the site as well. Along with many of the native plants originally seeded, other native plants have invaded the site. The cover on this land is probably sufficient to prevent erosion. INFORM was unable to determine whether the application of herbicides to this area could account for the small numbers of weeds or the proliferation of invading native plants.

AMAX has collected some data on productivity, plant diversity, plant cover and plant frequency on reclaimed lands. However, analysis was not complete as of November 1979.[77] The only attempt at reclaiming lands to grain production was an experimental planting of wheat on 35 acres in 1973. The land produced 20 bushels of wheat, compared with an average of 35 bushels produced on undisturbed lands.[78]

No grazing trials are planned at the site because, the company says, federal mine safety regulations will not allow cows to graze near haul roads. Yet, grazing is conducted at the Decker, Colstrip and Navajo mines. Futher investigation is needed.

The company does not plan to attempt to release any lands from bond in the near future.[79]

WILDLIFE

Attempts to create wildlife habitat at Belle Ayr have been fair to good. AMAX is constructing rock shelters for small wildlife, such as rabbits and rodents, and is attempting to grow trees and shrubs in selected areas. Plants are compatible with wildlife use, but more trees and shrubs will be necessary to provide adequate forage for larger animals.

Many animals--such as deer and rabbits--are already back on the site, especially on Hollies Hill and lands that have not yet been disturbed. Small mammals and insects were found on IN-FORM's visit in June 1979, and deer tracks and droppings indicate that these animals are also making use of the land.

RECLAMATION COSTS

TABLE B*

RECLAMATION COSTS AT BELLE AYR

Operation	Cost Per Acre	Percentage of Total Cost
Earthmoving		81%
Grading	$ 500	
Touch-up grading	100	
Ripping subsoil before topsoil replacement	100	
Topsoil removal and replacement	2,800	
Revegetation		19%
Revegetation (seeding, mulching, fertilizing, etc.)	800	
Total Cost:	$4,300	

*These costs and percentages are not strictly comparable to those of the same operation or activity at other mines. See the Cost of Reclamation chapter for a detailed explanation.

Source: AMAX Coal Company

    In January 1977 AMAX estimated that reclamation costs were approximately $4,597 per acre at Belle Ayr.[80] However, when asked for reclamation costs in 1979, the company told INFORM it preferred not to address the question because of changes in regulations and the highly competitive nature of the coal industry. AMAX also told INFORM it was unaware of any uniformly acceptable formula for determining reclamation costs.[81] But reports from David Ham in January 1980 revealed that the company has assembled reclamation costs for its annual report to the state regulatory authority. This information disclosed that total cost per acre was $4,300.

Colstrip/Western Energy/MT

Mine: Colstrip (Rosebud) Mine

Location: Colstrip (population 2,250), Montana, in Rosebud County (population 10,800)

Operating company: Western Energy Company

Parent company: Montana Power Company

Parent company revenues: 1977, $191 million; 1978, $229 million

Parent company net income: 1977, $24.4 million; 1978, $31.5 million

First year of production: After being mined by the Northern Pacific Railroad from 1924-1958, the mine was reopened by Western Energy in 1968.

Coal production in 1978: 10,576,000 tons

Acres for which a mining permit has been issued: 5,500 as of December 31, 1979

State regulatory authority: Montana Department of State Lands

Reclamation bond posted: $15,889,000 as of December 1978

Reclamation cost: reclamation cost per acre, $11,800; *reclamation cost per ton of coal,* $0.46

Expansion plans at mine: To mine 19.1 million tons of coal per year by 1983

Additional acres for which the company will seek a permit: Over the next 40 years (the expected life of the mine), Western Energy plans to permit approximately 20,000 acres.

Parent company U.S. coal reserves: Owns or leases 1 billion tons of coal reserves

Development and expansion plans for surface coal mines in western United States: No new mining is planned by Western Energy except for expansion at Colstrip to produce 19.1 million tons per year by 1983.

Sources: Western Energy Company; Montana Power Form 10-K, *1978;* 1977 *and* 1979 Keystone Coal Industry Manual; *County Planning Office, Forsyth, Montana; Montana Department of State Lands,* Final Environmental Impact Statement for the Proposed Expansion of Western Energy Company's Rosebud Mine into Area B, *July 26, 1976, p.* 2

SITE DESCRIPTION

Colstrip, Montana, in Rosebud County, is home to both the Western Energy (WECo) Colstrip Mine and the Colstrip Power Plant. Lying northeast of the Cheyenne Indian Reservation and east of the Little Wolf Mountains, the town, power plant and mine are almost completely owned and run by the parent Montana Power Co. Colstrip is only 30 miles east of Westmoreland Resource's Absaloka Mine (see Absaloka profile), but the environment differs considerably.

The topography is typical of much of south-central Montana. Rolling plains are broken occasionally by sandstone outcroppings, hills covered with ponderosa pine trees, and intermittent creeks and washes.[1] The Colstrip Mine lies in a generally broad, rolling valley of the East Fork of Armells Creek, which runs through the mine. The creek is considered intermittent--flowing only in peak periods of spring runoff of melting snow and in times of rain. Lately, however, the creek has been running almost constantly.[2] According to both the U.S. Geological Survey (USGS)[3] and the Montana Department of State Lands (DSL),[4] this fork of the creek may be an alluvial valley floor. WECo disagrees, and a final determination of its character remains to be made.

Precipitation averages 15.8 inches annually[5] but has fallen below 12 inches at a rate of one year in every five years since 1941.[6] Half the annual precipitation falls in April to mid-June; summers are hot and very dry. The frost-free season averages 115 days.[7] As rainfall can vary considerably--it was very high (28 inches) in 1978 and several inches below normal in 1979-- successful establishment of vegetation in one or two seasons may be ruined in a following dry season.[8]

Most of the mine area is grazing land. None of it could be considered prime agricultural land.[9] About 72-79 percent of the site was described as grassland, mixed shrub-grassland, or cultivated rangeland. The remainder is mixed forest--ponderosa pine and Rocky Mountain juniper--and grassland, or cultivated cropland.[10] Trees appear principally along ephemeral or intermittent streams and water drainage areas, or on rocky outcroppings. WECo has reported a high diversity of plant species on unmined lands, with perhaps 300 identifiable species in the mine area alone.[11]

OVERVIEW

In a very real sense the Colstrip Mine* in Rosebud County, Montana, is a microcosm of the problems and the promise of western surface coal mining and development. At no other mine can the total effect of coal development be found in such concentration. And at no other mine has the impact of local public opinion--and outrage--been so effective.

With the neighboring Colstrip Power Plant and the company town of Colstrip (built to house the families of workers at the mine and plant, as well as construction crews), the effects of coal prospecting, mining and combustion are being imposed on both the environmental and social fabric of the area. Western Energy (WECo) is the subsidiary of Montana Power, which runs the area.

Thanks to the concerted activities of ranchers, farmers and other citizens, reclamation practices have improved considerably at Colstrip in recent years. In the distant future, it may be viewed as the best reclamation anywhere--depending on many variables. At this date, however, the question whether reclamation can be guaranteed at Colstrip--as at all other western mines --remains wide open.

The blackest mark on the company's reclamation ledger--and the sore point that has raised the most objections to *any* mining in the area--is the Pit 6 and Pit 6 Extension, covering approximately 300 of the nearly 3,200 acres of land disturbed since 1968. Here steeply graded slopes, combined with poor initial reclamation practices, have left an area of unstabilized soils covered with mostly useless vegetation. This area shows the most severe erosion of any of the 15 mines surveyed by INFORM. The land remains testimony to the poverty of Colstrip's early reclamation program--at the same time providing a benchmark against which

---

*Although WECo refers to the mine as the Rosebud Mine, it is commonly called Colstrip to avoid confusion with Peter Kiewit's Rosebud Mine in Wyoming.

to measure the positive development of WECo's program in recent years.

Water control is another serious problem. The company attempts to hold all the water on the site instead of releasing it to the creeks that run through and around the mine. But the water retention system leaks, and although the effects on off-site surface waters and groundwaters are poorly documented, it is clear that they have caused several problems. Southeast of the mine, in the direction in which groundwater flows, water quality in the wells of at least one ranch has forced the rancher to dig a new well. To the north, several meadows have been flooded, and the level of neighboring creeks--which once ran intermittently and now run year-round--has posed serious problems for ranchers and farmers. The key to the water controversy appears to be that the coal seam being mined by WECo is the principal groundwater aquifer for the area. WECo has not sufficiently addressed the problem, with the result that opposition to mining has increased as the mine's neighbors find their most precious natural resource, water, being adversely affected.

Aggravating the groundwater/surface water problem is the erosion that still plagues the Pit 6 area and also shows up in other small areas. "Piping"--an underground erosion phenomenon in which holes suddenly open up on graded land--has occurred. This has increased the concern of local landowners over the slow settling of lands where mine pits have been filled with overburden--which expands in volume when mined and contracts and settles when replaced. The result can be more piping and erosion, making the land useless and increasing the threat of sedimentation onto previously undisturbed lands.

Perhaps the most positive feature of WECo's Colstrip operation is the continuing development of its revegetation program. Largely in response to pressure from ranchers and state regulators (who are themselves prodded by rancher groups), and as a result of studies conducted on-site, Colstrip's complex seeding and amendment program attempts the successful replacement of trees, shrubs, and native grass and forb communities. Even early results show improvements in revegetation. But the long-term stability and productivity of the land remain questionable.

Grading has improved dramatically, with lands being left much flatter and better approximating the original contour than before, reducing erosion potential. Drainage reconstruction is among the best-conceived in the West. Topsoil removal and replacement have been only fair, but reprimands by state authorities have sensitized WECo to the value of this natural resource, and such positive practices as direct haulback have been instituted.

State authorities have commended Colstrip's program for re-

creating the habitat for wildlife, which abounds in the area. Ultimate success will rest on the outcome of the company's tree and
shrub planting program.

Colstrip operates at all times under the watchful eyes of its
neighbors, who frequently express their concerns to WECo and
state regulators. Many observers believe it is the activities of
these ranchers and farmers that have caused the Montana Department of State Lands (DSL) to become the most stringent regulator in all the West. From October 1973 to December 1979,
WECo had to pay more than $18,000 in fines--typical for Montana
mines but higher than in other states of INFORM's sample--for a
variety of violations. The federal Office of Surface Mining (OSM)
has been less active, but its intervention may be expected to increase as questions of plant diversity and the effect of mining on
water resources are expressed more and more vehemently.

If reclamation can be successful in this region of the West,
Colstrip's may--as noted--turn out to be the best. If this develops, it will be because local landowners and groups like the
Rosebud Protective Association and Northern Plains Resource
Council have been vigilant, and have continued to question the
practices of corporate giants like Montana Power and its WECo
subsidiary. Such individuals and groups may set the pattern for
other areas of the West, if similar individuals and groups wish to
ensure that the ravages of surface mining are minimized.

EARTHMOVING

Grading at Colstrip follows close behind active pit operations, with few spoil ridges following mining. The Montana DSL
told INFORM that WECo keeps within the two spoil-ridge standard
established by the state--as do other mines in the state--and
site visits have confirmed this.[12] According to WECo, by January 1, 1980, of the 3,158 acres of land disturbed at the mine
since WECo began operations in 1968, 2,087--more than 65 per-
cent--have been graded.[13]

Premining terrain around Colstrip was made up of gently
rolling hills with low profiles. Current grading is producing
fairly flat, rolling slopes with most major drainages recreated.[14]
According to WECo, grading now produces slopes of between
12:1 and 14:1.[15] But early grading at the mine--especially
around Pit 6, Pit 6 Extension and Area A--have left steep slopes,
some steeper than 5:1, that have been especially prone to erosion.[16] (See Erosion.) Owing to the sandy nature of soils and
subsoils in the area, slopes that might not be too steep for other
regions in the West are much more liable to be eroded, as the
unconsolidated nature of soils disturbed by mining lends itself
to increased rilling and gullying. Although these older, steeper

slopes are not traversable by reclamation machinery, newer grading is easily traversable.

The Montana DSL told INFORM WECo was the "first [company] in the state to grade a drainage the way we wanted them to," albeit grudgingly.[17] Current drainage construction is some of the most advanced seed by INFORM. Drainages concentrate, disperse and reconcentrate flowing water in an effort to slow its velocity and minimize erosion. Drainages are now laid down according to the direction in which water runs after final contouring has been completed, thus letting nature define the channel.[18] (See Surface Water and Groundwater.)

Because the Colstrip Mine was once operated to supply coal for the Northern Pacific (now Burlington Northern) Railroad, some old spoils exist on the site. Although WECo has no legal responsibility for these eyesores, the company might consider seeking funding from the Office of Surface Mining (OSM)'s Abandoned Mined Land program to pay for the reclamation of some of these spoils. A few will be graded and seeded to blend with WECo mining in the future, but the bulk will remain, unless further action is demanded.[19]

Topsoil

Topsoil and subsoil are now stripped and segregated at Colstrip for use on graded lands once mining has been completed. However, WECo's topsoil removal practices have varied in the past, with numerous instances in 1977 and earlier years when insufficient topsoil was removed[20] from small areas, or was covered with spoils before being removed. A DSL memo of June 1976 stated that poor topsoil removal and protection at the mine made WECo stand out (adversely) from the rest of the coal mine operators in Montana, suggesting that the company had a low regard for the value of this soil resource.[21] This forced DSL to require that WECo leave topsoil "islands" (small areas approximately 2½ feet square) where the amount of topsoil removed can be ascertained by looking at the depth of the soil left exposed on the island. Earl Murray, one of the mine's reclamation engineers, told INFORM that the practice is expensive because the company must remove the islands once DSL has determined that sufficient topsoil removal has been done.[22] But the requirement has improved soil removal on the site.

When removed, 6-12 inches of topsoil and up to 4 feet of subsoil are taken from lands to be disturbed.[23] A good practice being followed at the mine is the direct haulback of topsoil and subsoil wherever possible. According to reports by OSM, DSL and the U.S. Geological Survey, some direct haulback was employed in 1977, 1978 and 1979.[24] WECo says it prefers the direct haulback technique[25] and is continuing to use it whenever possible in 1980.[26] However, Murray could not tell INFORM exactly how much direct haulback is actually being conducted.[27]

INFORM saw few stockpiles on its visit to the site in April 1980, and those seen were sufficiently vegetated. Past reports by OSM and DSL have noted that stockpiles have eroded, or were improperly seeded.[28] The more direct haulback that is practiced at Colstrip, the less this problem will occur. When replaced on graded lands, soils are laid down evenly.[29]

Toxic Material

Although few salts have been reported as appearing on seeded lands at Colstrip, there is great concern that salt problems may arise in the future because of the loosening of soils and the loss of soil structure caused by mining. DSL has reported that the upper 10-30 feet of overburden is very saline,[30] and in a draft environmental statement it warns that high levels of molybdenum, cadmium, zinc, and salt may affect reclamation success and subsequent land use if they accumulate near the surface.[31] WECo also plans to fill some pits on the mine with ash from the nearby power plant. This could allow toxic minerals to contaminate groundwater supplies.[32] These issues concern long-term implications of mining in this area and require thorough investigation (see Surface Water and Groundwater).

WATER QUALITY

Surface Water and Groundwater

Because WECo has attempted to contain all water on the mine site instead of discharging it, the company has built ponds to hold the water, allowing it either to evaporate or to be used for spraying dusty roads at the mine. The company did not receive a permit to discharge waters from the site until the spring of 1980 and had yet to do so as of INFORM's visit in April.[33]

The system for holding water has generally been well protected from erosion, but some problems have occurred. Montana DSL inspectors have urged that more rip-rap (stone or other objects used to slow the flow of water and prevent erosion and sedimentation) be used in some of the drainage areas, as well as near culverts that transport water over parts of the mine.[34] This is an important consideration because soils disturbed by mining have become especially prone to erosion, which could result in sedimentation both on and off the site (see Erosion).

Although not permitted, some discharge has occurred from Colstrip. One pond was undersized and discharged from the mine twice. It is now being rebuilt.[35] DSL also reported in March 1978 that one drainage ditch had discharged directly into the East

Fork of Armells Creek.[36] However, this was not considered a
violation by DSL because there was inconclusive proof as to exactly where water was leaking from the ditch.[37]
    This raises a problem that has made local landowners stand
up and begin challenging the company over leaks from water
storage ponds. For example, the East Fork of Armells Creek has
always been considered an intermittently flowing water course.
It now flows virtually all year, often overflowing, and several
ranchers in the area blame this phenomenon not only on the
mine's water holding ponds, but also on the water storage (or
surge) pond for the power plant, and on sewage disposal from
the town of Colstrip.[38]
    The problem appears to be that DSL refuses to classify water
that leaks from ponds on the site as discharges that must meet
the effluent standards of the Environmental Protection Agency
(EPA).[39] This problem has been argued extensively by the
Kluver family. With the use of home-made movies, the family has
blamed the loss of good-quality well water found on its ranch on
leakage from Colstrip into Cow Creek, which feeds some subsurface water supplies. OSM concedes that water is leaking from
ponds on the site, but neither the state nor OSM has determined
exactly how much water is flowing, where it is going, its quality,
and its effect on surface and groundwater systems in the area.[40]
While the investigation continues, the Kluver ranch has had to
drill a new well to replace the old one, whose water had become
salty, discolored and totally unusable.
    The groundwater-surface water issue shows WECo off to its
worst advantage at Colstrip. Several significant questions have
been raised--and have produced wholly unsatisfactory answers.
In the West, where water is a scarce necessity, concern for its
protection should be paramount. WECo has more than 50 monitoring wells around the mine area--yet it has never fully grasped
the value of groundwater systems to local ranchers.[41] The crux
of the problem is that the 25-foot-thick Rosebud coal seam is
tapped by area residents--indeed, it is critical to them--for domestic and stock water. According to the Bureau of Land Management, *the aquifer in the coal seam is one of the most important sources of water for livestock in the Colstrip area.*[42] Yet,
when asked whether mining at Colstrip was disturbing any important aquifers, WECo responded that: *"The only seam mined pres-
ently is called the Rosebud and it can be a water-bearing unit.
However, it is a very disrupted and unreliable ground water
system. It is only of economic importance as a supplemental
source of stock water."*[43] In the face of this statment, WECo itself directed INFORM to contradictory reports by hydrogeologists
of the Montana Bureau of Mines and Geology and the Montana
College of Mineral Science and Technology.[44] In a December 1978

report, these researchers stated that *"Ground water in the Rose-bud coal bed...provides generally adequate water supplies to stock and domestic wells and is most heavily utilized where it lies within about 60 meters below land surface. Logically, these areas nearly coincide with areas where the coal bed is economically mineable."*[45] They further report that in five years of monitoring water levels near active mining operations, no changes have been found in water levels.[46] But they make no mention of water studies conducted further from the mine, where changes might be much more obvious. The researchers do admit that waters from mine areas are more highly mineralized than regional coal-bed waters and waters used locally for stock and domestic supplies--which (they state) are vital to the area's agricultural economy.[47] According to Dennis Hemmer, a DSL reclamation specialist, further water monitoring is critically needed.[48]

The groundwater controversy is being especially pressed by the Rosebud Protective Association, made up of ranchers and farmers in the Rosebud County area, and the Northern Plains Resource Council, a group of ranchers, farmers and concerned citizens from the entire Northern Plains region. The basic fear is that replacement of the Rosebud coal seam with spoils--pre-dominantly shale, clay and sandy soils--will provide neither the pathway nor the cleansing filter that nature provided with the coal seam, for groundwater supplies so crucial to the area. Already, fields to the north of the mine are being flooded by rising water tables, and the increasing level of waters flowing through the East Fork of Armells Creek has also caused some overflowing onto pasture and croplands.[49] As mentioned earlier, water supplies to the southeast have been affected. Moreover, DSL has warned WECo that salty overburden in graded areas could affect the quality of groundwater,[50] and the U.S. Bureau of Land Management has stated that "any discharge causing or adding to a constant flow downstream from Colstrip would contribute to water logging."[51] Although the damage may already have been done, further study of groundwater resources and their interaction with surface waters is clearly needed. The concerns of the area's ranchers and farmers must be answered before further expansion of mining can be considered.

Alluvial Valley Floors

Both the Montana DSL and the U.S. Geological Survey believe that the East Fork of Armells Creek, which flows through the mine, may qualify as an alluvial valley floor (AVF) under the OSM and DSL criteria.[52] WECo contends that the creek is at best a "borderline case" and is not subirrigated, and that riparian vegetation occurs in broken patches along its length.[53] Yet, according to WECo, the East Fork does meet the preliminary cri-

teria for classification as an alluvial valley floor.[54] DSL has
stated that if East Armells is found to be an AVF, disturbance of
the alluvial aquifer directly underneath and to the sides of the
creek might render mining unacceptable.[55] Again, further study
is necessary. At present, WECo mines to within 100 feet of the
creek, leaving the buffer as a protective measure. However, the
company has disturbed the buffer zone in the past.[56]

Erosion

        Colstrip's Pit 6 and Pit 6 Extension reveal some of the most
severe erosion found by INFORM in the West. It is so serious
that it raises basic questions of the suitability of the area for
mining. Inspections from 1976 through 1978 turned up severe
problems, which have been repaired periodically but with little
success.[57] On its visit to the site in April 1980, INFORM found
the area--although covered in many areas with vegetation--still
eroding badly, with many gullies deeper than 6 feet. The soil is
very loose, and vegetation has done little to stabilize the area.
Huge (2-3 foot-wide) furrows in the side of the slopes are breaking and creating further erosion and sedimentation, and the 300-
acre area is useless for grazing. As far back as 1977, Michael
Grende, the mine's permit supervisor and former reclamation supervisor, asserted that, although the erosion was a problem, it
was nothing WECo could not handle.[58] But in at least one instance, more than one year went by with little repair work done
on the land as WECo explained that it could not find materials to
repair the erosion. DSL refused to accept the excuse, writing
WECo that "The lack of a key element in reclamation such as
seed, mulch, or equipment should be brought to the attention of
the Department [DSL] as other methods can be worked out. Delay of work for a year because of lack of material is not a valid
excuse...."[59]
        Much of the damage in the Pit 6 and Pit 6 Extension area is
due to poor reclamation practices, including steep grading and
the use of large "dozer-basins" instead of small contour furrows
on the land. The fine, sandy nature of soils in this part of
Montana creates the potential for severe erosion, as this area
demonstrates. In response to pressure from local ranchers and
state regulators, WECo now grades its slopes much flatter, and
it may be hoped that the Pit 6 experience will not be repeated
(see Grading).
        Because of the nature of the soils, and the amount of erosion
that has occurred at Colstrip, sedimentation has also resulted.
It has repeatedly affected an area where a haul road crosses a
creek,[60] and in several instances sedimentation has occurred on
unmined lands.[61]
        Chris Cull, WECo's senior reclamation engineer, told INFORM

Eroded areas of Colstrip's Pit 6 still remain. Photograph taken by INFORM in April, 1980.

that the company, learning from experience, now employs several
new techniques (along with flatter grading) to combat both sedimentation and erosion. These include straw bales, netting and
regrading. But only good, stable vegetative cover will hold the
soils together and control this problem permanently.[62] The results of this work won't be known for a few years, but indica-
tions of subsurface erosion--piping--have left area residents
convinced that the situation is irreparable. It may be that the
nature of the soils, rocks and other overburden materials is such
that destruction of geologic layers by surface mining will have
the same effect, regardless of the stability of the surface of the
land. Whether eroding from the top or subsiding from within,
the land at Colstrip is perhaps one of those areas that never
should have been disturbed in the first place.

SEEDING AND AMENDMENTS

Seeding Procedures

From 1968, when WECo resumed mining at Colstrip, to January 1980, 3,158 acres of land had been disturbed, and 1,537
acres had been seeded and growing for at least one year.[63] This
includes the approximately 300 acres of land in the Pit 6 and Pit
6 Extension area (see Erosion). The Montana DSL told INFORM
that WECo is current with its seeding of graded lands.[64] Although the company seeds in the spring, late summer and fall, it
has been increasingly favoring fall seeding for dormancy, so
that germination and growth will begin early in the following
spring.[65] This preference resulted from on-site experiments by
Montana State University researchers, who found that fall seeding produced a more diverse vegetative cover.[66] Many other
mines in the INFORM sample also conduct their seeding in the
fall.

WECo has one of the most diverse seed mixtures of any company in the INFORM sample, although its 21 plant species do not
even approach the 300 or so found on the site before mining began. With both cool- and warm-season plants, and some forbs
and shrubs, represented, the effort is a good one. Two variations of the mixture are designed for slopes that face to the
north and east, where evaporation is less, or to the south and
west. A third variation is used in drainages. Much of the development of this seed mixture is tied to research by WECo and
Montana State University.[67]

WECo says this seed mixture has been further refined for
the 1980 planting season. Under the new revegetation plan,
reclamation will fit into five postmining management units, with
each represented by a plant community or combination of plant

TABLE A

1979 SEED MIXTURE
FOR LANDS BEING RECLAIMED AT COLSTRIP

Plant Species	Intro-duced/ Native	Cool/ Warm	*Pounds Applied Per Acre* North/ East Slopes	South/ West Slopes	Drainages
Grasses					
Western wheatgrass	N	C	2.2	2.6	3.0
Green needlegrass	N	C	3.0	2.0	2.0
Slender wheatgrass	N	C	2.0	2.0	3.0
Indian ricegrass	N	C	2.6	2.0	---
Prairie sandreed	N	W	0.8	1.4	2.0
Crested wheatgrass	N	C	0.6	0.4	---
Blue grama	N	W	1.0	1.0	---
Altai wildrye	I	C	1.4	---	1.0
Thickspike wheat-grass	N	C	1.0	1.2	1.0
Sideoats grama	N	W	0.8	1.0	---
Big bluestem	N	W	---	1.0	---
Little bluestem	N	W	---	1.4	---
Switchgrass	N	W	---	---	2.0
Canadian bluegrass	N	C	---	---	2.0
Prairie cordgrass	N	W	---	---	2.0
			15.4	16.0	17.0
Legumes					
Alfalfa	I	C/W	1.0	1.0	1.0
Cicer milkvetch	I	W	1.0	1.0	1.0
			2.0	2.0	2.0
Forbs and Shrubs					
Fourwing saltbush	N	W	1.0	1.0	1.0
Lewis flax	I	C	0.6	---	---
Shadscale	N	W	0.8	0.8	---
Big sagebrush	N	C	0.2	0.2	---
			2.6	2.0	1.0
Total			20.0	20.0	20.0 Pure live seed

Source: *Western Energy Company*

communities. Trees, shrubs and forbs will be planted or transplanted first to allow them to establish. The ground will be stabilized with a sterile hybrid of wheat. Later (up to three years),
the permanent mixture of plants will be seeded.[68] This may
prove to be the best approach to establishing a variety of plants
on lands disturbed by mining. The new revegetation plan is
commendable.

Amendments and Maintenance

Now that grading of lands at Colstrip is generally flat and
rolling, all reclamation equipment is operated along the contour.
Inability to do so in the past--as in the Pit 6 and Pit 6 Extension
--may have contributed to erosion.[69] Furrows are not placed on
graded lands to help control erosion and sedimentation; these
might enhance the WECo reclamation program.[70] In the Pit 6 and
Pit 6 Extension areas, large "dozer basins" were used for this
purpose, but their size proved to be their undoing. Once filled
with water, the basins broke apart, allowing water to cascade
down the slope, breaking up other basins and creating the mess
that remains today (see Erosion).

When planting seed, WECo loosens the soil with a chisel plow
and broadcasts fertilizer (discussed below), then seeds with a
broadcast seeder. The seeder is designed to protect the bed
from blowing winds. (The company switched to this type of
seeding machine because of dissatisfaction with the uniform rows
of grasses that result from the use of a drill-seeding machine.)
The seed is then covered with soil.[71] This is a fine practice for
seeding as long as seed are evenly distributed, as they appear
to be at Colstrip.

Besides seeding from 5-10 pounds of wintergraze--a sterile
grain hybrid--for a standing mulch, either wheat straw or grass
hay is applied to the land, and some steep slopes have been
mulched with wood-fiber.[72] The combination of these two techniques provides a standing cover for protection, and the native
grasses in the hay provide an extra source of seeds to the land.
(If the hay is made up of undesirable plants, this could back-
fire. Care must be taken that only native hay is used.)

WECo conducts no irrigation at Colstrip, but it applies fertilizer when the soil is tilled prior to seeding. WECo says it samples the soil to determine how much fertilizer to use. As in any
operation where fertilizer is used, it should be fully ascertained
that the plant communities will continue to grow and flourish
without the use of such additives.

Trees and Shrubs

Although WECo has been transplanting trees and shrubs for

many years, it has only recently determined that a new approach
is needed for successful establishment and growth. According
to Earl Murray, a reclamation engineer at the mine, the grasses
planted by WECo were out-competing and killing the trees and
shrubs. The company now transplants and stabilizes the soil
with annual grains, planting the regular seed mixture two or
three years later.[74] (See Seeding Procedures.) Tree spades are
used to remove trees from areas that are to be mined; these are
then planted on graded and topsoiled lands. Small amounts of
fertilizer and irrigation are used to promote growth.

WECo directs much of its work at areas where trees are usually found, such as drainages. The long-term results remain
uncertain, but the program deserves praise for the moment. Reports on the survival rates of past plantings vary from 98 percent to 27 percent.[75] Further investigation is needed, but Dennis
Hemmer of the DSL notes that the tree and shrub transplant
program at the mine is "very impressive."[76]

VEGETATION

As a ranching area, the lands around Colstrip are in better
condition than many of the areas studied by INFORM. Productivity of undisturbed land allows for the grazing of one animal unit
(the equivalent of one cow and calf, or five sheep) per month on
from less than two acres of land up to five acres of land.[77] In
areas where wheat, barley, or oats are grown, average productivity is from 20 bushels per acre to 60 bushels per acre.[78]

Revegetation success at Colstrip varies greatly with the location and year of initial seeding--primarily because of improved
practices in recent years. DSL and the U.S. Geological Survey
concluded in 1979 that:

> Early reclamation efforts [at Northern Powder
> River Basin mines] have met with variable initial
> success, ranging from impressively poor to im-
> pressively good. Both extremes are exemplified
> at the...mine in Colstrip.[79]

At the time of INFORM's first visit to Colstrip in June 1979,
the seasonal plant growth was at its peak, affording a view of
the best that could be offered on reclaimed lands. In general,
vegetation is made up predominantly of cool-season grasses,
many of which are introduced.

Revegetation has failed in lands seeded through 1976 in the
Pit 6 and Pit 6 Extension areas (see Erosion). Although there is
some plant cover, it has utterly failed to stabilize the soil and
prevent erosion. Bare patches and gullies abound. Besides
other undesirable grasses, a large amount of yellow sweetclover

--an introduced legume--has grown, and only a few desirable
forbs can be seen. WECo told INFORM in 1979 that the growth
on this area was becoming stagnated.[80] And on a follow-up visit
to the mine in April 1980, virtually no new growth could be
found. Of course, Pit 6 and Pit 6 Extension are the worst areas
to be found at the mine, but given that they encompass about 300
acres of former rangeland, the impact is considerable. Chris
Cull told INFORM that he would like to return to Pit 6 and try to
completely rework the reclamation there, using improved techniques developed by WECo in the last five years. He estimates
that this would cost $750,000.[81] This may be an optimistic view,
as the total would break down to $2,500 per acre (assuming 300
acres), when total reclamation costs reported by WECo are
$11,800 per acre. (Of course, not all the costs encountered on
newly disturbed lands would apply, but the company would do
better to remain pessimistic in this regard.)

On lands where revegetation has been more successful,
plants most commonly encountered include several wheatgrasses
(many introduced species), smooth brome, legumes like cicer
milkvetch, alfalfa and yellow sweetclover, and some forbs and
weeds. In 1977 the U.S. Fish and Wildlife Service found that although there was a fairly good cover on much of the land, at
least 40 percent of the cover was made up of the cool-season, introduced smooth brome. Moreover, plants such as tall and
crested wheatgrass (both introduced species) and some native
Indian ricegrass could be found.[82] The best results cited earlier
by USGS and DSL include lands where healthy covers, primarily
of grasses--and few weeds--can be found. But the cover is still
dominated by introduced species, including several wheatgrasses
and legumes. WECo reports that test clippings have found up to
88 percent of the cover on its best lands are made up of native
plant species, but INFORM could not verify this data.[83]

The Montana State University research team that has conducted most of the tests on Colstrip lands found that diversity
of plants is a problem when introduced species are included in
the seeding mixture used. On some of their test plots they
found that "although a variety of native and introduced plant
species were originally seeded, the stands were ultimately nearly
completely dominated by a relatively small number of introduced
legumes and naturalized grasses." Although diversity was increasing in the fourth and fifth years after seeding, with invasion of native grasses and forbs, MSU said diversity "in an ab-
solute sense was still very low."[84]

The question of establishing native species--both cool- and
warm-season varieties--worries local ranchers. The most vocal
critics are the Northern Plains Resource Council and the Rosebud
Protective Association. According to Bill Gillin--a native of the

area and member of both groups--the problem is one of establishment. He says that once native plants are established, they will
take care of themselves. But the growth of native species in the
first few years after mining and grading is not an accurate rep-
resentation of what will come later. Gillin likens the growth of
plants after mining to the growth of plants after fertilization, as
mining has released significantly larger amounts of minerals to
be used by the plants. But when that source runs out, the
plant must then fend for itself. Diverse, native plant communities have developed over many years and are well protected by
their diversity against total devastation. "The last 10 years
have been better [more rain] than the previous 40. What happens when the real drought hits in the next 10?"[85] This question
applies to all lands being reclaimed in the West, but will probably
be asked more often, and with greater conviction, by the ranchers and farmers of Colstrip.

The final measure of productivity is the ability to use the
land for grazing once reclamation is complete. To keep up the
company's reclamation efforts, WECo's Cull has conducted some
grazing trials as part of his reclamation management plan. He
rightly feels that grazing is the natural maintenance scheme that
produced the plant communities now found in the area--buffalo,
deer and other animals grazed the land for many years before
man ever settled in the region.[86] Results of grazing trials have
been inconclusive, but they tend to show that reclaimed lands
are not as productive as native, undisturbed areas. In fact, the
results of these trials are still argued by critics of both WECo
and MSU because of some of the techniques used. Results reported by Meyn et al. found that steers grazed on native range
gained an average of 1.9 pounds per day, while those grazed on
introduced grasses on reclaimed lands gained only 1.3 pounds
per day.[87] Yet even these data are not sufficient for ranchers
in the area. They argue that it is ridiculous to conduct studies
using steers (emasculated cattle raised solely for beef production) when almost all the animals grazed in the area are cows
(raised first for their ability to reproduce, and secondarily for
beef). This is like comparing apples with sea cucumbers.[88]
Thus, reclamation success is still far from proven at Colstrip, as
at the other mines in the INFORM sample. However, WECo is one
of two companies (Peter Kiewit at its Decker Mine is the other)
that have attempted to measure the success of reclamation with a
regular grazing program, for which it should be commended.
But again, it is most likely the pressure from local landowners
that has caused the company to take such action. The jury remains out.

WILDLIFE

The Montana DSL and the U.S. Geological Survey have noted that Colstrip is an important wildlife area, affording habitat for several species, including deer, antelope, grouse, and coyote. The habitat along the banks of the East Fork of Armells Creek and Cow Creek are also important sources of food and shelter for wildlife.[89] DSL's Hemmer told INFORM that the wildlife habitat re-creation program at Colstrip is "one of the better programs" in the state.[90] A couple of sandstone outcrops have been avoided during mining in an effort to leave natural habitat, and riparian zones have been created in drainage bottoms.[91] When WECo re-constructs water drainages, corridors of trees and shrubs are created; their continuous nature is attractive to animals that will not use isolated trees or shrubs.[92] Again, the success of the company's tree and shrub planting program (see Trees and Shrubs) will determine the ultimate success of the wildlife pro-gram.

Rock shelters--similar to those built at other mines in the INFORM sample--and the trunks of dead trees have been used on some lands to attract wildlife. Local ranchers question the value of these structures, and one, Wallace McRae, told INFORM that no self-respecting bird would ever be caught nesting in an iso-lated, dead tree trunk like the one researchers have stuck in the ground on one section of the mine.[93] In fact, aside from their concern for the grazing potential of reclaimed lands, ranchers are especially aware of the value of wildlife in the area to main-tain the correct balance between animal and plant communities. Disturbance of this balance could easily change the entire area's usefulness. But WECo is making an effort.

RECLAMATION COSTS

TABLE B*

RECLAMATION COSTS AT COLSTRIP

Operation	Cost Per Acre	Percentage of Total Cost
Planning and Design		5%
Planning, design, permitting, bonding	$ 600	
Earthmoving		73%
Topsoil salvage and redistribution	1,000	
Regrading and contouring	1,500	
Final pit reclamation	6,100	
Water Quality		17%
Drainage reconstruction	2,000	
Revegetation		5%
Revegetation	500	
Management (long-term)	100	
Total Cost:	$11,800	

*These costs and percentages are not strictly comparable to those of the same operation or activity at other mines. See the Cost of Reclamation chapter for a detailed explanation.

Source: Western Energy Company

Decker/Peter Kiewit/MT

Mine: Decker Mine

Location: Decker (population 420), Montana, in Big Horn County (population 10,688)

Operating company: Peter Kiewit Sons' Company

Parent company: Decker Coal (Peter Kiewit Sons' Company and Pacific Power & Light Company)

Parent company revenues: 1977, $404.6 million*; 1978, $506.5 million*

Parent company net income: 1977, $88.6 million*; 1978, $105.8 million*

First year of production: 1972

Coal production in 1978: 9,073,592 tons

Acres for which a mining permit has been issued: 7,514.4

State regulatory authority: Montana Department of State Lands

Reclamation bond posted: $15.4 million

Reclamation cost: reclamation cost per acre, $4,334; reclamation cost per ton of coal, NA

Expansion plans at mine: Of 17,530 acres of land under lease, more than 12,000 are surface-mineable

Additional acres for which the company will seek a permit: NA

Parent company U.S. coal reserves: 1.7 billion tons*

ern United States: In 1980 Northern Energy Resources Company
(a subsidiary of Pacific Power & Light) opened the Spring Creek
Mine in Montana to produce 10 million tons per year by 1982. By
1984, the company plans to open the Cherokee Mine in Wyoming
to produce 6 million tons annually by 1985, and the Antelope
Mine in Wyoming (in 1983) to produce 10 million tons per year by
1985. Peter Kiewit is developing three mines in Wyoming--as a
joint venture with the Rocky Mountain Energy Company--and one
mine in Montana. The three Wyoming mines are scheduled to
produce 12.1 million tons of coal per year by 1981.

NA = Not available

*Data refer to Pacific Power & Light only.

Sources: Peter Kiewit Sons' Company; Pacific Power & Light
Form 10-K, *1978;* 1979 Keystone Coal Industry Manual; *"New
Coal Mine Development Survey 1978-1987,"* Coal Age, *February
1979; U.S. Department of the Interior, USGS, and Montana De-
partment of State Lands,* Final Environmental Impact Statement,
Proposed Plan of Mining and Reclamation, East Decker and North
Extension Mines, Decker Coal Co., Big Horn County, Montana,
1976, *Vol. I, pp. 16-23; Computer printout from H. Robert
Moore, Assistant to the Director for Coal Management, Bureau of
Land Management, U.S. Department of the Interior, November
20, 1978; INFORM interview with Dennis Hemmer, April 28, 1980*

SITE DESCRIPTION

    The Decker Mine is in the southeast corner of Big Horn
County, five miles from the Wyoming state line in southeast Montana. Actually two mines, the West Decker operation lies west of
the Tongue River Reservoir on the Tongue River; the East
Decker operation lies to the east. The Tongue, a tributary of
the Yellowstone River to the north, is fed by ephemeral streams
that drain the permit area.
    The land is fairly flat, with slopes seldom exceeding 15 per-
cent.[1] The topography of the mine site rolls down to the reservoir, with a few of the cliffs and rock outcroppings that are
more common to the lands surrounding the mine. To the west,
the Bighorn Mountains create a "rain shadow" that limits site
precipitation.[2] The climate is characterized by wide variations in
annual and seasonal precipitation and temperature. Total yearly
precipitation averages 11.8 inches, but has ranged from 6.5 - 20
inches since 1949, with one year in four below 10 inches. June
is the peak month for rain, and 45 percent of the yearly precipi-

tation falls in April, May and June. Another 30 percent comes as snow in the winter; the balance falls in brief summer storms.[3] In 1978 the precipitation was very high, nearly 20 inches, but 1979 was dry: much of the reclamation work done that year will have to be repeated.[4] The period when the ground is not frozen and the growing season range from 90 - 130 days.[5]

Several vegetative communities are identifiable at Decker. On most of the land, grasses and sagebrush communities are plentiful. More than 100 species of plants have been identified, according to the U.S. Geological Survey (USGS) and the Montana Department of State Lands (DSL).[6] Sam Scott, Peter Kiewit's reclamation supervisor, says several hundred plant types have been found, though not all in the mine area itself.[7] Decker was ranchland prior to mining, used for grazing and hay production. Peter Kiewit plans to return the land to those uses.

OVERVIEW

Operated by Peter Kiewit Sons--in a joint venture with Pacific Power & Light--the Decker Mine has long been one of the largest coal producing facilities in the United States since it opened early in the 1970s, and one of the most widely studied. Its production, coupled with that from the smaller Rosebud and Big Horn Mines in Wyoming, has made Peter Kiewit Sons--a large, privately held construction firm--one of the 15 largest coal producers in the country. With the volume of coal that is produced, a great deal of land is disturbed. Peter Kiewit has made strong efforts to reclaim it--but the company's practices leave much room for improvement and only 9 percent of all lands disturbed at the site have been seeded permanently.

Perhaps the most acute concern at Decker--although the company professes unconcern--is salt. Very little has yet been found on reclaimed land, but the topsoil and subsoils are somewhat sodic, and the overburden considerably more so. The long-term effects of salt seepage on groundwater resources and on vegetative growth and productivity remain unknown. They should be cause for concern, but Kiewit officials say they doubt the problem will arise--and if it should, they say, they will be able to correct it, however expensive the effort. But no known cure for salt seepage exists. Decker's potential salt problem calls for close and continued scrutiny.

The mine's seeding and amendment program is fair to good. It would benefit from the use of a more rounded seed mixture--incorporating more warm-season grasses and forb species--and from direct replacement of topsoil. But five-year-old stands of grasses appear well stabilized, and limited grazing has not harmed the plant communities established so far.

Earthmoving and grading are performed well but lag considerably behind disturbance.   The foremost reason is the size of
the operation, soon to encompass from three to four separate
mining areas.   Moreover, the nature of the overburden is such
that the company and the state have decided to allow more time
for overburden to settle before final grading, replacement of
topsoil and seeding.   This has been done to prevent land subsidence (and extensive post-seeding surgery) caused by settling
and compacting of the overburden after only a few years.   The
results of this practice will become more evident in the next few
years.   As for topsoiling practices, the best thing Peter Kiewit
could do for its land reclamation program would be to haul more
soils directly back onto graded lands instead of stockpiling.
The advantages for plant growth and diversity have been demonstrated on small areas at the mine where direct haulback has
been practiced.   But Kiewit says operating procedures and the
mine's design restrict this practice.

Water issues at Decker leave many questions unanswered.
Numerous creeks feed into the Tongue River and its reservoir--
which separate the East Decker and West Decker operations.   Diversions were built to allow mining under the original creek beds
without preventing water flow.   But they have been heavily
eroded because of poor stabilization, and Peter Kiewit has had to
work hard to restabilize the diversions and prevent more erosion.   Eventually the company will return the surface waters to
their original paths, and the quality of that work--including the
stabilization of the recreated permanent creek beds--will require
careful monitoring.   Mining is affecting groundwater, too, and
documentation of lower water tables and increased mineralization
of some groundwater supplies has made the issue more visible.
As at so many mines where aquifers used by local residents for
water coincide with the coal seams being mined, the long-term
impact on local water users is hard to assess.   Preliminary data
appear to indicate that both quality and quantity will be jeopar-
dized.

The revegetative effort has been successful at establishing
grasses, but many are introduced species, and diversity is poor.
Much more work is required to establish forbs and shrubs in the
area.   Direct replacement of topsoil, in lieu of stockpiling, would
help in this regard.   Diversity of native plants growing from
seeds transported with the topsoil has been good, and more of
this practice would help revegetation.   Peter Kiewit should be
commended for conducting grazing studies on its best reclaimed
lands at Decker, but--like Western Energy Company at its Colstrip Mine--the studies are inconclusive and do not directly address the critical concerns of affected ranchers.   Further work
is recommended.   The company plans to continue testing.

Peter Kiewit has been engaged in research with federal agencies to attempt to define the more beneficial practices it might
incorporate into its reclamation program. Using research results
from on-site experiments is always a good practice, and much of
the Decker program is based on the work done when the mine
first opened. An innovative sediment pond design has been introduced recently in the East Decker operation; it should be
studied by all mining companies attempting to control surface
water discharges in the West.

OSM has conducted few inspections at Decker. The agency
seems still to be hampered by having too few inspectors for too
many mines in the West. But the Montana regulators have visited
Decker often, and their concerns are similar to those set forth
above. Since July 1974, only $6,700 in fines has been levied
upon Peter Kiewit at this site, less than at either of the other
two Montana mines studied by INFORM. This is not the sole indicator of a company's ability to perform within the boundaries
of the law, but it at least speaks well for the company's inten-
tions. And with the company's Sheridan, Wyoming offices only
20 miles away, better control over the entire complex appears to
have been achieved.

EARTHMOVING

Grading

Grading progress is as good at Decker as at other mines in
Montana. Strict state regulations require currency, and according to the Montana DSL and the U.S. Geological Survey, the
company keeps grading within one to two spoil ridges of the active mine pits.[9] The Office of Surface Mining (OSM) found
Decker operations to be within two spoil ridges of the active
pit.[10] Yet of the 2,540.7 acres of land disturbed by mining and
related activities through March 1980, only 221.3 have been
graded.[11]

Premining slopes are generally being recreated or made
somewhat flatter after grading. No lands have been graded to
more than a 5:1 slope, except for spoils from the initial mine cuts
for the East Decker operation, which have slopes as steep as
3:1.[12] However, these steep slopes will eventually be regraded
by Peter Kiewit, unless the vegetation that the company can establish on the land while mining proceeds is sufficient to prevent
erosion and permit grazing. The state and federal regulators
would have to agree to allowing these slopes to remain.[13]

It appears to both INFORM and the DSL that Peter Kiewit is
attempting to strengthen itself for a battle with state regulators
in the future over these lands. By concentrating revegetation
activities and wildlife habitat construction on these slopes, Kiewit

is going beyond the necessary amount of work to stabilize the land until regrading can be conducted. In the future, when the DSL demands that the land be flattened, Kiewit may well point to the work it has done and try to gain a waiver over the state's regulations.[14]

Drainage patterns are generally being restored on graded lands at Decker. However, according to a DSL inspection in March 1979, portions of the first area ever reclaimed at West Decker will need some corrective grading to improve drainage.[15] Slopes created by grading are all traversable by reclamation machinery, except on the 3:1 and 4:1 slopes mentioned above.

Numerous instances of subsidence on graded lands have been noted by state inspectors at the West Decker operations.[16] Very few have been found at East Decker. Peter Kiewit argues that the reports are exaggerated,[17] but in August 1977 DSL reported that a crack had appeared in graded lands (owing to subsidence); the crack was 3 feet deep, 100 feet long and 1 foot wide.[18] In March 1979 the state found another crack in graded lands to be 30 feet wide, 150 feet long and 4 feet deep.[19] The only explanation for the subsidence is that spoil materials are compacting after having been loosened and expanded by mining. Subsidence is found mainly on the west side, and Peter Kiewit told INFORM in June 1979 that it has begun grading spoil piles on the west side somewhat differently to help mitigate the problem. The DSL has allowed additional time to let overburden settle before final grading is done.[20] The issue of subsidence and the potential for erosion--as well as the uselessness of lands that have cracked and faulted--must be monitored closely so that persistent problems can be dealt with, if the technology exists to do so. Otherwise, the long-term use of the land will be severely limited (see Erosion).

Topsoil

Topsoil and subsoil are salvaged at Decker in two separate operations. The two uppermost layers of soil--termed the A-horizon and B-horizon--are removed together and considered topsoil. From 6-12 inches is taken, and, according to Dale Johnson, a reclamation specialist at the mine, an average of 8 inches comes from all the land being disturbed.[21] Further, the C-horizon is removed as subsoil, averaging 5 feet in depth.[22] On occasion, insufficient topsoil has been removed,[23] but topsoil salvage operations are generally reported by regulators to be good.[24] The company has not been cited for missing whole areas of topsoil as often as have the operators of both Absaloka and Colstrip--the other Montana mines studied by INFORM (see Absaloka and Colstrip profiles). When replaced, topsoil and subsoil are left to a uniform depth on graded lands.[25]

Little of Decker's topsoil is directly hauled back onto graded lands. Dale Johnson believes this practice would be an important asset to reclamation (see Vegetation), but he says operating procedures and the design of the mine limit this practice.[26] As a result, most topsoil and subsoil are stockpiled.[27] Stockpiles observed by INFORM in April 1980 appeared fairly well vegetated and stabilized, but DSL has noted occasional erosion similar to that found at most large surface mining operations.[28] Peter Kiewit has been quick to repair problems noted by state inspectors.[29] More direct haulback is essential. Stockpiles should be avoided.

Toxic Material

The sodic or salty nature of the soils and overburden is the biggest potential threat to plant growth at Decker. According to the U.S. Department of Agriculture's Agricultural Research Service, topsoil and subsoil are only slightly to moderately sodic, but overburden is moderately to highly sodic.[30] Peter Kiewit refuses to acknowledge that a major problem may exist; the company simply says that if salt becomes a problem, the company will solve it.[31] INFORM is unaware of any technique to repair problems caused by sodic soils--or, at least, any economical technique. Some salts already have been reported on the land. In 1977, salty seepage was found by DSL inspectors on reclaimed areas[32]; salts are also appearing on graded spoils that have yet to be topsoiled.[33] Some land near a pond next to Pond Creek-- which flows through the mine area--also showed salts on the surface.[34] This may indicate that where there is more water-- which acts to transport salts both downward and upward through the soil--the salts will appear sooner. Although reclamation results to date show no appreciable problems attributable to sodic seepage, it may be some time before the true impact can be discerned. It is mining--loosening of soils and destruction of earthen strata--that not only increases the amount of salt available for transport, but also allows the salty material to be carried through the overburden more readily. Montana DSL officials speculate that upward migration of sodium may take a long time to show up.[35] The lack of apparent problems to date should not be cause for complacency. The final judgment on the effect of sodic materials on reclamation--and on surface and groundwater supplies--should be reserved for a number of years. Peter Kiewit should not be released prematurely from responsibility for reclaimed lands.

Surface Water and Groundwater

Most water on the Decker site flows in, toward the mining area. Very little water flows off the permit area, and all waters are consolidated into settling ponds.[36] Although the company is permitted to discharge water from five points at the edge of the mine, only one discharge point (Point #2) discharges regularly.[37] The system itself is fairly well protected by vegetation and the use of rocks--scoria--to protect the ground from erosion and slow the scouring flow of water.[38] In December 1979 OSM reported that the system was well protected.[39] However, in several instances erosion has occurred where surface creek diversions--Pond Creek and Pearson Creek--have been built. (Peter Kiewit has diverted intermittent creeks around sections of the mine where it has mined through the creek beds. The original paths are being re-established once mining and other reclamation activities are completed.)[40] These areas have presented the biggest problem for the company in the past. Although they were first cited in July 1977 for being severely eroded,[41] temporary repairs--which were only minimally helpful--were not completed until April 1978.[42] Erosion continued; by May, repair operations had tripled the size of the affected area, according to the DSL.[43] By June 1978 the size of the disturbance--covered by rocks and other attempts to control erosion--spanned an area 900 feet long, 150 feet wide, and 9 feet deep.[44] In October--more than a year after it was cited for severe erosion problems by DSL--the area was seeded to allow vegetation to help stabilize the land. Current creek diversions seen by INFORM in April 1980 are heavily covered with scoria rock to aid in preventing erosion, but according to the Bureau of Land Management, any attempts to create diversions away from natural creek beds will be met with inherently less-stable conditions.[45] As at AMAX's Belle Ayr Mine in Wyoming (see Belle Ayr profile), attempts to redirect creeks are often met with erosion and require extensive maintenance.

Peter Kiewit has generally been able to meet the requirements of its National Pollutant Discharge Elimination System permit for waters discharged from the mine into a reservoir. (The reservoir is part of the Tongue River, which flows between the East and West Decker operations.) However, a careless incident in March 1978--when water was pumped onto a frozen sediment pond, running off and disturbing surrounding lands--resulted in a violation of standards for sediment (Total Suspended Solids), that reached more than three times the allowed levels.[46] Other discharges have been within limits or slightly above the limits for Total Suspended Solids. No fines have been levied.[47] Not

Innovative sediment pond design at Decker. Note earthen fingers projecting into water body.

only is more care now being given to the retention of water on
the site before discharge, but the company has implemented a
very innovative technique for the construction of a sediment
pond in the East Decker area. Large earthen baffles extend into
the pond from both sides, so that water entering from one end
must flow in a circuitous route before reaching the point of discharge at the far end. This allows more time for sediment to
drop from the water and remain in the pond. In February 1980
DSL reported the pond was functioning well, with dirty water
being pumped into the pond from active mine pits and clean water exiting from the other end.[48] Peter Kiewit merits praise for
this design, which INFORM believes is unique to the West.

Mercury is one element that is not generally monitored in
waters discharged from surface coal mines. Several years ago,
it was found that mercury concentrations were increasing in the
Tongue River Reservoir, next to the mine. However, early fears
that mining was responsible have been all but completely quelled:
it was determined that the entire watershed area for the reservoir was contributing mercury to the water,[49] and that less than
2 percent of all the mercury found was coming from mine-related
activities.[50] These mercury concentrations, quickly diluted in
the reservoir, are not considered a problem,[51] as the waters are
impounded for use in irrigation in surrounding areas.[52]

Surface water monitoring consists basically of monitoring
discharges from the mine, and ongoing studies of the quality of
water in the Tongue River and its reservoir, by state and federal agencies. Groundwater monitoring is another matter.
Decker is one of the few sites where effects of mining coal aquifers have been substantially quantified. Groundwater levels in
1975 were found to have dropped by 40 feet on the mine site and
10 feet within $1\frac{1}{2}$ miles of the site. Even groundwaters in a coal
seam not being mined at Decker were found to have dropped--
some guess this is due to upward leakage of water to replace that
lost from the destroyed aquifer above.[53] Furthermore, the Bureau of Land Management estimates that groundwater quality will
probably change, with a 300 percent increase in mineralization
(Total Dissolved Solids), thus making the water unfit for domestic use because of bicarbonates and sodium, though still useful
for livestock watering.[54] Sam Scott, Peter Kiewit's reclamation
supervisor, feels that dewatering of coal aquifers is not a problem because deeper aquifers may be tapped. As for the decreasing quality of some groundwaters, he believes little can be done
beyond replacing overburden selectively, placing the most be-
nign spoils in the ground at the approximate position of the destroyed aquifer.[55] But according to the Bureau of Land Management, residents rely heavily on groundwater--especially the aquifers in coal seams--for both stock watering and domestic use.[56]

Flowing from the west to the east and south,[57] groundwater--its quality and quantity--will have to be continuously monitored for a long time, as mining operations will continue at West Decker and East Decker for many years. To assume that deeper aquifers may be tapped to replace shallower waters lost to mining is not the answer; not only is it expensive to drill deep wells, but the cost of bringing deeper groundwater to the surface is prohibitive for many ranching operations. Some groundwater may return in quantity, owing to the presence of the Tongue River and the nearby reservoir, which may allow large quantities of water to soak into subsurface systems. Yet the quality of this water is still unknown. And what of the effect of increased salt availability? (See Toxic Material.) These questions remain to be answered.

Alluvial Valley Floors

It once appeared that some lands around Spring Creek and valleys around lower portions of Pearson Creek might qualify as alluvial valley floors (AVFs),[58] but the closest any land came to being an AVF was the area surrounding Spring Creek. However, because of the lack of sub-irrigation (irrigation from shallow aquifers) the lands are not protected under the alluvial valley floor criteria of either Montana or OSM regulations.[59] According to Sam Scott, no other areas are considered AVFs, but Deer Creek (which might be an alluvial valley floor[60]) in the East Decker portion of the mine was deemed a unique wildlife habitat (see Wildlife) and ordered preserved. Peter Kiewit reports it is not mining this land.[61]

Erosion

Erosion occurs only occasionally at Decker. Except for the problems associated with creek diversions on the site (see Surface Water and Groundwater) and periodic erosion of stockpiles (see Topsoil), the mine is relatively free of common rilling and gullying. When erosion of stockpiles or diversions does occur, some sediment is carried from the mine.[62] DSL inspectors told INFORM in April 1979 that Decker had no major sedimentation problems.[63] If sedimentation were excessive, it might be difficult to find, as most surface drainage is directed into either the Tongue River or Tongue River Reservoir. However, recent stabilization of water diversions on the mine--along with innovative sediment pond construction--reduces this potential (see Vegetation).

SEEDING AND AMENDMENTS

Seeding Procedures

Seeding at both West Decker--where the mine pits are an expanding semicircle--and East Decker--where the newer operation
is only now beginning to grade lands and make them available for
seeding--lags behind disturbance for the total amount of land affected. By April 1980 only 221.3 of the 2,540.7 acres of land
disturbed at the entire Decker Mine had been seeded.[64] Nonetheless, DSL told INFORM that Peter Kiewit is always prompt to
seed lands once graded,[65] yet the shape of the mine leaves much
land in a disturbed state.

As at most mines in the West--except those in the Southwest
--seeding generally takes place in the fall, late enough so that
seeds remain dormant until the very early spring, then germinating and allowing for the longest possible growing season in
which to establish.[66] A commendable basis for much of the company's reclamation plan has been the U.S. Forest Service's on-
site research.[67]

TABLE A

1979 SEED MIXTURE
FOR LANDS BEING RECLAIMED AT DECKER

Plant Species	Introduced/ Native	Cool/ Warm	Pounds Applied Per Acre
Grasses			
Western wheatgrass	N	C	6.0
Thickspike wheatgrass	N	C	3.0
Slender wheatgrass	N	C	3.0
Beardless wheatgrass	N	C	3.0
Green needlegrass	N	C	4.0
Pubescent wheatgrass	I	C	1.0
Smooth brome	I	C	1.0
Sideoats grama	N	W	2.0
Indian ricegrass	N	C	2.0
Legumes and Forbs			
Sanfoin	I	W	3.0
Fourwing saltbush	N	W	0.5
			28.5 Pure live seed

Source: Peter Kiewit Sons' Company

Surprisingly, with all the evidence Peter Kiewit has accumulated documenting the number of forb species common to the area, few are being seeded. In fact, of 28.5 pounds of seed being applied per acre, only 0.5 pounds of either forbs or shrubs are being used. Although some introduced species are being used, they are wisely kept to a minimum—comprising only 5 of the total 28.5 pounds. Too few warm-season seeds or species are being used, and the resultant grass cover will tend to be dominated by cool-season plants. The mixture contrasts sharply to that used at Colstrip to the north, where numerous warm-season plants and forbs are planted (see Colstrip profile). The seed mixture, developed with DSL and the U.S. Soil Conservation Service,[68] should be further refined.

Amendments and Maintenance

Because the land being graded is fairly flat, and no slopes are steeper than approximately 5:1, reclamation machinery can operate along the contour of the slopes being reclaimed. No furrowing is conducted along the contours; although the terrain is somewhat flat, furrows would help prevent erosion and otherwise benefit plant growth. Dale Johnson feels contour furrowing would be an asset to the Decker reclamation program.[69] Peter Kiewit should investigate adding this practice to its operation.

In seeding, topsoil is first disced. The seeds are planted and covered with soil. Mulch is applied with a hydromulcher. (Wood fiber mulch cannot be tacked or crimped into the ground.) Peter Kiewit is one of the few companies in the INFORM sample that prefers using a wood-fiber mulch to a straw, hay or standing mulch. Approximately 2,000 pounds of the mulch is applied per acre, and experience at the site has shown it to be very effective in holding soils together, and not being lost in high winds.[70] In June of the first season of growth, weeds are mowed to provide additional mulch, as well as to prevent seeds from forming. Again, research by the U.S. Forest Service helped to determine the type and amount of mulch used.[71]

No irrigation is conducted at Decker. Fertilizer is applied twice in the first year of growth. The first application is made before the seeds are sown, the second within the first year of growth—to offset the nitrogen deficiency caused by decomposition of the wood fiber mulch.[72] Because no fertilizer is added later, the long-term growth of plants on the land is most likely not being affected by the early applications, which aid in establishing vegetation in the earliest years only.

Trees and Shrubs

The program for establishing trees and shrubs at Decker is
very limited. A few trees grew on the site before mining, but
no tree replacement is being attempted. As for shrubs, little
has been done, and success has been minimal. Peter Kiewit told
INFORM that it is conducting extensive trial plantings of shrubs
at the site, both on its own and with the U.S. Forest Service.[73]
But little of this work has been transferred to the actual reclamation program. According to company reports, more than 4,000
shrubs were planted in October 1976 in water drainage areas,
and some trees were planted in the spring of 1977.[74] Reports by
DSL in April 1978 indicate that survival has not been good.[75]
Peter Kiewit disputes this assessment, but on the best reclamation the company had to show during a visit to the site in April
1980, no shrubs were present.[76] Although it is important to
place shrubs along drainages, it is also important to create a
more diverse plant community on other lands at the mine (see
Vegetation).

An intriguing experiment being conducted by the company
uses containerized shrubs on some newly seeded lands. Black
plastic sheets surrounding the shrub planting act to direct dew
and other moisture toward the plant, thus providing natural irrigation.[77] The only problems so far have been caused by
browsing by wildlife, which tends to destroy these young and
still poorly established shrubs. The work and experimentation
are encouraging and should be expanded to many other parts of
the mine where land is being reclaimed. The Office of Surface
Mining,[78] DSL[79] and the Bureau of Land Management have all
suggested that more shrub plantings be undertaken.[80]

VEGETATION

The land in the Decker area is moderately productive for the
grazing of cattle. According to Sam Scott, the Soil Conservation
Service recommends stocking rates for the area at about 7.7
acres per month per animal unit (an animal unit is the equivalent
of five sheep or one cow and calf).[81] But Scott feels that the
land is more productive than that, and that it requires only 5.5-
5.8 acres of land to graze an animal unit for one month.[82] According to an environmental impact statement for part of the
mine, the land produces forage ranging from a low of 685 pounds
per acre in ponderosa pine communities, to a high of 4,000
pounds per acre in riparian-shrub communities near water drainage bottoms.[83] The Bureau of Land Management found that
stocking rates on 440 acres of land proposed for mining actually
ranged from 4-10 acres per animal unit per month, averaging 6.9

acres.[84] Thus, it seems likely that the productivity of the land
does vary considerably, and that the best measure of productivity on reclaimed areas would be actual grazing trials.

Trials have been conducted at Decker, but their results are
somewhat inconclusive. Twenty cattle were grazed on 75 acres
of land for approximately two months in the "scraper pit" reclamation area. Results--according to Dale Johnson--showed that
at the end of the period, weight gain was approximately 1.7
pounds per day for each animal, and only 25 percent of the
available ground cover had been removed. (Most grazing under
proper management removes 50 percent of the available forage.)[85]
With these good weight gains, the company feels that twice as
many cattle could have been productively grazed on the land.[86]
However, INFORM feels that the controls on this experimental
grazing favored weight gain excessively. First, a high level of
management prevailed: the 75-acre grazing area was split into
four smaller grazing pastures, and the cattle were rotated among
them over the two-month period. Second, wooden shelters were
constructed to allow cattle to escape the hot afternoon sun--
which causes dehydration and weight loss--and a water supply
was available on the 75-acre plot. Thus, the cattle had very
small distances to travel for either shelter or water, and vegetative production was near a peak when weight gains were being
recorded. Peter Kiewit admits that further experimentation is
needed to ascertain the productivity of the land over a full
growing season, as well as to test the productivity of the land
not just for weight gain, but for reproductive potential for the
grazing animals. Of the 20 head grazed, only one animal was
capable of reproducing, the others having been spayed or emasculated.[87] When incapable of reproducing, more weight is gained
since no energy is needed to fuel the reproductive processes in
the animals. This situation is similar to the grazing trials conducted at Colstrip (see Colstrip profile). Nevertheless, Peter
Kiewit should be commended for at least attempting to make accurate studies of the productive potential of lands it has attempted to reclaim.

As for the plants growing on reclaimed lands, almost all the
desirable vegetation is made up of grasses.[88] Although vegetative cover is good on the "scraper pit"--the area of best reclamation so far[89]--many introduced species are prevalent, and few
forbs or shrubs are growing; thus diversity is poor.[90] Sam Scott
concedes that most vegetation is made up of cool-season plants--
primarily grasses[91]--but claims that native forbs and warm-season grasses are coming in.[92] Dale Johnson noted that the best
diversity of both grasses and forbs could be found on small
areas where topsoil had been immediately replaced instead of being stockpiled.[93] Again, more direct placement is needed (see

Topsoil). Much more work and time are required before the land
can be considered to have a good, diverse cover of native plant
species, including grasses, forbs and shrubs.

Regarding the stability of the land, DSL has reported subsidence causing cracks in the land (see Erosion) and also piping—
a phenomenon where the land erodes from below, creating deep
holes in the ground.[94] The Bureau of Land Management says it
expects piping to occur on lands reclaimed at the mine, but only
in localized areas.[95] If subsidence and/or piping become prevalent, the land may become useless for grazing or farming. Coupled with the possibility that salts may begin to accumulate in
surface soils and inhibit vegetative growth (see Toxic Material),
these stability problems indicate that the final word on reclamation success is far from known. Members of the Rosebud (County) Protective Association—a group of ranchers and farmers
concerned with agricultural issues in Montana—told INFORM that
they were impressed with the results of reclamation so far at
Decker, especially given the circumstances under which it is being conducted. But they feel that the short-term achievements
and positive results should not be cause for optimism, as so
many factors—subsidence, salts, etc.—must be observed for
years before a clean bill of health may be given to reclaimed
areas at Decker.[96]

WILDLIFE

Sam Scott, a wildlife specialist, is concerned about wildlife
issues at Decker. A sage grouse strutting area was established
on some lands at the mine in order to attract birds away from an
area where the company plans to mine. According to Peter Kiewit, the experiment has been successful, and mining will continue.[97] Nesting structures have been tried, and rock pile
shelters—common to most of the mines in the INFORM sample—
have been built.[98] Experimentation is a positive factor in the
Decker wildlife program, but more shrubs and forbs would benefit animals as well. A variety of wildlife can be found in the
area, including antelope, deer, game and nongame birds, and
many mammals. Much of the diversity can be attributed to the
varied riparian habitat along the Tongue River Reservoir and
small creeks in the area. Peter Kiewit also is working with local
and state officials to maintain a pike spawning pond near the
reservoir for stocking the river.

According to officials at the Colstrip Mine, representatives
from Peter Kiewit visited Colstrip at the suggestion of the Department of State Lands to view the work that Western Energy
was doing in recreating water drainages and riparian habitats.[99]
The outcome is not known, but communication between mine op-

erators is always helpful and should benefit wildlife habitat restoration at Decker.

RECLAMATION COSTS

TABLE B*

RECLAMATION COSTS AT DECKER

Operation	Cost Per Acre	Percentage of Total Cost
Earthmoving		75%
Backfilling and grading	$1,450	
Topsoil replacement	1,810	
Revegetation		25%
Seeding	373	
Fertilizing	24	
Maintenance	677	
Total Cost:	$4,334	

*These costs and percentages are not strictly comparable to those of the same operation or activity at other mines. See the Cost of Reclamation chapter for a detailed explanation.

Source: Decker Coal

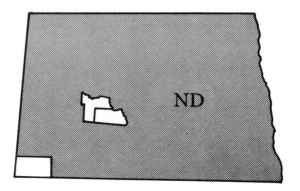

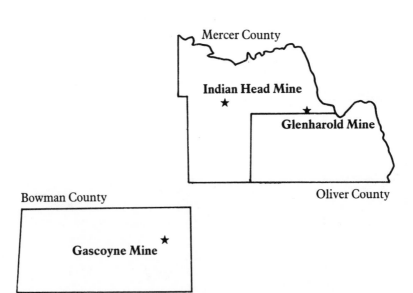

Mercer County

Indian Head Mine
★

Glenharold Mine
★

Oliver County

Bowman County

Gascoyne Mine ★

North Dakota

With plentiful, rich soils and annual precipitation of between 15 and 17 inches, reclamation in North Dakota should be the easiest in the West. But salty overburden can wreak havoc on both land and water resources if not carefully monitored. With a fairly good regulatory program conducted by the Public Service Commission, reclamation has been improving in the area, and work by groups such as the Dakota Resource Council in Dickinson--which centers its efforts on energy development issues--keeps the public involved in decisions that will affect the state, such as the call for increased coal production and utilization--especially in the manufacture of synthetic fuels. North Dakota is the grain center of the West, and conflicts between energy development and farming interests are pronounced.

Gascoyne/Knife River/ND

Mine: Gascoyne Mine

Location: Scranton (population 360), North Dakota, in Bowman County (population 4,100)

Operating company: Knife River Coal Mining Company

Parent company: Montana-Dakota Utilities Company

Parent company revenues: *1977,* $110.5 million; *1978,* $135.7 million

Parent company net income: *1977,* $10.5 million; *1978,* $14.8 million

First year of production: Opened in 1946 by Helm Brothers; purchased by Knife River in 1950

Coal production in 1978: 2,871,839 tons

Acres for which a mining permit has been issued: 1,228 as of December 31, 1978

State regulatory authority: North Dakota Public Service Commission

Reclamation bond posted: $1,006,621

Reclamation cost: reclamation cost per acre, $2,500-$3,000; reclamation cost per ton of coal, NA

Expansion plans at mine: None

Additional acres for which the company will seek a permit: 2,000

Parent company U.S. coal reserves: Owns or leases 600 million tons of coal in the United States

Development and expansion plans for surface coal mines in western United States: Aside from the Gascoyne Mine, Knife River operates the Beulah Mine in North Dakota, which produced 1.9 million tons of coal in 1978. This mine is slated to increase production to 2.5 million tons per year by 1981. The company also plans to open the Sprecher Mine in North Dakota to produce 2.0 million tons of coal per year by 1983.

NA = Not available

Sources: Knife River Coal Mining Company; Montana-Dakota Utilities Company Form 10-K, *1978;* 1977 *and* 1979 Keystone Coal Industry Manual; *"New Coal Mine Development Survey 1978-1987,"* Coal Age, *February 1979;* Rand McNally Road Atlas, *54th Edition*

SITE DESCRIPTION

In 1978 the Gascoyne Mine in the southwestern corner of North Dakota produced more coal than all but 16 other coal mines in the western United States. Located in the Missouri Plateau area of northeastern Bowman County, the land rolls from flat plains to a gentle easterly slope.[1] Although in a few small areas slopes could be measured up to 9 degrees,[2] the relief at the mine is minimal and very few slopes are steeper than even one-half of one degree.[3] Bordering on the western part of the mine is the Gascoyne Lake, used by townspeople for recreation.

Average annual precipitation is high compared with other surface mining sites in Montana, Wyoming and the Southwest-- 15-16 inches,[4] with about 70 percent of the rain falling between May and September.[5] Like other sites in North Dakota, precipitation levels are higher than most other portions of the West, greatly increasing the chances for reclamation success. Precipitation over the last 30 years has ranged from 10.4-23 inches annually.[6] Most rain generally falls in June. The area is frost-free for about 100 days from mid-May to late September.[7]

Knife River Coal reports that the land being mined in Bowman County is used as cropland, hayland and rangeland. About 70 percent of the land is used to raise crops such as wheat, barley and oats.[8] About half of this cropland lies uncultivated (fallow) each year, a standard farming practice in the area. Hayland covers 20 percent of the land, and the remaining 10 percent is used as rangeland.[9] Most of the land the company will mine in the next 30 years is now cropland and hayland; it will be reclaimed to these land uses,[10] according to David Jordan, conservationist for Knife River.

The Public Service Commission (PSC) told INFORM it considers none of the company's lands to be prime agricultural lands. As a result of water runoff patterns and the quality of some soils, certain productive soils are prime agricultural soils. But after making soil surveys in the area, PSC representatives concluded that because the highly productive soils occur only in small patches, they do not constitute agricultural units of sufficient size to meet prime agricultural land classification.[11]

OVERVIEW

Given the high quality of soils and adequate supply of water found at Gascoyne, one would expect that reclamation would be some of the best found in the West. But little has been done, primarily because Knife River Coal Mining Company is more interested in coal production than land reclamation. So reclamation at the mine must be classified as poor. Although plans presented by the company's reclamation staff since 1974 incorporate thoughtful methods of revegetation, the effectiveness of the plans is impossible to analyze because virtually nothing has been done. Strong federal regulatory authority, beginning in 1978 over this site, may result in speeding land reclamation in the early 1980s. The company is already responding to regulatory suggestions that it grade lands and prepare for seeding.

For the last several years, the story of earthmoving and grading at the Gascoyne Mine has been a story of neglect. Spoil ridges were left untouched, and not until the fall of 1979 did Knife River begin to plan to grade lands so that it could begin replacing topsoil for seeding. Topsoil and subsoil have been removed in two separate layers and may be stockpiled for as long as 15-20 years,[12] diminishing their revegetative potential. No problems with toxic substances have been discovered, but again, this may be because so little land has yet been fully reclaimed. The backlog in grading and the prolonged storage of topsoil and subsoil are not positive steps toward a comprehensive and successful reclamation program.

More water is to be found on this mine site than any other in INFORM's study. The Gascoyne mining operations are affecting water quality and quantity in several ways. First, mining interrupts a highly conductive layer of coal that provides a pathway for water--an aquifer--used by farms for stock watering and limited domestic purposes, and well-water levels have dropped. Second, the interaction between groundwater and surface water, coupled with the fineness of the overburden, creates the potential for sedimentation in water discharged from the site. A chem-

ical treatment program is being used to combat this problem.
Finally, some erosion problems arise because of the fine nature
of the overburden.

The seeding and amendment plan—still only a plan—is
unique in the INFORM sample in its attempt to establish warm-
season plants first (because they are harder to establish than
the cool-season plants generally seeded by most companies) and
to protect the land in cooler seasons with annual grains used as
mulch. (See Glenharold profile for a different approach to this
problem.) However, as no land has been available for seeding,
this plan has not been put into practice, and David Jordan, the
company conservationist who drafted the plan, is unsure whether
Knife River management will allow his staff to implement it even
when lands are available, because of the cost and time involved.[13]
Thus evaluation of the plan can only be speculative.

Because so little land has been revegetated in the last few
years, little can be said, either, about the productivity of reclaimed lands. The company has no plans to enhance wildlife
refuge, and no work is being done toward this end, since little
of the land supported this use prior to mining.

EARTHMOVING

Grading

About 77 percent of the land that has been disturbed at
Gascoyne remains ungraded. As of January 1980, only 257 acres
had been graded of the 1,105 disturbed.[14] In 1978 Knife River
disturbed 90 acres of land and graded none.[15] Although the
company stated in its 1977 mining and reclamation plan that "leveling of the overburden piles will be conducted at all times on
the second peak nearest the [mine] pit"[16] (which would be a
very good practice), this has not been done. Inspections by the
Office of Surface Mining (OSM) and the state Public Service
Commission in August 1978 found that grading was from one to
three years behind mining operations.[17] The PSC found that
Knife River was "behind in many areas...where grading work
should be completed [that] fall. There are large areas in the
Red, Blue and Yellow Pit area where grading work should be
done."[18] It determined that, if grading was not done in the fall
of 1978, no areas would be left for respreading topsoil and subsoil, and stockpiling would have to be increased. "It is *strongly
recommended* that a considerable amount of *grading work be
completed this fall* [all emphasis PSC's] to prevent the stockpiling problems at the mine."[19]

Areas mined in 1975 were still being graded during INFORM's
tour of the Gascoyne Mine in June 1979. Curtis Blohm, manager

of reclamation for Knife River, says many of these ungraded lands are now being rough-graded, and that the final leveling and smoothing prior to replacement of subsoil and topsoil should take place in 1980.

Why the prolonged delay? Part of the problem in keeping current with grading, Blohm says, stems from an expansion in coal production operations that occurred in 1974-1975.[20] Coal from four different pits at the mine is blended to produce a more constant level of sodium in the coal. Blohm says this attempt to produce a consistent quality for its utility customers "affects the ability of Knife River to undertake timely reclamation activities."[21] David Jordan faults production operations, arguing that the use of only one dragline to mine in all four coal pits wastes time--as the machine is constantly in motion, moving from one pit to the next instead of digging--and noting that, to ensure that coal has been exposed and can be removed in case the dragline breaks down or must be stopped for maintenance, the company exposes coal four to six months before removal.[22]

Grading, when it is done, approximates the original contour of the flat lands of the Gascoyne Mine area. Machinery can easily traverse any of the slopes created.

Topsoil

Topsoil and subsoil abound at Gascoyne, but although removal is performed well, the subsequent use of these soils has been poor. Topsoil and subsoil are removed to a depth commensurate with state law.[23] Since July 1975, Knife River has been removing topsoil and subsoil in separate operations and stockpiling them separately.[24] These are sound practices (required by North Dakota law), but others that are important for efficient handling of soils are not employed. Immediate haulback of topsoil and subsoil is not practiced--even though the company stated in one of its mining and reclamation plans that "ordinarily, the operator will haul soil materials [topsoil and subsoil] that are removed directly from the area to an area where the overburden has been leveled and thus spread the soil materials in the same operation as when it is removed."[25]

As of June 1979, no topsoil or subsoil had been replaced on any graded lands at the Gascoyne Mine. David Jordan reported that some stockpiles are likely to remain in place for 15-20 years before the material is respread on graded areas.[26] The stockpiles are generally well vegetated and resist erosion. However, the PSC observed in April 1979 that newly placed stockpiles were in need of stabilization.[27] Curtis Blohm reports that these stockpiles were seeded and stabilized in the fall of 1979 and that North Dakota laws allowed this delay between placement of the stockpile

and stabilization because the company was still placing soils in
the stockpiles.[28]

Toxic Materials

The only potential toxic or sodic problem would be posed by
the occurrence of salts in the overburden. Sodium levels are
deemed "no serious problem"[29] by Paul Packer, a scientist with
the Forestry Sciences Laboratory in Logan, Utah, who has studied the site. But Knife River's Jordan observes that seeping
salt problems will probably occur when there is no vegetation on
the land to trap water at the surface. He says it is common in
North Dakota that when land is left fallow, or vegetation is just
being established, water will infiltrate to layers of salt below the
root zone and bring this salt to the surface.[30] This raises the
risk of harming vegetation that cannot live in a salty environment. This situation will require careful monitoring, especially
if irrigation is to be used at the mine in the future (see Amendments and Maintenance).

WATER QUALITY

Surface Water and Groundwater

Surface water and groundwater control are largely intercon-
nected at the Gascoyne Mine. As coal is being mined, the aquifer in the Harmon Lignite Bed coal seam discharges water into
the mine pit. It is pumped into sedimentation ponds after being
treated by an anion-cation chemical process, which speeds the
rate at which sediment settles out of the water. The mine site
has 18 sediment ponds and five anion-cation injection systems
available for use. These chemical treatment facilities are housed
in trailers that can be moved from one mine pit to the next if
needed. Water pumped from the mine pit flows into the trailers,
where first an anion solution, then a cation solution, are injected
into the water pipe. The water then flows into a sediment pond.
The process by which sediment drops is completed within hours.[31]
A great quantity of water passes through the sedimentation
system at the Gascoyne Mine each day. Knife River reports that
maximum discharge from the mine pits could exceed 50,000 gallons per day at certain times of the year.[32] The Public Service
Commission reports that water flows from the coal seam aquifer
all year.[33] The 18 sediment ponds have a total holding capacity
of 215 acre-feet of water,[34] and they discharge fairly constantly.
The sediment ponds at the Gascoyne Mine are quite well
maintained. They show few signs of erosion, and when problems
occur, the company is quick to fix them.[35] The OSM has found
that "sediment ponds exhibit a consistent growth of herbacious

vegetation above the water,"[36] although minor problems have
been witnessed by the Public Service Commission and the U.S.
Geological Survey. Winds occasionally whip the water in the
pond against the sides, causing erosion,[37] and sometimes the
ponds overflow and erosion occurs along the sides.[38] In general
the company is working hard to control the excessive amounts of
water on the mine site.

Water discharged from the mine site occasionally exceeds the
limit for total suspended solids (TSS) allowed in the mine's discharge permit. Overload of the chemical treatment facilities,
above-average snow or rainfalls, and rapid melting of snow in
the spring affect the quality of water discharged from the site.[39]
Since the anion-cation system was established in the fall of 1977,
TSS discharge limits were exceeded in every quarter from January 1978 through October 1978.[40] The North Dakota Health Department has issued violation notices. However, INFORM has
not found that any fines were levied by the state. It appears
that when the anion-cation system is working, most water discharge problems are controlled, but any one of the contingencies
mentioned above can upset the delicate balance struck by this
system.

The U.S. Geological Survey monitors four observation wells
in the Gascoyne Mine area.[41] Four private wells also exist within
a half-mile of the mine, but the company reports that all four tap
aquifers below the Harmon Lignite Bed.[42] The company also reports that USGS monitoring has found a lowering of well-water
levels in the area. Curtis Blohm says, "The quality of water, to
our knowledge, has not been diminished....The quantity of
groundwater in the area did diminish to some degree...however,
information from USGS hydrographs indicated that the static
level of the shallow groundwater aquifer has stabilized during
the last $2\frac{1}{2}$ to 3 years."[43] This does not mean that the level has
returned to its previous height--only that it has stopped dropping.

Alluvial Valley Floors

No alluvial valley floors have been identified in the area, according to the U.S. Geological Survey and Knife River.[44]

Erosion

Erosion is not a major problem at Gascoyne. Although the
overburden is made up of very fine materials and can be easily
carried away in flowing water, this usually occurs only in the
spring, when snow melts and runs off into mine pits or sediment
ponds.[45] A Public Service Commission inspector told INFORM
the mine had few erosion problems, and these were largely con-

fined to mine pits and some topsoil and subsoil piles.[46] (In fact, all production in the White mining pit had ceased indefinitely during INFORM's visit in June 1979 because of the slumping of spoils into the pit from the cuts made in the earth.) However, there has been no sedimentation off the site.

SEEDING AND AMENDMENTS

Seeding Procedures

Seeding cannot be done at the Gascoyne Mine because so little land has been prepared for revegetation. As of January 1980 only 4 percent of the 1,105 acres of land that had been disturbed by mining operations had been planted.[47] In 1977, 76 acres of land were disturbed and no land was seeded. In 1978, 90 acres of land were disturbed and no land seeded.[48] Grading is now under way at the mine, and final grading before replacement of subsoil and topsoil was to be finished early in 1980 (see Earthmoving).

As Jordan explains it, he and his staff would like to seed annual grains, which would serve as a standing mulch, before the seeding of grasses. After seeding a mixture dominated by warm-season grasses in the fall, Jordan would allow the grain mulch and warm-season grasses to grow for about two years. Then the company would seed more cool-season plants, crowding out the grain and establishing both warm- and cool-season plants on the same piece of land. His reasoning is that it is difficult to establish both grasses simultaneously because cool-season grasses germinate sooner and outcompete warm-season grasses, so efforts should be directed toward establishing warm-season plants first.[49] The plan is innovative; many revegetation programs have problems establishing warm-season varieties. Using a grain mulch to protect the land in the cooler seasons would prevent erosion. But Jordan says the company management is unwilling to try this plan because of the expense and time involved in two seedings.[50]

Knife River--when it seeds at all--uses two types of seed mixtures. If land is to be returned to grazing or hay production, the seed mixture consists of five native plants (two warm-season grasses, three cool-season grasses). If the land is to be returned to crop production, under North Dakota law it must first be revegetated to grass for two years. For this purpose, Knife River seeds two introduced grasses and two introduced legume species. The grasses provide a vegetative cover sooner and are then displaced by the grain that is seeded two years later. Legumes help by supplying nitrogen to the soil as they grow; the nitrogen remains available to the later crop. Stock-

TABLE A

1979 SEED MIXTURE
FOR LANDS BEING RECLAIMED AT GASCOYNE

Plant Species	Introduced/ Native	Cool/ Warm	Pounds Applied Per Acre
Grasslands*			
Grasses			
Sideoats grama	N	W	1.0
Little bluestem	N	W	1.0
Western wheatgrass	N	C	6.0
Needlegrass	N	C	5.0
Slender wheatgrass	N	C	4.0
			17.0 Pure live seed
Cropland†			
Grasses			
Tall wheatgrass	I	C	4.0
Pubescent wheatgrass	I	C	3.0
Legumes			
Alfalfa	I	C-W	2.0
Sweetclover	I	W	1.0- 1.5
			10.0-10.5 Pure live seed

*Seed mixture used when final land use is to be grassland graz-
ing

†Seed mixture used when final land use is to be production of
crops such as wheat, barley, oats, etc.

Source: Knife River Coal Mining Company

piles are seeded for protection with annual grains, such as oats
or rye.[51] Seed amounts are unvaried on all portions of the mine
site, as the topography is basically uniform, and no research is
being conducted on different seed mixtures.

At most mine sites, seed mixtures are constantly altered as
experience with them grows, and some companies conduct experiments with different plants. As little seeding is being done at
Gascoyne, not much is being added to the company's base of information to help determine the best mixture for the site.

Amendments and Maintenance

Topsoil is not prepared much before planting. Furrowing is
not practiced along the contour, as there is little need to slow or
trap the flow of water down slopes that are almost horizontal. All
seed is planted with a drill seeder operating along the contour,
and mulching--which has never been part of Knife River's reclamation program--will now be undertaken on lands being reclaimed
to comply with federal requirements.[52]
Fertilizer and irrigation tests have been conducted. Results
of the fertilizer experiments were inconclusive, according to Jordan, because application of different amounts of fertilizer yielded
identical results.[53] Irrigation tests are being conducted to determine whether the effect of water pumped from the mine pits
will be different from that of well-water on lands where it is used
for irrigation. Curtis Blohm says the company is awaiting release
of a report by the U.S. Department of the Interior's Science and
Education Administration Research Station on the use of irrigation. "Once the report is published, Knife River will probably
undertake a review to see if it would be feasible to incorporate
irrigation into its reclamation program."[54] Jordan fears irrigation
might cause increased movement of salts upward into root zones
from the overburden. He believes the long-term implications of
irrigating on the Gascoyne site must be measured before being
implemented on all lands to be reclaimed.[55]

Trees and Shrubs

Knife River is not planting trees and shrubs as part of its
reclamation program, which calls for re-establishing grazing and
croplands, as few trees and shrubs existed prior to mining.[56]
The company once attempted to transplant trees onto reclaimed
lands. That effort failed, because it was not carefully performed,[57] but the company will try again in the same areas because it wants to establish a golf course on the land. Knife River
calls this an "experiment," but David Jordan says it is really because many employees are avid golfers who would like to host the
annual Lignite Golf Classic on reclaimed lands at the Gascoyne
Mine.[58]

VEGETATION

Because so little seeding has been done at the Gascoyne Mine
in the last few years, there is little vegetative growth to judge,
either for productivity or for control of erosion. The Department
of the Interior's Fish and Wildlife Service says lands seeded in
1973 and 1974 exhibit growth dominated by alfalfa[59] (which is not
good for year-round grazing) and some patches of wheatgrasses.

No crops have yet been seeded onto reclaimed lands at the mine.[60]
The land, as noted above, is generally free of erosion, however.

Productivity in the area ranges from 27 bushels of wheat per
acre to 54 bushels of oats and 43 bushels of barley, under land
management regimes deemed typical by the Soil Conservation
Service.[61] When alfalfa is grown for hay, lands in the area produce 1.0-1.9 tons per acre.[62] The company feels that Soil Conservation Service data are accurate for its mine site, and it has
developed no other data.[63]

One area at the mine has been cut for hay in the past. Some
20-25 acres of land produced 2,760 pounds of hay in 1975 and
2,000 pounds of hay in 1976, according to Curtis Blohm. No
topsoil was replaced on this land, and it is producing hay at
rates approximating those found on surrounding undisturbed
lands.[64] No further measurements have been taken.[65]

Owing to its lagging reclamation efforts at Gascoyne, the
company has been in no position to seek release of any of its reclaimed lands from reclamation bond. However, some lands have
been released from bond under North Dakota reclamation laws
from the early 1970s after three growing seasons at the company's
Beulah Mine.[66] Lands have also been released at the other two
North Dakota mines in the INFORM sample (see Glenharold, Indian Head profiles).

WILDLIFE

Knife River is not attempting to reclaim any of its land to
wildlife habitat because the area is generally used for crop production and only the 10 percent of land used for grazing might
have a secondary use as wildlife habitat. In the future the company may place a few shelter belts of trees along some reclaimed
areas for animal use.[67]

RECLAMATION COSTS

TABLE B*

RECLAMATION COSTS AT GASCOYNE

Operation	Cost Per Acre		Percentage of Total Cost
Planning and Design			3%
Bonding†	$ 82		
Earthmoving, Revegetation			97%
Topsoil removal∇	218–	718	
Grading, seedbed preparation, seeding, and fertilizing	2,200		
Total Cost:	$2,500–$3,000		

These costs and percentages are not strictly comparable to those of the same operation or activity at other mines. See the Cost of Reclamation chapter for a detailed explanation.

†*Calculated by INFORM based on $1,006,621 reclamation bond for 1,228 acres under permit.*

∇*Though Knife River reported this cost to be $0.52 per cubic yard of topsoil removed, INFORM calculated the figure shown based on the total reclamation cost provided by the company.*

Source: Knife River Coal Mining Company

Glenharold/CONSOL/ND

Mine: Glenharold Mine

Location: Stanton (population 520), North Dakota, in Mercer (population 6,800) and Oliver (population 2,320) counties

Operating company: Consolidation Coal Company (CONSOL)

Parent company: Continental Oil Company

Parent company revenues: 1977, $4.05 billion; 1978, $9.87 billion

Parent company net income: 1977, $381 million; 1978, $451 million

First year of production: 1966

Coal production in 1978: 3,686,094 tons

Acres for which a mining permit has been issued: 2,700 as of December 31, 1978

State regulatory authority: North Dakota Public Service Commission

Reclamation bond posted: NA

Reclamation cost: reclamation cost per acre, NA*; reclamation cost per ton of coal, NA*

Expansion plans at mine: By 1980, CONSOL plans to expand the mine from two to eight separate working pits, each considered a separate mining operation.

Additional acres for which the company will seek a permit: 5,300 over the life of the mine

Parent company U.S. coal reserves: Owns or leases 13.7 billion tons of coal reserves

Development and expansion plans for surface coal mines in western United States: By 1981, CONSOL plans to produce 5 million tons of coal per year at one surface mine in Wyoming; by 1985, one mine in New Mexico should provide 6.5 million tons of coal annually, and by 1986, four surface mines in North Dakota should produce 14.1 million tons of coal per year.

NA = Not available

*CONSOL considers this proprietary information.

Sources: Continental Oil Company, 1978 Annual Report; *1977 and 1979 Keystone Coal Industry Manual; "New Coal Mine Development Survey 1978-1987," Coal Age, February 1979;* Rand McNally Road Atlas, *54th Edition; Written communication from Gary Slagel, Manager of Reclamation, Consolidation Coal Company to INFORM, June 8, 1979 and October 10, 1979*

SITE DESCRIPTION

The Glenharold Mine is in central western North Dakota on the southern border of Mercer County, along the north border of Oliver County. In 1978 this mine produced more coal than all but ten other coal mines in the United States. Gently to steeply rolling hills are dissected by water drainages. Some of these are called hardwood draws--areas into which rain or melting snow drains. More topsoil accumulates in these draws than on surrounding lands. Because water and topsoil collect in these draws, they usually contain dense stands of trees and bushes, including many hardwoods. Hardwood draws provide excellent wildlife habitats, affording food and shelter for animals.

Average annual precipitation in Mercer and Oliver counties has ranged from 10.97 - 21.74 inches over the past 30 years. Average precipitation is 16.9 inches yearly. About 25 percent of the yearly precipitation is snow; the rest is rain.[1] The land is frozen from about September 20 through May 20,[2] but for the rest of the year, it can be either seeded or scraped of its topsoil.

Of the 2,700 acres of land now permitted for mining at Glenharold, approximately 30 percent was agricultural before mining began; 70 percent was rangeland. Before mining began, about 25 to 30 percent of land at the mine consisted of wooded areas.[3] Most of the land mined today is rangeland, but more cropland will be disturbed in the future.[4]

Most land in this part of North Dakota is not prime agricultural land as defined by the Department of Agriculture's Soil Conservation Service. The U.S. Bureau of Land Management

considers none of the land at the Glenharold Mine to be prime
agricultural land.[5] However, many of the soils usually found in
small patches of this part of North Dakota do produce large
amounts of grains and grasses relative to the rest of the West.
An inspector for the North Dakota Public Service Commission
labeled these prime agricultural soils but indicated that in most
cases they do not exist in sufficient quantity over a large
enough area to constitute a "logical farming unit"; that is, they
could not be economically farmed by themselves.[6]

OVERVIEW

Consolidation Coal Company (CONSOL), owner and operator
of the Glenharold Mine, has recently begun a good reclamation
program at the site--where reclamation has lagged far behind
mining, and erosion and water discharge problems have persisted.

CONSOL's earthmoving practices are fair to good. However,
the time between coal production/ grading and respreading of
topsoil and subsoil prevents revegetation from following mining
more closely. By 1979 59 percent of all the lands disturbed at
the mine site had been graded but only 30 percent of disturbed
lands had been seeded.[7] Unreclaimed lands are susceptible to
erosion because they lack vegetative cover. Water infiltrating
the soils of such land has a greater chance of bringing salts from
the overburden to the surface, where they may cause problems
in establishing vegetation when seeding is completed. Topsoil
and subsoil quality declines when these soils remain stockpiled.

Although most grading at this mine blends graded lands with
surrounding landscape, CONSOL has radically changed the topography of at least one 20-acre area, altering the patterns for
water drainage on that land. The state Public Service Commission is now assessing this change.

Integrated planning between production and reclamation
operations could allow reclamation to follow land disturbance more
closely and minimize the impact of the expanding mining at the
Glenharold Mine.

CONSOL has improved its seeding and amendment program at
the Glenharold Mine in the last 2 years. Responding to legisla-
tive requirements, the company hired a range scientist in 1979 to
coordinate much of the revegetation program at Glenharold and
to evaluate its success. CONSOL's plans to use four different
mixtures of seeds, along with fertilization and irrigation, reflect
a strong company commitment to land reclamation. The company
has responded quickly to federal regulations by hiring technical
staff experienced in western reclamation techniques. CONSOL is
now growing warm-season plants on irrigated lands, and cool-
season plants on non-irrigated lands, both good procedures.

However, the program is young; little land has actually been reclaimed by these techniques, and it is too early to assess the results. They could provide the basis of a successful seeding and
amendment program. CONSOL is one of four companies in the
INFORM sample that is varying its seed mixtures based on special characteristics in its seeding program such as soil type,
slope aspect or application of amendments.

In the past CONSOL has not kept the quality of water discharged from the mine site within state and federal requirements.
Erosion on the site has increased the amount of sediment carried
by water flowing across the mine to levels that exceed the requirements of regulatory agencies. Waters containing high sediment levels have occasionally been discharged at the mine and in
at least one instance have disturbed the property of adjacent
landowners. CONSOL has also failed to seek permits (for discharging water) within allotted time limits, and the company has
reported water discharges and precipitation levels inaccurately.
In 1979 however, staff at the mine prepared and presented a
new water management program to the federal Office of Surface
Mining (OSM) and the North Dakota Public Service Commission,
agencies that regulate reclamation activities at the mine, outlin-
ing how it plans to avoid further problems.

The company has only recently begun to tackle its persistent
erosion problems. A visit to the mine in June 1979 confirmed
that erosion problems still exist at the site and on adjacent lands
where water is discharged from the mine's sediment ponds. Erosion of haul roads, graded and topsoiled areas, spoilpiles, and
topsoil piles have caused many minor sedimentation problems in
the past. The company's attempts to use terraces to slow the
flow of water downhill have met with little success, but according
to a recent OSM evaluation of areas reclaimed in the last few
years, "conscientious revegetation operations can establish consistent stands of vegetative cover"[8] at the mine; thus, ultimate
control of erosion by vegetative cover should be achievable.

The long-term effect of mining at Glenharold on surface
water and groundwater is unknown. Work at the mine already
disrupts a local aquifer. But CONSOL has collected little data
on the effects of this aspect of its mining operations; neither
have state or federal regulators.

Actions of state and federal regulators have led to improved
reclamation practices at the mine. The federal Office of Surface
Mining (OSM), created in 1978, inspected the Glenharold Mine
four times between June 1978 and October 1979. As a result,
CONSOL upgraded the mine's sedimentation-pond system in 1979.
Since 1977, the North Dakota Public Service Commission (PSC), the
state regulator, has inspected the mine site about twice a month.
The Commission has been generally effective in monitoring mining
and reclamation practices at the mine, but until recently it has

had little impact in controlling the amount of sediment that is
sometimes discharged in waters leaving the mine site. A PSC
regulation prohibiting placement of topsoil taken from one surface owner's land on the surface of another owner's land has
slowed reclamation. The Commission is writing new regulations
in an effort to gain regulatory authority under the federal Surface Mining Control and Reclamation Act. The Commission plans
to submit these regulations to OSM in 1980. In 1978 and 1979,
CONSOL paid $12,000 in fines to OSM and the Public Service
Commission for discharging water from the mine without a permit
and for discharging water containing sediment levels above those
allowed by law.[9]

Although CONSOL is upgrading its revegetation and water-
control programs, the company still disturbs many more acres of
land annually than it reclaims. As the mine expands to eight
separate mining pits operating simultaneously, more lands will be
disturbed and will remain unreclaimed for several years if present procedures continue. This situation could present even
greater problems for CONSOL in maintaining water quality both
on and off the mine site, as more unvegetated soils will be exposed to the erosive forces of rain and melting snow. Mining at
Glenharold can accelerate environmental degradation by its sheer
size. Only future performance will determine whether CONSOL
will cope well with its growing responsibility.

EARTHMOVING

Grading

Grading does not follow mining and land disturbance closely
at the Glenharold Mine. The company reports that mining operations disturb 253 acres per year on the average.[10] In 1977, 232
acres were mined and 200 acres were graded. In 1978, 277 acres
were mined and 145 acres were graded.[11] The more soil that
remains unvegetated and exposed to the erosive forces of rain
and water flow, the greater the chances that water quality problems will continue. CONSOL has experienced such problems in
the past. At this mine, as at North American Coal's Indian Head
Mine, after mining begins in an area, all coal in that area must
be removed before grading begins. As North Dakota's regulations require that all the land be graded to original contours,
spoil from the first cut of the operation must be graded over all
the land being graded so that no hill remains, and so that the
last pit in a series of mine pits is filled with spoil from preceding
spoil ridges. Many mining areas are not simple rows of cuts
made in the earth to remove coal. The pits are shaped in varying patterns--sometimes in a horseshoe shape, for example. As
a result, CONSOL grades the entire series of pits to blend with
the topography of the surrounding area. This usually means

that up to 18 months can elapse between the end of mining and
the end of grading.[12]

Although most grading at the mine approximates the rolling
contour of the surrounding land, a 20-acre area graded in the
last two years is much steeper than the original pre-mining topography. This graded land is 30-35 feet lower than the surrounding land and is like a large crater. The company wishes to
use this land, initially farm and grazing land, as a drainage area
for surrounding lands and as grassland.[13] Water draining from
lands surrounding this section would flow through this lower
elevation to create a new pattern of surface water flow in the
area. The impact of this change is not yet measurable. The
North Dakota Public Service Commission and/or OSM will have to
approve a modification to the mine plan for the Glenharold Mine
before such a change in topography and land use can become
part of the final reclamation plan. CONSOL says it plans to submit a request to modify its reclamation plan in early 1980.[14] But
the company told INFORM it could grade this area to conform to
the surrounding lands, if necessary.[15]

Topsoil

As required by North Dakota regulations, topsoil and subsoil are removed in two separate operations at the Glenharold
Mine, similar to practices at the Indian Head Mine and Knife
River Coal's Gascoyne Mine.[16] Although state regulations require that at least five feet of topsoil and subsoil be removed--and
that it be either stockpiled or replaced on graded lands--before
mining begins, in some areas of the Glenharold Mine, the company states that up to 15 feet of this material may be removed.[17]
At most of the areas INFORM visited in the summer of 1979, only
five feet of topsoil and subsoil had been saved, but the company
reports that it salvages all material suitable for plant growth to
improve the chances for successful revegetation.[18]

In 1977 CONSOL placed spoil material without first salvaging the topsoil or receiving permission from the Public Service
Commission to leave it in place. CONSOL removed both the spoil
and topsoil within five months of receiving notice of the problem
from state regulators. But the company was not cited for these
practices.[19]

Topsoil and subsoil stockpiles are carefully maintained on
the site and are generally well vegetated and free of erosion.
Some of the older stockpiles have 45-degree side slopes, which
make revegetation difficult. Newer, less steep stockpiles are
being built, which should prevent erosion and simplify the establishment of plant cover.[20] However, when these stockpiles
are removed so that topsoil can be replaced on graded lands,
some mixing of subsoil and topsoil may occur because of the

proximity of one pile to another. This might defeat the original purpose of segregating the soils.

The long period between mining and replacing topsoil at the Glenharold Mine may reduce the reclamation program's chances of success. Most of the topsoil and subsoil has been stockpiled at the mine for four to five years before being replaced on graded lands.[21] As noted, because of the way the mining operations proceed, a whole area is mined before grading begins. In some years, no topsoil is respread; in others, the company replaces topsoil over many more acres than were disturbed.[22].

Replacing topsoil is made more difficult at Glenharold because of the Public Service Commission regulations. North Dakota is the only western state with a regulation prohibiting topsoil and subsoil removed from the surface of one individual's land from being replaced on another's.[23] Mining at Glenharold cuts across the lands of many owners, and reclamation produces graded areas where topsoil recently removed from other areas could be placed.[24]

The overburden atop the coal seam expands in volume when it is removed. As a result, once spoil material is replaced and graded, the company allows it to settle for up to one year before topsoil is replaced. Upon settling, cracks may appear in the ground, hampering not only seeding and mulching operations but farming as well. The cracks may be large enough to swallow a tractor or a cow, and they must be filled and regraded before topsoil and subsoil are respread.[25] The Public Service Commission said that most coal operators in North Dakota wait one year after they have graded an area before replacing subsoil and topsoil and beginning seeding operations, due to special characteristics of the overburden, which cause it to expand during mining and then settle after replacement.[26]

Settling overburden has also produced small depressions in graded lands. These depressions, if left unfilled, sometimes collect pools of water, which can hamper planting and harvesting and pose problems for operators of agricultural machinery. Machinery cannot operate in wet areas, since it may either become immobilized or damage the land. Water may also cause salts and contaminants from spoil under the topsoil and subsoil to rise to the surface. In June 1979 INFORM observed lands graded in 1973 and 1974 containing water-filled depressions.

Toxic Material

Sodium in the overburden and the use of irrigation could hurt CONSOL's chances for reclaiming land at the mine. High levels of sodium salts that could harm vegetation unable to tolerate salty environments occur in much of the overburden. However, INFORM found no evidence that these salts were migrating

to the surface of reclaimed lands or into the root zones of plants
in revegetated areas. Irrigation of lands where vegetation is not
dense may increase the likelihood of these salts' reaching the
upper layers of topsoil and subsoil, as sufficient plant materials
will not be available to hold the water in the root zone, allowing
the water to penetrate deeper to salty layers.

WATER QUALITY

Surface Water and Groundwater

The Glenharold Mine in June 1979 had an extensive water-
control system consisting of 24 sediment ponds. The company
plans to build six additional ponds in each of the next two years
to provide water-retention facilities for the eight mine pits
scheduled to be opened by 1980.[27] The mine has also received a
permit from the U.S. Environmental Protection Agency (EPA)
allowing it to discharge pond water through 32 different points,
under the National Pollutant Discharge Elimination System
(NPDES) program.[28] Under the terms of the permit, discharges
must be monitored for any precipitation event which is less than
a 10-year/24-hour storm (the maximum amount of rain or the
equivalent of this quantity of rain expected to fall in a 24-hour
period once every 10 years) defined for the mine as 3.1 inches
of rain.[29] Because the main coal seam at Glenharold is an aquifer, water seeps into the mine pits through the highwall or
working face. As a result, water must constantly be pumped
from the mine pits into sediment ponds, keeping the ponds filled
for many months.[30]

In April 1979 all sediment ponds at the Glenharold Mine
were being rebuilt or replaced by new ponds that comply with
OSM specifications.[31] Ken Wangerud, manager of reclamation at
Glenharold, says the new ponds have been designed to hold sediment collected up to 3 years before they must be cleaned.[32] An
August 1978 OSM inspection report noted that no vegetation was
present on the banks of sediment-control ponds, and that the
company needed to add vegetation to stabilize the surface areas
of the ponds to prevent erosion.[33] In June 1979 INFORM saw no
vegetation growing on the banks.

The company plans to improve its water control system by
installing a piece of equipment called a "drawdown." Acting like
a pump floating in the middle of the sediment ponds, this equipment will draw water from the pond's surface, where it is clean-
est, and discharge it through a hose. Hence, pure water can
leave even when the overall level of water in the ponds is not
high enough to reach the discharge exit pipes normally used.
Before rainstorms or in thaws, drawdowns would increase the
pond capacity to hold newly entering water, as water levels can
be lowered before they occur.[34] No drawdowns had been built

as of June 1979, but in the fall of 1979, the company reported
they were in place.[35]

CONSOL has experienced numerous problems with water being discharged from the Glenharold Mine, including sediment
levels much higher than the mine's permit allows. The problems
were most severe in the summer and fall of 1977 and 1978. A discharge during a storm in September 1977 far exceeded all levels
for Total Suspended Solids (TSS) allowable for the mine. In that
storm, EPA and Public Service Commission representatives took
water samples downstream from NPDES discharge points as well
as at the discharge points themselves. At one of these points,
TSS levels were 45,925 milligrams per liter (mg/liter) and 122,075
mg/liter. One and one-quarter miles downstream from these discharge points, sediment levels dropped to 42,756 mg/liter and
48,275 mg/liter.[36] Under the mine's NPDES permit, the maximum
permissible TSS level at the mine is 45 mg/liter.[37] Yet the Discharge Monitoring Report prepared by CONSOL for September
through November 1977 stated that no discharge occurred in that
period even though climatological data showed that over 4 days
in September 2.65 inches of rain had fallen.[38] In a June-August
1977 report, the company erroneously referred to this storm as a
3-inch rainfall occurring over 2 days.[39]

Again, in 1978, the Discharge Monitoring Report (DMR)
CONSOL filed with the EPA for June 1-August 31 said no discharge had occurred at the mine except during a storm considered by the company to be in excess of the 10-year/24-hour criterion of 3.1 inches. Climatological data attached to the DMR
showed that over 5 days from June 24 to June 28, only 1.58
inches of rain fell. Other storms in the 3-month period included
less than 1 inch of rain over 2 days in July, 1.4 inches over another 2 days in July, and 1 inch over 2 days in August.[40] A
Public Service Commission inspection report in July 1978 said,
"The series pond system was receiving inflow and was also discharging at the time of inspection,"[41] almost 2 weeks after the
1.4-inch storm. In June 1978 a member of the Public Service
Commission's staff sampled the water discharged at that time and
found TSS above the permitted limits.[42] He also found that
water was being discharged at points that were not included in
the NPDES permit. According to the North Dakota Department
of Health, water had been discharged from these points for at
least a month before the company requested permission to use
them.[43]

On June 28, 1978, CONSOL paid a $1,000 fine for violating
water discharge limits.[44] In November 1978 OSM also fined the
company $3,000 for discharging water without a permit.[45] That
same month, the North Dakota Health Department fined CONSOL
$18,000 for violating the TSS discharge limits. The Health De-

Discharge point off the mine site has eroded because of poor stabilization of the land where CONSOL has allowed water to flow onto neighboring field. (CONSOL did not want INFORM to photograph this scene.)

partment suspended $6,000 of this fine but said it would reassess
this decision if violations occurred at the same discharge points
in the future.[46] In the summer of 1979 the company paid the
fine after court proceedings.[47]

The company disagrees with the EPA guidelines. Both Steve
Young, vice president for government affairs at CONSOL, and
Gary Slagel, the company's manager of reclamation, said that
the effluent guidelines established for Total Suspended Solids
are too strict. They contend that TSS levels in waters around
the mine and in waters flowing from undisturbed lands in the
area are much higher than these effluent limits.[48] In late 1979
the U.S. EPA was re-evaluating its effluent guidelines. However, according to Frank Beaver, an inspector for the EPA in
North Dakota, the limits that the federal agency has established
are based on the capabilities of the "best available control technology" to protect water quality.[49] Robert Walline, a chemical
engineer with the EPA's regional office in Denver, said that the
EPA bases its standards on a "well-run" rather than exemplary
operation's ability to control water runoff.[50] Beaver added that
in North Dakota, "most of the companies are complying with the
law."[51]

The mine pits at Glenharold Mine cut through major water
drainages and hardwood draws. More of these areas will be
mined.[52] The company is reclaiming hardwood draws, and CON-
SOL reports that although some stream channels may be altered,
proper grading will restore the same basic drainage pattern to
the mine area.[53] If the company can re-establish vegetation in
the hardwood draws that it will eliminate and avoid disruption of
water drainage patterns, mining operations should create no
long-term environmental problems. This would represent a
strong reclamation success, as a well developed plant and animal
community of grasses, shrubs and trees will have been created.
Moreover, lands where mining is being prevented because of the
value of these hardwood draws might then be opened for development.

Little information is available on the overall effect of mining
operations at Glenharold upon either groundwater or surface
water in the area. However, state and federal inspectors have
noted problems. Mining disrupts an aquifer that provides water
to local farmers. Yet, as of fall 1979, CONSOL had not yet
analyzed available information on the quality and quantity of
either on- or off-site water in the area.[54] Gary Slagel reports
that the company is trying to collect groundwater data.[55] As of
October 1979, the company had sunk 30 monitoring wells at the
mine and planned 24 more. By late 1979 Slagel said the company
had gathered preliminary data and was beginning to reach conclusions based on information collected at the wells.[56]

Alluvial Valley Floors

The company and the Public Service Commission agree that there are no alluvial valleys on the mine site.[57]

Erosion

In April 1979, at the request of the Public Service Commission and OSM, CONSOL presented its new water management program to reduce the sedimentation produced by discharging water.[58] This program, which includes the newly built sediment ponds and drawdown devices mentioned earlier, offers no strategies for controlling erosion on the land, a logical first step in controlling the quality of water on the mine site.

The company's past record of correcting erosion problems has been unimpressive. A memorandum from the Public Service Commission's Reclamation Division to Ray Walton, the Commission's legal counsel, noted nine incidents of sedimentation and erosion at the mine in June, August, and September 1977. These incidents included erosion and sedimentation of ephemeral streams running along a mine haul road, sedimentation of a major drainage (the natural path of overland water flow), and wind and water erosion of stockpiles, which carried sediment onto undisturbed lands and lands that had been graded and topsoiled.[59] During a September 1977 rainstorm, a Public Service Commission inspector noted "gross runoff was found coming off haul roads and nonactive mining area spoil piles and top soil storage piles."[60] This runoff raised sediment levels in water discharged from the mine site far above permissible limits. According to the memorandum, the company had taken no corrective action up to 4 months after state inspectors had identified the problem.[61]

The Public Service Commission has also criticized CONSOL's delay in clearing away spoil materials that had eroded and were carried onto stockpiles in a section of the mine. In a letter of July 15, 1977, the Commission told the company to correct the problem. CONSOL failed to act, and the Commission sent a followup reminder three months later. The company then tried to remedy the problem. CONSOL said the delay was caused by management's inability to decide whether the reclamation or production divisions would incur the cost of the cleanup.[62] In October 1979 Gary Slagel told INFORM that all problems described above had been rectified.[63]

Sediment flowing from the Glenharold Mine has damaged the lands of local ranchers. After a storm in September 1978, one rancher found up to 3 feet of sand washed over prime river-bottom land on which alfalfa was growing.[64] A spring used to water livestock was almost buried by sediment after the storm. The Public Service Commission found that sediment from the mine

had accumulated along the banks of the stream over many months
--up to 5 feet deep--in one quarter-mile stretch. During the
rainstorm, this sediment was deposited on at least one acre of
the rancher's fields.[65]

CONSOL says that since January 1978, no more than 3 weeks
have elapsed between the detection of erosion damage and attempts to repair it.[66] In July 1978 the Public Service Commission found that graded slopes south of a haul road in one section
of the mine had eroded, but that the company had repaired them
by regrading, mulching, and installing a netting over the area
to hold mulch in place.[67] In August 1978 OSM said stockpiles
were protected from wind and water erosion by vegetative cover.[68] However, new water management strategies have not yet
solved all CONSOL's water discharge and erosion problems. In
a tour of the mine site in June 1979, an INFORM researcher was
shown one of the new sediment ponds built as part of the new
water management plan. The pond was discharging at the time;
water was traveling along a U-shaped piece of an old conveyor
belt to the edge of the mine site, where it was discharged onto
land adjacent to the mine site. A hole approximately 6 feet in
diameter and 4 feet deep had been eroded into the ground where
the water was discharged. Numerous small rocks in the center
of the hole did not prevent the runoff from gouging the land.
This may be an isolated incident, but it does not speak well for
a company that has acknowledged its problems with controlling
water discharges and is trying to prevent their recurrence.
CONSOL representatives would not allow INFORM to photograph
this hole.

In most areas of the mine the company has established
enough vegetation to prevent erosion. Topsoil stockpiles are
also generally free of erosion. In one instance, steep-sided
stockpiles were stabilized by using a chemical glue to hold seeds
and mulch in place. This technique prevented loss of seed or
mulch from the fall of one year to the spring of the next, allowing the seeds to germinate and grow.[69] However, steeper,
graded slopes are largely barren. CONSOL is trying to use terraces to slow the flow of water down slopes, but in April 1979,
the Public Service Commission reported that these efforts had
met with very little success.[70]

SEEDING AND AMENDMENTS

Seeding Procedures

In 1977, 232 acres of land were mined at the Glenharold
Mine, 200 acres were graded and 75 acres were seeded. In 1978,
277 acres were mined, 145 acres were graded and no lands were
seeded.[71] These disparities illustrate CONSOL's fluctuating rec-

lamation schedule. Mining disturbs an average of 253 acres[72]
annually while reclamation proceeds at an erratic pace: in 1977
and 1978, CONSOL seeded only 22 percent of all the lands graded
in those years.[73]

The seed mix at Glenharold consists of 7 species of grass,
which are almost equally divided between cool- and warm-season
varieties (see Table A). From 21-25 pounds of this seed mix is
spread on each acre of graded land. If the land seeded is to be
cropland, a half-pound of alfalfa or sweetclover is added. Both
are introduced legumes. Their ability to fix nitrogen in the soil
is an asset, as most soils lack this nutrient. The use of only
one-half pound of these seeds in a total application of 21-25
pounds per acre should not overwhelm other grasses more suitable for grazing. North Dakota regulations require that reclaimed lands returned to crop production be seeded initially
with grasses and legumes. After 2 years, the crop to be grown
can be seeded.[74]

The Glenharold reclamation program relies primarily on 5
grasses: sideoats grama, little bluestem, blue grama, western
wheatgrass and green needlegrass. Western wheatgrass, blue
grama and green needlegrass are native cool-and warm-season
plants that dominate the seed mix used on lands not intended for
irrigation. On lands to be irrigated, sideoats grama and little
bluestem, both warm-season, native plants, dominate.[75] Because
cool-season grasses grow first, they tend to use most of the
available water. Thus CONSOL's reclamation program will seed
higher concentrations of warm-season grasses and irrigate them
to grow both cool- and warm-season grasses on reclaimed grasslands.[76]

The company uses introduced plant species most heavily on
temporary structures such as the sides of sediment ponds, along
haul roads and in water diversions that direct water from the
mine into sediment ponds. These introduced plants grow quickly
and stabilize the soil sooner than native species.[77]

CONSOL employs a drill seeder to create small furrows and
cover seeds along the contour of all reclaimed land at the mine
except topsoil and subsoil stockpiles, where the broadcast
method is employed.[78] Broadcasting is generally by hand.[79]

Amendments and Maintenance

Since 1977, CONSOL has spread mulches on land at the mine
to control erosion in areas where it is trying to establish permanent vegetation. Straw mulch is sprayed on the ground, or annual plants such as wheat or rye grass are seeded and mowed
once they have grown.[80]

The company began irrigation at the mine site in the summer
of 1979. CONSOL irrigated lands, providing 6 inches of water in

TABLE A

1979 SEED MIXTURE
FOR LANDS BEING RECLAIMED AT GLENHAROLD

(Sweetclover and Slender wheatgrass included in all plans)

Plant Species	Introduced/ Native	Cool/ Warm	Pounds Applied Per Acre	
Sweetclover	I	W	1.0	
Slender wheatgrass	N	C	2.0	
			3.0	Pure live seed
Plan 1 (lands irrigated during reclamation)				
Sideoats grama	N	W	9.0	
Little bluestem	N	W	4.0	
Blue grama	N	W	1.0	
Western wheatgrass	N	C	3.0	
Green needlegrass	N	C	2.0	
			19.0	Pure live seed
Plan 2 (lands irrigated during reclamation)				
Sideoats grama	N	W	5.5	
Little bluestem	N	W	6.5	
Blue grama	N	W	1.0	
Western wheatgrass	N	C	3.0	
Green needlegrass	N	C	2.0	
			18.0	Pure live seed
Plan 3 (lands not irrigated during reclamation)				
Sideoats grama	N	W	1.0	
Little bluestem	N	W	4.0	
Blue grama	N	W	5.0	
Western wheatgrass	N	C	9.0	
Green needlegrass	N	C	3.0	
			22.0	Pure live seed
Plan 4 (lands not irrigated during reclamation)				
Sideoats grama	N	W	1.0	
Little bluestem	N	W	4.0	
Blue grama	N	W	5.0	
Western wheatgrass	N	C	4.5	
Green needlegrass	N	C	6.0	
			20.5	Pure live seed

Source: CONSOL

the growing season in some areas where previous revegetation
attempts had failed. This added over 30 percent more water
than is available naturally in an average year. The company was
to irrigate other lands using amounts of water determined by the
success of initial experiments at the mine site.[81]

Until the summer of 1979, CONSOL had never fertilized as a
regular part of its reclamation program. However, in this period, the company fertilized all seeded lands. Using soil data and
the experience of local farmers to determine the application and
amount of fertilizer, CONSOL applied 25 pounds of organic nitrogen, 55 pounds of synthetic nitrogen, 20 pounds of organic
phosphates, and 105 pounds of super phosphates per acre.[82]
The company said these amounts are higher than the standard
amounts dictated by soil surveys and higher than amounts commonly used in this part of North Dakota.[83]

This artificial supply of nutrients and water may create a
vegetative cover that cannot sustain itself once the supply of
fertilizer ceases. In June 1979 Ken Wangerud, reclamation supervisor at the Glenharold Mine, said CONSOL's reclamation program was "shooting in the dark" in trying to determine how long
irrigation and fertilization must continue.[84] According to Rick
Williamson, the mine's range scientist, the company wants to establish vegetation, then cease maintenance of the area, but he
fears that the areas where this maintenance program is in effect
will require careful management for quite a while.[85]

CONSOL conducts no experiments in small test plots to develop its seeding and amendment program. Tests are made over
larger areas. The company determines whether to use mulch and
fertilizer based on experience, not experiment.[86] It terraces
some graded slopes. This procedure slows the flow of water down
the hillside and minimizes erosion. CONSOL usually seeds in the
fall, the generally accepted seeding time at all western mines in
this study (except those in the Southwest).[87]

Trees and Shrubs

CONSOL is transplanting trees and shrubs to re-create
woody vegetation in hardwood draws. The mine's reclamation
staff manually transplanted 40 trees from the banks of the Missouri River, which flows within 5 miles of the mine. It also
transplants trees and shrubs from nursery stock, a procedure
inferior to transplanting from the surrounding area because the
latter stock has already proven its ability to survive.[88] However,
the species being planted are compatible with plans for the site
and are similar to plants that have grown there.

VEGETATION

Short-term revegetation at Glenharold has been good. But
data on the vegetation and productivity of the mine's reclaimed
lands is insufficient to determine whether reclamation will be
successful over the long term. Seeded lands at the Glenharold
Mine range from those with enough plant cover to prevent erosion, to those with virtually none. The company has yet to analyze covered areas for density of vegetation, amount of land
covered or amount of plant forage produced. In August 1978
OSM inspectors reported that "Revegetated areas...attest to the
fact that conscientious revegetation operations can establish consistent stands of vegetative cover."[89] Reclaimed lands on which
wheat has been planted appear dense with it. CONSOL has collected little data on the productivity of these croplands.

Lands reclaimed since 1975 (over 300 acres) have enough
vegetation to control erosion but are not as densely vegetated as
surrounding, undisturbed lands.[90]

Lands seeded in 1973 and 1974 suffer from the company's
failure to respread topsoil or subsoil, an operation not then required by law. Thin soil surface is hard and difficult for water
to penetrate. Some vegetation exists on these lands; however,
the salty nature of the spoils, hardness of the soil surface and
lack of topsoil make it difficult for plants to root, making these
areas extremely unproductive and more susceptible to erosion
than any other reclaimed or undisturbed lands on the site.

The results of CONSOL's attempt to re-create hardwood
draws are still inconclusive. These areas are well maintained
and covered with enough grasses to prevent erosion, but so far
the company has not documented the growth of any shrubs or
trees, although in late summer, 1979, CONSOL reported a 75
percent survival rate.[91]

CONSOL has no information on pre-mining land productivity
or on the productivity of reclaimed lands except for a 16.5-acre
area used to grow wheat.[92] In 1979, CONSOL established four
one-hectare (10,000-square meter) plots on undisturbed lands as
reference areas on the productivity of undisturbed land.[93] The
company is collecting data on the variety of species present on
undisturbed and reclaimed lands. Information on both productivity and plant diversity should be available in the fall of 1979.[94]

CONSOL said that the 16.5 acres of reclaimed wheat lands,
seeded in 1976, produced 17 bushels per acre in its first yield.
The company considered this a good yield as the year was especially dry.[95] CONSOL has also reported yields of 20 and 30 bushels per acre from wheat lands that have been harvested in 2
years when precipitation was at least average.[96] Company officials say CONSOL should consistently be able to produce 31
bushels of wheat per acre on the land beginning in 1979,[97] well

in excess of the 22 bushels the U.S. Soil Conservation Service
has determined could be grown in the area using standard farming practices.[98] However, Rick Williamson, CONSOL's range scientist at the mine, says the Soil Conservation Service's figures
are low, and that most farms produce 29-31 bushels of wheat per
acre under a management scheme similar to that practiced at the
Glenharold Mine. In fact, 1979 production at the mine rose to
32.9 bushels per acre,[99] a good yield.

The productivity of mined land depends on the quality of its
topsoils. Rick Williamson says each acre annually produces from
700 to 2,000 pounds of dry-forage (plant materials that are dried
before they are weighed in order to assess accurately the
amounts of organic material produced).[100] A Soil Conservation
Service survey of the soils of nearby off-site lands found that
under a land maintenance program similar to that practiced in
central-western North Dakota, land could annually produce about
22 bushels of wheat, 35 bushels of barley or 39 bushels of oats
per acre. The Soil Conservation Service also found that if alfalfa were grown for hay, the land could produce 1.3 tons of this
hay per acre each year.[101] These levels will probably be used
to assess the success of CONSOL's reclamation program at the
mine.

CONSOL is studying the productivity of the mine's reclaimed
grasslands through cattle grazing studies on two 25-acre plots
with their own water supply in tubs. The weight-gain of cattle
in these experiments will be difficult to compare with that of cattle grazing on lands where the water is more difficult to obtain.
Most rangelands consist of hundreds of acres with only a handful
of watering ponds. The company plans to compare weight-gain
data on reclaimed land with that on unreclaimed land to indicate
the reclaimed lands' productivity when seeking release of reclamation bond.[102] So far, CONSOL has not applied for release of
bond.[103]

WILDLIFE

If the program to re-establish hardwood draws is successful,
the vegetation in these draws should be well suited to wildlife.
CONSOL told INFORM that it is conducting extensive research to
re-create habitats for birds and small mammals.[104] The company
concedes that much of its effort involves re-establishing these
hardwood draws, which should in itself provide shelter and food
for wildlife in the area.[105]

RECLAMATION COSTS

TABLE B*

RECLAMATION COSTS AT GLENHAROLD

Operation	Cost Per Acre	Approximate Percentage of Cost†
Planning and Design		8%-9%
Planning, permit application, reclamation bond	$ 200	
Earthmoving		82%-83%
Removing topsoil	1,000	
Replacing topsoil	1,000	
Revegetation		8%-9%
Seeding	100	
Fertilizing and adding other amendments	100	
Subtotal:	$2,400	

*These costs and percentages are not strictly comparable to those of the same operation or activity at other mines. See the Cost of Reclamation chapter for a detailed explanation.

†Because Glenharold did not disclose a total reclamation cost, approximate percentages have been calculated based on a subtotal of defined costs.

Source: Consolidation Coal Company

CONSOL says it does not collect data on the cost of grading as it relates to reclamation. The company did not disclose the total cost of reclamation per acre or its cost per ton of coal mined at Glenharold.[106]

Indian Head/NACCO/ND

Mine: Indian Head Mine

Location: Zap (population 271), North Dakota in Mercer County (population 6,800)

Operating company: North American Coal Corporation (NACCO)

Parent company: NACCO

Parent company revenues: 1977, $256,230,000; 1978, $257,194,000

Parent company net income: 1977, $7,034,000; 1978, $9,380,000

First year of production: Purchased mine in 1957 when already producing 250,000 tons per year. Increased to 1,000,000 tons per year in 1967.

Coal production in 1978: 914,207 tons

Acres for which a mining permit has been issued: 1,826 since 1969

State regulatory authority: North Dakota Public Service Commission

Reclamation bond posted: $2,205,396

Reclamation cost: reclamation cost per acre, $200-$20,000; *reclamation cost per ton of coal,* NA

Expansion plans at mine: Production to increase to 1.2 million tons per year by 1984

Additional acres for which the company will seek a permit: Approximately 2,000 more will be needed to meet current contracts.

Parent company U.S. coal reserves: Owns or leases 5.4 billion
tons of coal reserves

*Development and expansion plans for surface coal mines in west-
ern United States:* By 1985, NACCO plans to have three mines
producing 16.5 million tons of coal from North Dakota. Another
mine will be opened in Texas, but its projected production is unknown.

NA = Not available

Sources: NACCO; North American Coal Corporation Form 10-K,
1978; Rand McNally Road Atlas, *54th Edition; Interview with Dr.
Edward Englerth, Director, Reclamation Division, North Dakota
Public Service Commission*

SITE DESCRIPTION

    North American Coal Corporation (NACCO)'s Indian Head
Mine is near Zap, North Dakota, in the middle of the western
half of the state. The area is sparsely populated; the nearest
sizeable town is Beulah, with 1,344 residents, some five miles
east of the mine, and the little town of Zap--population 271--is
two miles northwest of the mine. The mine is about 20 miles
west of CONSOL's Glenharold Mine, profiled elsewhere in this
book.
    The lands surrounding Indian Head are like those at Glen-
harold. Gently to steeply rolling hills are dissected by water
drainages, some of which are highly productive areas known as
wooded draws. These draws accumulate topsoils that flow into
them in spring when snows melt, or when rains carry them in,
and the abundant soils and water combine to create a lush community of trees, shrubs and grasses. Wooded draws provide excellent wildlife habitat, affording food and shelter to the area's
denizens.
    Like the climate at the Glenharold Mine, annual precipitation
has ranged from 10.97-21.7 inches over the last 30 years[1]; the
average is 16.9 inches, which is more than most regions in the
West. Approximately 25 percent of this precipitation falls as
snow.[2] The growing season at the mine lasts approximately 125
days from late May through early September.[3]
    Of the 1,826 acres for which NACCO has received permits to
mine, about 80 percent is used for grazing as range and pastureland. The remaining 20 percent is used for the production of
small grains such as wheat, oats and barley.[4] NACCO will be
reclaiming Indian Head to these uses, as well as assuring a wildlife habitat on lands reclaimed to rangeland. Unlike the opera-

tions at the Glenharold Mine, according to James Brown, NACCO's
director of environmental control, the company will not have to
mine through any wooded draws.[5]

Although the soils in this region of North Dakota are among
the most fertile in the states studied by INFORM, not enough of
this land has been sufficiently productive to be termed "prime-
agricultural." Although some "prime" soils may exist, they are
insufficient to create a logical farming unit on prime farmland.[6]
Given the range of soil and climatic characteristics, reclaiming
mined lands in North Dakota should be easier and more successful than in any other part of the West.

OVERVIEW

The Indian Head Mine, operated by the North American Coal
Company, produced only 914,207 tons of coal in 1978--less than
any other mine in the INFORM sample. The highlight of the
company's reclamation work is its excellent grading. Surface-
water control at the mine has been generally good, and a new
system incorporating three sediment ponds was built in the fall
of 1979. Beyond these factors, the basic reclamation program
falls apart. A primary problem, which the company has failed to
acknowledge or address, is that the overburden is highly sodic.
As operations at Indian Head are contaminating the groundwater
and allowing the salts to migrate toward the surface, the long-
term growth of plants may be hampered. But the seeding and
amendment program at the mine is also poor, reflecting little concern for creating a viable, long-term productive ground cover,
relying instead on a cover made up of just a few cool-season introduced species. Moreover, the company has made no attempt
to collect information on the productivity of lands reclaimed at
the mine in the past, and is only now collecting data on the productivity of unmined, surrounding lands against which to compare its own reclamation results. Many of the companies in IN-
FORM's sample have also been lax in collecting this data.

Grading is both well done and current. By the end of 1979,
the company had graded more than 85 percent of the lands disturbed between 1970 and 1979. Contouring of graded lands is
also good despite the fact that the company must mine irregular
coal seams that underlie the area, and despite a law unique to
North Dakota that prohibits the replacement of topsoil belonging
to one surface owner on the land of another surface owner. Thus
the company cannot immediately use topsoil and subsoil stripped
from the land prior to mining, and it must stockpile.

Control of surface water at the mine, using sediment ponds,
is good and being upgraded, but control of groundwater is poor.
And sodic materials in soils on the Indian Head site, now affect-

ing groundwater quality, may soon also create problems for vegetation on land as these salts migrate up through the soils into
plant root-zones. Gerold Groenewold, a hydrogeologist for the
North Dakota Geological Survey, told INFORM, "this mine stands
out as one of the worst [in North Dakota]"[7] for the groundwater
contamination and the potential migration of salts into root-zones.
He questions whether the mine should be allowed to remain open.

NACCO's program for re-establishing vegetation on lands
disturbed by mining can be rated only fair to poor. Using a
seed mixture consisting primarily of introduced plant species,
which would provide sufficient forage only in the cooler months,
the company told INFORM it would rely on nature to invade reclaimed lands with native plant species. This is time-consuming
and unwise. The Indian Head program does not reflect a desire
to help accelerate natural processes once mining and related disturbance have been completed in an area. Moreover, in four
years NACCO's only experiment with trees and shrubs has been
to plant bare-root seedlings on less than 20 acres at the mine,
with no plans to continue this work.

Although land has been released from reclamation bond
under state laws in effect in 1969, 1971 and 1973, NACCO has no
data on the productivity of these lands. (The 1969 and 1971 laws
consisted of minimal reclamation requirements.) Nor has it collected data on the productivity, plant composition or cover on
any other reclaimed lands at the mine. Only in late 1979 did the
company begin to collect information on undisturbed lands surrounding the mine, and it had yet to begin gathering data on
land that it had disturbed and then revegetated.

The North Dakota Public Service Commission inspects Indian
Head once or twice a month, but it has not had as great an effect upon reclamation practice as has the federal Office of Surface Mining. OSM visited the mine in August 1978, November
1978 and October 1979. As a result, new sediment ponds have
been built and the company has begun to collect productivity
data on lands yet undisturbed by mining. However, both state
and federal regulators will be busy in the future as questions
are raised about the long-range effects of increased mineralization in groundwater and migrations of salts into plant root-zones
--the effects of which are not yet visible--from sodic materials
that are exposed, by mining, to increased water infiltration.

The company has established an island within a water impoundment to provide forage and habitat for wildlife. It is questionable, however, whether NACCO's reclamation program will
meet the year-round forage needs of wildlife.

EARTHMOVING

Grading

Grading is remarkably up-to-date at Indian Head. According to Peter Neilson, NACCO's director of special projects, only 160 acres of land remained to be graded at the end of 1979 of the 1,392 acres of land mined from 1970 through 1979.[8] The company caught up to disturbance quickly in the 16 months following an August 1978 OSM inspection that found that "grading and backfilling operations do not appear to be as current as they should possibly be. An area mined in 1975 was the most recent area to have grading operations completed on it."[9]

However, grading does not always follow mining directly. Grading can be from one to many spoil piles behind the active mine pit. Sometimes mine pits remain open so that different quality coals can be mixed, and, as with CONSOL at Glenharold, mining may be under way in an area for up to one and one-half years before grading begins. Because the configuration of the mining pits is oddly-shaped to follow the coal deposit, instead of in straight rows, mining must advance farther ahead of grading. This allows the company to grade to a contour that better replicates the pre-mining contours, but creates a delay in revegetation and in subsequent use of the land for grazing and crop production. In winter months, to avoid burying snow and ice which can create depressions in the overburden after it melts, grading is slowed at the request of the Public Service Commission.[10] Yet this problem is not as serious as at Glenharold, because control of sedimentation from ungraded lands off the mine site is better at Indian Head. Grading for the most part has approximated pre-mining contours of the land, and it is traversable by reclamation machinery along the contours. Slopes on approximately 22 acres of land surrounding a water impoundment and "island" are steeper than original contours, and the Public Service Commission and OSM are carefully watching this area for "success in establishing vegetation."[11] The water impoundment has been created to enhance wildlife and grazing uses of the land and contains a small island in its center which should serve as a waterfowl habitat.

Topsoil

Topsoil removal and stockpiling practices at Indian Head are generally good, as are those at other North Dakota surface mines. All topsoil is removed from the land before mining, and subsoil is also segregated to provide suitable amounts of topsoil and subsoil when respread on lands being reclaimed, as required by North Dakota law. At Indian Head, 6-24 inches of topsoil lies

on the land, depending on whether it is at the top of a hill where
soil erosion will have been greater over many years (less topsoil), or in valleys where years of eroded soils accumulate, providing more topsoil.[12] An average of 12-15 inches of topsoil is
removed and stockpiled separately from subsoils scraped from
the land.[13] The company buried some topsoil on about 200
square feet of land during operations to add subsoils to a stockpile in June 1977. However, after realizing this error the company removed the subsoil and salvaged the topsoil beneath it.
No violation was found by the Public Service Commission because
of NACCO's quick and successful action to correct its mistake.[14]

NACCO stockpiles all the topsoil and subsoil taken from the
land before mining. Like the operations at Glenharold, whole
sections of land within the mine are completely mined out before
grading and topsoil replacement operations are begun. After
grading, the company waits six months to one year before
spreading topsoil and subsoil to allow the graded overburden to
settle.[15] The soil and rock characteristics at both Glenharold
and Indian Head require this delay in order that unevenly settled
material may be regraded before topsoil replacement and revegetation are begun. As a result of stockpiling, the soil is not as
good for revegetation, but the circumstances described above
appear to preclude NACCO's immediate use of all salvaged subsoil
and topsoil.

About 50 topsoil and subsoil stockpiles, each covering an
average of two or three acres of land, dotted the mine site as of
November 1979.[16] One innovative practice used by the company
in the placement of soils into stockpiles was noted by the Office
of Surface Mining in November 1978. By approaching and leaving the stockpile area from various directions, the scrapers that
dump the soils can create a more rounded, less steep stockpile.
This relegates more land to use as topsoil and subsoil storage
areas, but these rounder piles can be seeded with drill seeders
operating on the contours, unlike other steeper stockpile config-
urations that generally require that seed be broadcast by hand
or machine on all slopes.[17] OSM stated:

> In that the entire stockpile area has slopes which
> are traversable by machinery, the entire surface
> area can be seeded/drilled, utilizing conventional
> farm machinery. This should help reduce the po-
> tential of erosion from the stockpile....Although
> the existence of this new technique has not yet
> been in operation long enough to evaluate the de-
> gree of success which it may have on controlling
> stockpile erosion, the company personnel respon-
> sible should certainly be commended for their
> originality in trying to develop new and practical

techniques to mitigate adverse environmental effects due to mining.[18]

Topsoil and subsoil stockpiles are vegetated, thus are generally
free of erosion. When topsoils and subsoils are replaced on
graded lands, they are respread to a uniform depth appropriate
to the post-mining land use.[19]

Toxic Material

Though a common characteristic of many mining areas in the
state, Indian Head stands out in North Dakota for its highly sodic overburden. The effects of mining on the quality of groundwater and salt migration through these spoils is well documented.[20]
Many of the soils in the area of North Dakota studied by INFORM
have high levels of sodium sulfates (salts). These salts can migrate through the soils into plant root zones and can contaminate
groundwater. When asked whether salts had been found migrating up into graded areas at the mine, NACCO told INFORM that
it had:

> not experienced any instances of salt migration
> up into graded and revegetated lands at Indian
> Head Mine. This is due to the deep water table
> and to the fact that the precipitation tends to
> carry the salts downward.[21]

However, the company suggested that INFORM contact hydrogeologist Gerold Groenewold of the North Dakota Geological Survey. He told INFORM that mining at Indian Head, in fact, showed
a "particularly strong potential" for the migration problem. He
said measurements at the mine revealed that sodium sulfate salts
were definitely moving up through the soils on graded lands
and that, although the mechanism for this movement was not understood, the measurements showed that salt would reach the
upper layers of soil. These salts could prevent desirable vegetation from growing on areas where migration occurs. But he
said there was little NACCO could economically do to prevent
this. Special earthmoving practices would have to be incorporated, he said, so that clay-like soils could be used to seal off
sodic materials before final grading and replacement of subsoil
and topsoil. This would be costly.[22] (See Surface Water and
Groundwater.)

But NACCO refuses even to recognize the problem, or at
least to admit that it exists. James Brown, director of environmental control for NACCO, repeatedly told INFORM that salts
simply are not a problem because they move *downward* through
soils instead of upward.[23] In fact, according to Groenewold, it
appears that both upward and downward movement are occurring. (See Surface Water and Groundwater.)

Surface Water and Groundwater

Control of surface water on the Indian Head mine site has been only fair, but is being significantly upgraded. In late 1979 the Public Service Commission approved a new water management plan for the mine to bring the company into compliance with federal water control standards. By December 1979 all three sediment ponds planned had been built.[24] Before installing these new ponds, NACCO had relied on small stock-watering ponds on adjacent ranch property for water retention to control sediment. On a visit to the mine in the summer of 1979, INFORM observed the mine's "main sediment pond" on property adjacent to an active mining area. It was actually a rancher's stock-watering pond, about the size of a swimming pool, which the rancher had allowed NACCO to use. Discharge from the pond occurred whenever water levels reached the lower end of this depression, and water then flowed into neighboring fields. The new sediment ponds will allow NACCO to better monitor the quality of water being discharged from the site. They will also retain water longer, allowing sediment to settle before water is discharged. Some water is also impounded so that it is not discharged at all but evaporates, soaks into the soil or is used. Because sediment ponds were still under construction in late 1979, INFORM was not able to assess whether they were stabilized with vegetation, as planned. However, the "borrowed" sediment pond described above was well vegetated and would not allow additional sediment to be carried into water flowing from the mine. Rocks have been used as rip-rap to slow the flow of water leaving this sediment pond, and the company has used a rock filter dam to slow water flow and help settle suspended solids on another section of the mine.

The quality of water discharged from Indian Head has been within National Pollutant Discharge Elimination System (NPDES) limits since October 1977.[25] However, in March 1979 the company exceeded total suspended solids (TSS) limits and total iron content limits.[26] NACCO explained that because sediment had been cleaned from the bottom of the pond where this discharge occurred, "a significant amount of particulate matter [was made] available for transport during the initial stages of spring runoff."[27] The company said it would investigate the high levels of iron in the water, a situation it deemed "atypical."[28] With the construction of new sediment ponds, this problem should not recur, as the ponds will provide increased capacity for initial runoff in the spring and will not immediately discharge the water.

NACCO does not monitor groundwater on or near Indian

Head. (There is no surface water to monitor except for actual
discharges from the mine, which are measured.)[29] However, as
described above (see Toxic Material), the North Dakota Geological Survey has studied the problems of sodic materials as they
relate to groundwater and soils at the mine; hydrogeologist Gerold Groenewold has established approximately 100 groundwater
measuring devices on and off the Indian Head site. His measurements show that groundwater quality has changed as a result of
mining at Indian Head. Normal groundwater levels for total dissolved solids (TDS) range from 1,000-2,500 parts per million.
However, mineralization--primarily by sodium sulfates, bicarbonates and calcium--has raised these levels to an average of 5,000-
7,000 parts per million and has reached peaks of approximately
16,000 parts per million. This is caused by percolation of water
down through mined and graded spoils. Groenewold points out
that "this mine [Indian Head] stands out as one of the worst [in
North Dakota] for this problem."[30] The ultimate issue, Groenewold says, is whether OSM will ban further mining at the Indian
Head site, or the company will be willing to trade off some coal
production for protection of groundwater and soil productivity.

In a relatively minor instance, groundwater supplies to a
neighboring rancher's well dropped as a result of mining activities near his ranch. NACCO drilled another well for him, and
after mining in the area was completed, water in the original well
returned to normal levels.[31]

Alluvial Valley Floors

The U.S. Geological Survey and NACCO both report that no
alluvial valley floors have been identified on the Indian Head
permit area.[32]

Erosion

INFORM is unaware of any disturbances caused by erosion
and sedimentation off the Indian Head mine site. Erosion is well
controlled by vegetation where vegetation exists. On lands
lacking a plant cover, erosion is sometimes found to depths of
four or five inches, but NACCO has implemented measures to
control any resulting disturbance. The company uses bales of
straw, anchored to the ground with stakes, as dikes to slow water flow and trap sediment. And operation of seeding machinery
along the contour creates small troughs, which trap water and
sediment running downhill.

One area at the mine may present special erosion problems.
The graded area containing the "wildlife island" has slopes of
12-14 degrees, which drain toward the island. These steeper
slopes are more prone to erosion, so NACCO has created terraces

Straw bales staked to the ground are used at this mine to attempt to catch sediment before it is carried into streams on the site.

on these slopes to help reduce this potential. The land in this 22-acre area has also been roughened to retain more water until subsoil and topsoil are replaced and revegetation efforts begin.[33] Although erosion from these slopes would not cause offsite disturbances, it could eventually deposit an excess of soil into the water being impounded, thus reducing the capacity and utility of the impoundment.

SEEDING AND AMENDMENTS

Seeding Procedures

For many of the same reasons that grading does not follow mining promptly, seeding also does not take place on a constant basis. However, it has been accelerated. In 1977 mining disturbance totaled 49 acres. That same year NACCO seeded 225 acres. In 1978, 67 acres were disturbed and 24 acres seeded. By the end of 1979, approximately 70 percent of the land mined since 1970 had been seeded.[34] Seeding generally takes place in the fall or early spring. This allows seeds to germinate before and during the spring, when moisture is readily available to help ensure plant establishment. Fall seeding is a generally accepted practice in the region.

The seed mix used is poor. NACCO uses three introduced grass species, one introduced legume, and only two native grass species. All are cool-season varieties that do not grow in warmer seasons, except alfalfa, which is not desirable as a sole source of forage. The company feels that native plant species will eventually establish by themselves, and thus it relies on natural invasion for native plants.[35] This practice does not adequately deal with the need to create--after mining--a native plant cover suitable to the proposed land use, grazing and wildlife habitat. Only when the postmining land use is to be crop production would this mixture be adequate, as after two years the land would be seeded to crops, pursuant to North Dakota law. Initial cover on topsoiled lands is created by the seeding of oats, which are used as mulch. Stockpiles receive a mixture of rye, bromegrass and intermediate wheatgrass at a heavy rate of 30 pounds per acre to produce a quick vegetative cover.[36]

It is difficult to evaluate the application of NACCO's seed mixture at Indian Head. The company told INFORM that:

> Seeding levels of grasses vary on a site-specific basis. Soil quality, quantity, season of planting and post-mining land use data are all incorporated into the grass mixtures to determine the species and application rates.[37]

TABLE A

1979 SEED MIXTURE
FOR LANDS BEING RECLAIMED AT INDIAN HEAD

Plant Species	Introduced/ Native	Cool/ Warm	Pounds Applied Per Acre
Grasses			
Smooth brome	I	C	*
Western wheatgrass	N	C	*
Tall wheatgrass	I	C	*
Crested wheatgrass	I	C	*
Needlegrass	N	C	*
Legumes			
Alfalfa	I	C-W	*
			17 Pure live seed

*Unavailable (see text)

Source: NACCO

(NACCO later said that an average of 17 pounds of the mixture is now applied to each acre.[38]) However, the only seed testing plot of which INFORM was made aware was a test plot to evaluate the growth of seeds, developed by Northrup-King Company, which are resistant to alkaline environments.[39] (See Toxic Materials.) The company does no experimental work of its own, except for limited tree and shrub planting.[40]

Seeding is accomplished primarily with a drill seeder, operated on the contour of slopes being reclaimed. The tracks created by these machines help to retain water running downhill and provide more water for establishing vegetation. The tracks also help to control erosion. Furrows are not created as a separate operation. When drill seeding is not used, a broadcast seeder is employed. Land is plowed ahead of the machine to prepare the seed bed, and a drag bar covers the seed after dispersal.[41]

Amendments and Maintenance

As mentioned above, oats are used at Indian Head as initial cover on lands that have been graded and topsoiled. This provides a mulch that needs not be tacked to the surface. This cover of oats is mowed periodically to allow it to break down and return minerals and nutrients to the soil.[42] The company does not experiment with other types of mulch, but use of a standing mulch is a growing and acceptable practice in the West.

Indian Head does not irrigate, as moisture on the site is adequate to establish the plants being seeded.[43] But it does use fertilizer to make up deficiencies of nitrogen and phosphates in the soil. The company does not experiment with fertilizer use on the mine site.[44] The types and amounts of fertilizer are determined by standard soil analyses and rates suggested for North Dakota by the U.S. Department of Agriculture. The use of legumes also provides some enrichment for the soil, as these plants create organic nitrogen from the air and return it to the soil. According to Peter Neilson, director of special projects for NACCO, fertilizer is commonly used on most croplands in this area of North Dakota, but not on grazing land.[45] Thus if fertilizer is used on lands being reclaimed to grazing use, the question arises whether the vegetation will continue to grow after this maintenance is ended. OSM requirements should help define the answer to this question by holding bond for ten years after maintenance ends.

Trees and Shrubs

NACCO has planted few trees and shrubs at Indian Head. In cooperation with the U.S. Soil Conservation Service, portions of 19.7 acres of land at the mine were planted from 1974 to 1978 with various bare-root seedlings of green ash, Hansen hedgerose, caragana, Russian olive, buffaloberry, Siberian elm, American plum, ponderosa pine, eastern red cedar, laurel leaf willow, and hawthorn.[46] That is the extent of Indian Head's tree and shrub experimentation.[47] Depending on the proposed post-mining land use on any given piece of land, trees and shrubs may not be the most desirable form of vegetation. Yet so little work has been done with trees and shrubs, and so much of the surrounding area contains them, that NACCO's program cannot be deemed sufficient. Clumps of trees and shrubs do exist around the old, ungraded spoil piles from previous mining operations. These rough spoils have been invaded by many trees and other plants, but ground cover is sparse on the steep slopes of these piles. The plants that do exist, however, provide good habitat and some forage for wildlife.

VEGETATION

NACCO knows little about the type of plant cover or productivity found on lands to be reclaimed at Indian Head. Not until the fall of 1979 did NACCO begin to collect data on the productivity of native, undisturbed lands and of lands the company has been reclaiming--because, it says, "Reclamation laws covering

the lands reclaimed during [1974 to 1977] do not require establishment of productivity levels. As a result, this information is
not available."[48] Data are also being collected on the plant
cover and diversity on land reclaimed at the site.[49]
      Lands with vegetation show little if any erosion. Lands
lacking cover show some erosion. The composition of reclaimed
lands at the site is most likely dominated by introduced plant
species, but the company has not studied this area sufficiently
and has no information about it. Introduced plants prevent native, adapted species from growing and generally require
more maintenance for continued production. More information is
necessary to measure the extent of this dominance. Soil surveys
of Mercer County by the U.S. Soil Conservation Service
show that forage productivity in the Indian Head area is approximately one ton per acre, and crop production covers a wide
range depending upon the soils in which it is grown. SCS says
wheat production ranges toward 28 bushels per acre; barley
production to 47 bushels per acre; and oats production to 54
bushels per acre.[50] These productivity levels, along with the
data gathered by NACCO, will be used to assess the success of
the company's reclamation program.
      Under state laws of 1969, 1971 and 1973, slightly more than
42 percent of the 574 acres of land permitted for mining between
December 1969 and December 1972 have been released from reclamation bond.[51] North Dakota and Colorado are the only two
western states where lands have been released from reclamation
bond under any law. The criteria for bond release from earlier
state laws is much less strict than that for lands reclaimed under
newer state and federal regulations. In the fall of 1979, NACCO
requested that additional lands be released from bond.[52] Unlike
CONSOL, at Glenharold, NACCO has no data on the productivity
of lands released from bond at the site.[53] This information would
be helpful in assessing the changes that occur on the land when
compared to data on the land's pre-mining characteristics, or
those of nearby undisturbed lands.

                                                          WILDLIFE

      Wildlife habitat and forage are being enhanced by the devel-
opment of a water impoundment containing an island. The Public
Service Commission has allowed NACCO to create this impoundment and island, recognizing that it heightens the land use of
the area by providing a water body where none existed before.
The water will be useful for grazing animals as well as waterfowl
and other wildlife. The area was not seeded as of the fall of
1979; thus it is not certain whether the plants established will be
palatable to animals. If the mine's basic seed mixture is used,

the same problems associated with the dominance of cool-season,
introduced species may affect long-range use, because introduced
species are not well adapted to climatic extremes and require
more maintenance than natives. Moreover, without warm-season
plants, forage production will be reduced in the warmer seasons.
More tree and shrub establishment should be encouraged; NAC-
CO's experience with these plants, to date, has been minimal.
Thus, NACCO should develop a reclamation program that takes
into account the year-round needs of wildlife, if it expects them
to return to the area. Spoil piles left by the mine's previous
owner provide not only a good wildlife habitat, but forage in the
form of invading trees and shrubs and areas where water is impounded. While wildlife may benefit from and concentrate in
these spoil piles where their needs are being met, this could
cause an imbalance in the relationship between predators and
their prey--such as rodents--which, if not controlled, could significantly affect the quality of the plant communities on reclaimed
lands. Disturbing the balance established by nature is easy,
but knowing the results of this action, and the solutions if problems arise, is much more difficult.

COSTS

Reclamation costs vary so widely at Indian Head that NACCO
was unable to provide meaningful figures to INFORM.

> Due to the progression and implementation of in-
> creasingly stringent reclamation laws over the
> past several years, determining an average rec-
> lamation cost is next to impossible. Also, the
> type of land to be reclaimed and the land use
> during mining greatly affect these costs. Our
> current reclamation costs range from $200 to
> $12,000 per acre.[54]

However, Kent Horne, former executive assistant at NACCO,
told INFORM in June 1979 that reclamation costs average about
$4,000 per acre plus administrative overhead.[55]

Appendix

Costs of Reclamation

What does it cost a coal company to reclaim a surface mine and restore the land to its pre-mining condition? The answer to that central question was perhaps the most elusive of all in INFORM's survey of western surface mine reclamation. Even when answers could be gleaned--even from the most responsive and cooperative of the coal operators surveyed--the answers came in a complicated assortment of "packages," rendering even more difficult the task of presenting them to the public in a meaningful way.

The crux of the problem is that different companies measure reclamation costs differently. So do independent commentators in the field, including the U.S. Bureau of Mines. The basic distinction between *reclamation* costs and *production* costs is itself shadowy, and operators draw the line at different points.

But it is a question worthy of close scrutiny. The industry itself has made it so, having long (and loudly) questioned whether the environmental and agricultural gains derived from better reclamation are worth the dollars involved. We still don't know the answers--and won't for many years. Gary Grow, vice president of sales for Peabody Coal Company, which operates two mines in INFORM's sample, comments: "It's a shame the consumer has to pay so much for reclamation....for land that's not worth more than $50 an acre."[1] Although Grow's view may not be held throughout the industry, it points up the weakness of our knowledge of the long-term values the land provides. Coal may be considered a single harvest from the land, but plant and animal life will use productive lands for years. The environmental benefits of reclamation--including prevention of erosion and sedimentation on areas off the mine site, protection of water resources and long-term agricultural gains--must be taken into account.

The 1977 Surface Mining Control and Reclamation Act (SMCRA) brought the question of reclamation costs to the forefront. Consolidation Coal, according to Bowser-Morner Laboratories of Dayton, Ohio, a consultant group, will have to pay an average of $400 million annually to comply with SMCRA regulations.[2] Peabody assessed its costs of SMCRA compliance at $6.26/ton at one surface mine.[3] As the bill was being debated, companies argued that it would force cutbacks in production, layoffs of workers, and higher utility bills for consumers. According to Russell Boulding, a coal mining and reclamation consultant to the National Coal Policy Project, these figures many times represent "worst-case analyses."[4] Upon comparing the cost of reclamation under the SMCRA to the minimal cost for reclamation under poor, state reclamation laws, a wide margin can result, inflating the overall impact of federal legislation upon the cost of reclamation. Coal producers contend that the paperwork necessary to open new mines and to obtain new permits for existing mines is a threat to smaller operators, who cannot afford to comply with the new regulations. Defenders of the SMCRA counter that it will actually increase employment and protect lands vital to the national interest.

This section will not fully answer all of the questions about reclamation costs. However, it should clarify the various and conflicting methods by which such costs are measured and offer as much cost information as could be located in publicly available documents and in data obtained from companies that INFORM studied. Thus it should help explain why INFORM's attempt to provide meaningful figures has been only partially successful.

What will become evident from a study of the problem is the need for a uniform system of recording costs--including an accurate description of mining and reclamation conditions, itemized component costs, and a standardized accounting technique--if it is to be clear what costs are to be included in reclamation at any mining operation so that comparisons can be made. Why is such information important? If we are to develop a full picture of the costs of mining, transporting and burning this dirtiest of fossil fuels in the most environmentally sound fashion, each component cost must be better understood. Lacking this broad picture, figures tossed about among government, business and the public will mean little, and will be a poor foundation upon which to base regulation for environmental protection.

For example, consider the reclamation cost attributable to enenforcement of a federal surface mining and reclamation law. In 1975, on the assumption that such a law would pass (it was vetoed by President Ford), engineers for Gulf Energy & Minerals determined that federal regulation would yield reclamation costs from

$0.44 to $10.94 per ton of coal mined, depending on variations between western and eastern coal fields including topography and the nature of the coal seam being mined.[5] In 1980, with the federal Surface Mining Control and Reclamation Act firmly in place, the President's Commission on Coal found that the *increase* in reclamation costs would range from $0.10 to $17.77 per ton of coal.[6] According to Donald Koch, an analyst for the Arizona Public Service Commission, the highest cost that his office has seen that is due to the implementation of the OSM regulations is $2 per ton of coal delivered.[7] Comparisons, then, are difficult, if not impossible.

INFORM tried to pinpoint the cost of reclamation by analyzing what work is actually done for a given expense and by examining the component costs of reclamation to locate the most significant ones. (We did not assess the cost increase resulting from regulation because of the differences in state requirements as well as the lack of hard data on baseline reclamation costs.) Two questions had to be answered. First, what are the conditions under which mining and reclamation will take place? Second, how do the component costs correspond to the operations comprising overall reclamation?

Mining and Reclamation Conditions

At what levels of mining are reclamation costs being reported? The most general description used is simply "for the West." The West may be broken down into states and states into regions (often based on coal distribution). Finally, regions may be further divided into mine types, which may be specified in varying degrees. This variance in describing the mines themselves contributes greatly to the discrepancies in reported reclamation costs.

One method is to select a "model mine," which is probably among the least specific techniques to describe a mine type. N.B. Pundari and J.A. Coates, engineers for Gulf Energy & Minerals, take this approach, beginning with some fundamental assumptions about the mine's geophysical characteristics and its engineering design. Their model assumes topography with a uniformly increasing slope in the direction of mining, uniform topsoil thickness and composition. In terms of engineering, the model specifies the use of truck-and-shovel technology or bulldozers and scrapers for grading, a uniform width of mining cut, and a maximum mining cut depth of 200 feet.[8] Their findings--although not entirely accurate, owing to what the authors term "intangible costs"--are that reclamation will vary in cost from $9,854 per acre to as much as $76,905 per acre. The range itself suggests that studies of this nature are at best academic unless applied to site-specific conditions.[9]

The U.S. Bureau of Mines, providing background information on the circumstances surrounding reclamation costs, incorporates

a regional description along with key characteristics of the mine
type. In 1977, Franklin H. Persse and others, in *Coal Surface
Mining Reclamation Costs in the Western United States,* a Bureau
of Mines publication, describes the *region* in terms of coal type
(bituminous, sub-situminous, lignite), climate, vegetation, angle
of coal seam inclination, and land use. These factors are juxta-
posed against the mining and reclamation factors for a particular
mine type in the region. Persse et al. describe the *geophysical
characteristics* of the mine type in terms of the mine's production,
coal seam thickness, and annual acreage disturbed. They then
specify the overburden and topsoil thickness and composition,
and the maximum angle of highwall inclination. These authors
found that actual reclamation costs based on 1974 and 1975 information ranged from $1,670 per acre to $7,200 per acre.[10] They
did not calculate the increased cost due to federal reclamation
regulation.

A still more specific approach is taken by Kenneth L. Leathers. He specifies key characteristics for known stripping operations in the western region, identifying them by name, operator,
location, coal field and seams mined, and annual output.[11] Leathers includes both geophysical and technological characteristics,
assigning numerical values for coal seam thickness, overburden
depth and stripping ratio (the ratio of overburden volume, in
cubic yards, to coal volume, in tons removed). He refines his
approach further to specify the earthmoving technology in terms
of the type used for overburden removal, contouring and topsoiling. For example, he notes whether dragline or truck-and-shovel
technology is used for contouring. In general, the cost of removing overburden at a dragline operation is less than at a truck-and-
shovel operation. However, the grading costs at dragline operations are frequently higher. (See Techniques of Surface Mining
and Reclamation.)

Component Cost Approach

To what degree are reported costs divided into their components? The broadest (and least helpful) approach is simply to
state a total cost. The focus may be narrowed by specifying
component costs under five or more categories in up to three
different measuring units (cost per ton, per acre, or per Btu).
Even the use of these units as a cost measurement tends to leave
a large information gap, as will be discussed shortly.

The Bureau of Mines separates total reclamation cost into four
categories:

•Design, engineering and overhead. This category includes
development of the reclamation plan, preparation of environmental reports, supervision of reclamation work, engineering and surveying for environmental protection, water
quality protection and monitoring, overhead for environ-

mental activities, outside environmental consultants, surveying and mapping, dust control, and additional miscellaneous expense.

• Bonding and state permit or license fees.

• Backfilling and grading of mine pits. This category includes removal of vegetative cover from the area to be
mined when its removal is required for topsoil salvage;
removing and stockpiling topsoil; backfilling troughs,
haulage ramps and final cuts with soil; rough and fine
grading of spoil, and removing topsoil from stockpiles and
replacing it uniformly on graded spoil.

• Revegetation, including soil preparation, seeding and/or
planting, reseeding and/or replanting, and irrigation.[12]

Research conducted in 1975 by E.A. Nephew *et al.* under a National Science Foundation program employs four categories that
differ somewhat from those of the Bureau of Mines: topsoil removal and replacement, backfilling and grading, supplementary land-
scaping, and revegetation--in Appalachia. These divisions do not
break out from total costs those due to premining activities such
as design, permit and bond fees, and overhead.[13]

Other sources, citing the difficulty of differentiating costs,
opt for broad categories such as "tangible" and "intangible."
Gulf Energy's Pundari and Coates, who adopt this approach, define tangible costs to include those due to "moving of excess spoil
material to deficient areas in order to maintain original ground
surface elevation; elimination of highwall banks and last cuts;
grading of land to original contour level; topsoil removal, stockpiling and spreading on the mined out lands; (and) seeding and
fertilizing the ground for natural growth of vegetation within a
reasonable length of time...."[14] Intangible costs are described
as those of maintaining hydrologic integrity in the mining area;
analysis of waste strata for toxic compounds; fees for reclamation
bonds; continued maintenance of revegetated lands, overhead and
miscellaneous expenses.[15]

What "Unit Costs" Leave Out

Each of the measuring units--cost per ton, per acre, and
per Btu--has its intrinsic merits. But to employ any one is to
miss information contained in the others.

Thus the reclamation cost per acre says nothing about
the tonnage of coal produced for the cost, or about the market
value of the reclamation. The cost per ton reveals nothing
about the acreage of land reclaimed or the cost effectiveness of
reclamation. Cost per Btu provides information about the market value of reclamation but gives no clue as to the acreage of

land reclaimed. However, it is generally felt that when discuss-
ing the cost of reclamation, one is dealing with the land--not the
coal. In that light, the most relevant figure is the cost of reclamation per acre of land disturbed. As it is easy to translate
this into cost per ton (every mine operator knows how many tons
of coal are recovered from each acre of land mined) or cost per
Btu (adding the known Btu content of the coal), the cost per
acre is the best indicator for the general public of the cost of
reclaiming surface mines. Only later in the cycle is it essential
to translate this into the cost per Million Kilowatt Hours, or whatever unit is used to determine how much reclamation costs to the
consumer of electricity.

Other Cost Factors: Reclamation vs. Production

   Even after defining the mining and reclamation conditions
under which reclamation costs are incurred and breaking down
these costs, other significant elements influence the meaning of
reclamation cost figures. For example, when--for accounting
purposes--does production end and reclamation begin?
   This distinction is particularly ill-defined for those surface
mines that employ draglines--as do all but one in INFORM's sample.
Kenneth Leathers notes:

> "Depending on mining method, the types and sizes
> of stripping equipment used, and the particular
> engineering requirements of a given site configura-
> tion, production and reclamation procedures and
> costs are often jointly determined and therefore
> difficult to identify as separate activities."[16]

   Leathers cites backfilling at a mine using draglines as an
example. Except at the first and last cuts, as the dragline removes overburden it is placed in the preceding mined-out area.
This replacement of overburden is sometimes listed as a reclamation cost, under the assumption that it is necessary to return the
land to its approximate original contour. The *removal* of the
overburden, however, is clearly--to Leathers--a *production* cost.
Furthermore, the removal and replacement of the overburden are
one continuous operation: the dragline bucket is filled (overburden removal) and its boom is swung over the preceding pit
for backfilling (overburden replacement). How, then, should
backfilling be accounted for--as reclamation or production?
Leathers opts for production--as backfilling is an integral part
of removing the overburden. He points out that recontouring is
separate from backfilling at these mines--and therefore distinctly
a reclamation activity--because the dragline creates steep spoil
ridges, which later require extensive leveling.[17] (See Techniques
of Surface Mining and Reclamation.)

Leathers contrasts this to the situation at a truck-and-shovel mine (see Belle Ayr profile), where "the overburden is deliberately returned to the pit area rather than an adjacent dump site."[18] Because overburden must be hauled a significant distance rather than placed in area adjacent to the active mining region, the backfilling, Leathers contends, is a "genuine reclamation activity." Moreover, he notes, overburden at a truck-and-shovel operation is replaced in a level configuration," achieving maximum return to the appropriate original contour. This is recontouring, clearly an activity of reclamation, and accomplished jointly with backfilling. Therefore, backfilling becomes a reclamation cost.

Actual Cost vs. Market Cost

In a purely economic sense a company may account for its reclamation costs in two very different categories: actual costs or market costs. The actual cost of reclamation includes the capital investment in equipment, labor and overhead. It can differ significantly from market cost--commonly defined to mean the price of goods or services. Moreover, even within market costs, costs can vary widely because of the volatile nature of market conditions. These costs depend in part on contract agreements, which will vary widely from company to company and even within one company, depending on the year and conditions under which the contract was established.

Increased government regulation of the industry has dramatically affected both actual and market costs--and has made it even more difficult to compare reported reclamation costs meaningfully. INFORM asked the 15 companies in its survey to assess the increased costs of reclamation since implementation of the 1977 SMCRA. As of late 1979 their responses indicate that the incremental cost of reclamation under increased regulation varies from mine to mine. Five companies reported minor or no increases. Some companies report increased costs primarily in earthmoving activities, others in planning and design--two areas that vary greatly from each other in their component costs.

These, then, were some of the problems facing INFORM as it attempted to assess the costs of reclaiming western surface mines. We had hoped to be able to arrive at not only a meaningful analysis of reclamation costs, but an extrapolation of the impact of these costs on the ultimate price of coal to consumers. Leathers, too, attempted to assess this area--concluding that (based on 1976 data) reclamation costs in the West account for an average of less than 3 percent of the selling price of coal, with a maximum of 5 percent. At this early stage, however, INFORM believes the available information is too raw and unstandardized to yield a realistic assessment of its full impact on coal prices.

INFORM's Approach

INFORM requested itemized information--from the companies surveyed--on the direct and indirect costs per acre of the following:

- planning and design of reclamation program
- permitting and bonding
- removal, storage and stabilization of topsoil
- removal of overburden
- grading
- replacement of topsoil
- seeding (revegetation)
- fertilizing, mulching and irrigation
- continued maintenance after mining
- average cost per acre and per ton.

Not surprisingly, given the reclamation cost picture described throughout this chapter, even though most of the companies responded, they seldom itemized their answers according to INFORM's categories. Frequently several categories were grouped under one, or other categories were added.

For example, Peabody Coal Co., in accounting for its costs at the Black Mesa mine, took all of INFORM's categories and grouped 99 percent of its costs into two categories: "topsoil operations" and "grading." Others, such as Western Energy, added the categories of "final pit reclamation" and "drainage reconstruction" in accounting for the costs at the Colstrip Mine.

To understand more clearly where the costs of reclamation fall, INFORM grouped the reported component costs into four broad areas: Planning and Design, Earthmoving, Water Quality Protection, and Revegetation. We allocated the nine categories of information among the four broader categories as follows:

- Mine Planning and Design: planning and design, and permitting and bonding.

- Earthmoving: removal, storage and stabilization of topsoil, removal of overburden, grading, and replacement of topsoil.

- Revegetation: seeding, fertilizing, mulching, irrigation, and continued maintenance.

- Water Quality Protection: Where specified by the company, sediment pond construction, surface water monitoring, and groundwater monitoring were considered under this category.

This method of grouping assumes that any given company's reported component cost is comparable to another's reported component costs; for example, that the "grading" cost at the Decker
mine in Montana is comparable to the "grading" cost at the Navajo
mine in New Mexico. We recognize that the various geophysical,
technological and economic factors already described make each
of these figures unique to its own situation. Some generalization
remains possible, however. Using this technique, and information
from 11* mines, INFORM found that earthmoving costs ranged from
73-99 percent of the reported total reclamation cost for seven mines.
At the other four mines (where total reclamation costs could not be
used), only the percentage of the earthmoving costs to the subtotal of the reported reclamation costs (rather than the *total* reclamation cost) could be calculated--and the earthmoving cost fell
between 78-97 percent. Revegetation costs for these seven mines
ranged from negligible to 26 percent (over 25 percent at two mines)
of total reclamation costs. And for those four mines where only
the approximate percentage of revegetation costs could be calculated, the cost fell between 2 and 9 percent of the sub-total of
reported reclamation costs. The U.S. Bureau of Mines in 1977
estimated earthmoving costs to fall within 56-90 percent of the
total reclamation cost.[19] Leathers estimates 70-80 percent.[20]

For all of INFORM's efforts, the hard fact remains that it is
impossible today to compare further the component or total cost
figures of reclamation provided by the coal operators. The specific cost figures are reported here *solely* to permit researchers to
assess the incremental change in reclamation costs at a given
mine as reclamation practices change over time.

The conclusion that awaits the analyst several generations
away eludes us today: whether mining *and* reclamation succeed
at a cost that society is willing to pay.

---

* Four mines in INFORM's sample were not evaluated in this fashion, owing to a lack of information or extremely varied reporting
by the companies.

TABLE A

RECLAMATION COSTS PER ACRE

Mine	Planning and Design	Earthmoving	Revege-tation	Water Quality
Colstrip	$600	$ 8,600	$ 600	$2,000
Decker	---	3,260	1,074	---
Belle Ayr	---	3,500	800	---
Jim Bridger	---	7,529	407	---
Rosebud	---	1,919	142	---
Seminoe II	---	3,861- 7,160	200	---
Gascoyne	82	—— 2,418- 2,918 ——		---
Glenharold	200	2,000	200	---
Energy Fuels	331- 771	1,562- 3,765	92	---
Seneca II	---	5,200	230	---
Navajo	---	4,740	1,290	---
Black Mesa - Kayenta	---	12,000- 18,000	65- 85	---

Note: Reclamation costs for Absaloka, Indian Head, and McKinley Mines are not available.

TABLE B

PERCENTAGES OF RECLAMATION COSTS

Mine	Planning and Design	Earthmoving	Revege-tation	Water Quality
Colstrip	5%	73%	5%	17%
Decker	---	75%	25%	---
Belle Ayr	---	81%	19%	---
Jim Bridger	---	95%	5%	---
Rosebud	---	93%	7%	---
Seminoe II	---	95%-97%	3%-5%	---
Gascoyne	3%	——— 97% ———		---
Glenharold	8%-9%	82%-83%	8%-9%	---
Energy Fuels	20%	78%-81%	2%-5%	---
Seneca II	---	96%	4%	---
Navajo	---	74%	26%	---
Black Mesa-Kayenta	---	99%	negligible	---

Note: Reclamation cost percentages for Absaloka, Indian Head, and McKinley Mines are not available.

Glossary

NOTE: Italicized words or terms within definitions are defined elsewhere in the glossary.

acid soil: A soil that is acid throughout most or all of the part occupied by plant roots. Commonly, the term applies only to the surface-plowed layer or some other specific layer or horizon of a soil. It refers to a soil with a *pH value* of less than 7.0.

acre: A measure of land area equaling 43,560 square feet.

acre-foot: The volume of water that would cover one *acre* to a depth of one foot: 326,850 gallons.

aerial broadcast seeding: Scattering seeds on the surface of soil by air. This can be done by plane or helicopter. (It can also be done on the ground by machinery that throws or "broadcasts" the seed over the soil.) (See *hydroseeding* and *drill seeding*.)

alkaline soil: A soil that is alkaline throughout most or all of the parts occupied by plant roots; precisely, the term refers to any soil horizon having a *pH value* greater than 7.0.

alluvial valley floor: The floor of a valley in which materials such as clay, sand, silt, gravel or other rocks or minerals are deposited by running water. These areas are extremely valuable agricultural lands because their vegetative productivity is much higher than that of surrounding areas. (Criteria for defining alluvial valley floors are published by OSM in a paper by John Hardaway entitled "Subirrigated Alluvial Valley Floors.")

alluvium: Clay, sand, silt, gravel or other rock or mineral materials deposited by running water.

amendment: Any material, such as lime, sawdust or synthetic conditioners, that is worked into the soil. Technically, a fertilizer is also an amendment, but the term amendment is most commonly used in reference to added materials other than fertilizer.

animal unit month (AUM): The equivalent of the amount of forage consumed by one cow and calf, or five sheep, in one month. When an area of land has the forage productivity of five *acres* per AUM, five acres of land will support one animal unit (one cow and calf or five sheep) for one month.

annuals: Plants that complete their life cycles and die in one year or less.

aquifer: A subsurface stratum of rock, sand, gravel or coal that contains sufficient saturated permeable material to conduct *groundwater* and to yield significant quantities of water to wells and springs.

auger mining: A mining method generally practiced in, but not restricted to, hilly coal-bearing regions. It utilizes a machine designed on the principle of the drill, which bores into an exposed coal seam, conveying the coal to storage piles or bins for loading and transporting. Auger mining may be used alone or in combination with conventional surface mining. When it is used alone, a single cut is made sufficient to expose the coal seam and provide operating space for the machine. When it is used in combination with surface mining, the last-cut pit provides the operating space.

backfill: The entire operation of refilling an excavation. The term is occasionally used to identify the material placed in an excavation in the process of backfilling.

bare-root transplant: Tree or shrub stock, used for transplanting, which has no soil or other material surrounding and protecting the roots. (See Guideline 28.)

berm: A large pile of rock or other relatively heavy, stable material (dirt) placed at the outside bottom of the *spoil pile* or *stockpile* and/or placed higher on the pile to help hold the soils in position.

biennials: Plants that complete their life cycles in two years. In the first year they grow vegetatively, and in the second year they bear fruit and die.

boxcut: The first cut into the overburden layer during the opening of a new mining pit. The *spoil ridge* created is termed the boxcut spoils.

British thermal unit (Btu): Measure of energy content. More specifically, it is the quantity of heat required to raise the temperature of a pound of water one degree Fahrenheit.

broadcast: To scatter or sow seed over a broad area. (See *aerial broadcast seeding.*)

containerized transplant: Use of tree or shrub (sometimes grass) stock, which is transplanted with soil and/or other nutrients and materials surrounding the roots. (See Guideline 28.)

contouring: The *grading* of *spoil* banks to evenly sloped terrain features.

cool-season plants: Plants that have their major growth in the cool seasons of the year, primarily in spring and fall, but in some localities in winter. (See *warm-season plants.*)

crimper: Machinery used to tack or push mulch into soil so it will not blow or wash away.

deep mining: Mining method where the miner goes underground to extract coal.

depth band: An implement used in *drill seeding* to set the depth at which seed will be planted.

direct haulback: The process of removing topsoil from land to be disturbed and spreading it immediately over graded land, without first placing it in a *stockpile.*

discharging: The controlled or uncontrolled emission of water from a *holding pond.*

discing: Cultivation with an implement, such as a plow, that turns and loosens the soil with a series of discs.

disturbed land: Lands where excavation or other disturbance (such as road or building construction) has occurred and/or upon which *overburden* or other materials have been deposited.

DMR: Discharge Monitoring Report filed by the water discharger under conditions of that discharger's *NPDES* permit.

dragline: A large, crane-like device used for earthmoving.

drawdown: To deplete or lower a water level (in a reservoir or *aquifer*).

drill seeding: Selectively planting seed by placing it in rows in the soil.

ecosystem: The network of relationships between living things and their non-living but supporting environment.

effluent: Any water flowing from an enclosure or out of the ground to the surface flow network.

Environmental Protection Agency (EPA): The federal agency responsible for formulating and implementing policies under national environmental laws concerning human activity as it relates to the protection of land, air and water systems.

erosion: The deterioration of land formations by the action of water, wind, or construction activities.

flume: Concrete or steel structure used for measuring water flow and quality in surface streams and creeks.

forage: Plants that can be used as feed by animals. Forage may be grazed or cut for hay.

forb: An *herb* other than a *grass*.

germination: Sprouting; the beginning growth of a seed.

grading: Establishing a profile by *backfilling* and *bulldozing*.

grass: Herbage that is suitable or used for grazing animals.

groundwater: Subsurface water that supplies wells and springs.

gully: A trench formed in the earth by running water after rains. (See *erosion*.)

harrowing: Covering seed with soil.

hectare: 10,000 square meters or approximately 2.47 *acres*.

herb: Seed-producing *annual*, *biennial*, or *perennial* that does not develop woody persistent tissue, but dies down at the end of a growing season.

highwall: The active face of the mining pit. The forward or frontal face of the pit toward which mining proceeds. The highwall is constantly being mined into. (See Techniques of Surrace Mining and Reclamation.)

holding pond: A basin used primarily to detain runoff from *impervious surfaces*.

horizon: A distinct layer (usually refers to soil layers).

hot spots: Areas where highly *acidic* or *sodic* material leaches to the surface and can create conditions preventing vegetative growth.

hydrologic balance: The quantity and quality of water flowing into, held in, and flowing out of a given area.

hydroseeding: The process of planting by spraying seeds in a mixture of water and sometimes *mulch* and/or fertilizer.

impervious surface: A surface such as paving or compacted clay that does not allow water to penetrate.

interburden: Material found between coal seams.

irrigation: Supplying land with water by artificial means.

leasehold: Property leased by the mine operator, generally ranging from 1,000 acres to 30,000 acres or more.

legume: One of the most important and widely distributed plant families, which includes many valuable food and forage species. Some of these are peas, beans, peanuts, clovers, alfalfas and kudzu. Practically all of the legumes are *nitrogen-fixing* plants.

micro-habitat: The local environment of individuals or a small group of plants or animals—characterized by differences in relief, exposure or cover—in which life needs are supplied.

morphology: The external structure of rocks with regard to the development of forms caused by *erosion,* or other topographic features.

mulch: A protective covering spread on ground to reduce evaporation, maintain even soil temperature, prevent *erosion* and enrich the soil.

National Pollutant Discharge Elimination System (NPDES): This program was created by section 402 of the Federal Water Pollution Control Act Amendments of 1972. Since then the majority of States have been approved by EPA to administer the program within their boundaries. NPDES permits regulate the discharge of pollutants into U.S. waters. Further amendments were passed in 1977, and the Act is now known as the Clean Water Act.

nitrogen fixation: The conversion of atmospheric or free nitrogen to nitrogen compounds. In soils nitrogen fixation is the assimilation of free nitrogen from the air by soil organisms, which makes the nitrogen eventually available to plants. Nitrogen-fixing organisms associated with plants, such as *legumes,* are called symbiotic; those not definitely associated with plants are called nonsymbiotic.

nutrients: Elements taken into a plant that are essential to its growth.

Office of Surface Mining (OSM): Established under *SMCRA* to administer this Act. Region V office, in Denver, Colorado, has authority over all mines included in INFORM's sample.

outslope: The slope of a spoil ridge which extends away from the active mining operation.

overburden: Material, earth and rock, under the *topsoil* and above the coal *seam.*

packer wheel: Implement used in *drill seeding* to smooth the soil after seeding.

pasture: Grazing lands that receive more intensive management than *rangeland:* close control of the grazing animals and frequent use of cultivating treatments such as tillage, mowing, weed and brush control and fertilization.

perennials: Plants that survive for several years with continuous new growth.

performance bond: Similar to *reclamation bond.*

permit: Usually a mining permit. Regulatory allowance to conduct mining and related operations on a given piece of land. The company is allowed to engage in activities only on that land, and not outside the boundaries defined in the permit. A *reclamation bond* is usually required for lands included under the mining permit.

pH: An indicator used in expressing both *acidity* and *alkalinity* on a scale whose values run from 0 to 14 with 7 representing neutrality; numbers less than 7 indicate increasing acidity and numbers greater than 7 indicate increasing alkalinity.

porosity: The state of being permeable to liquids.

rangeland: Uncultivated lands, in particular those producing *forage* for animal consumption. Rangeland supports native herbaceous or shrubby plants or introduced forage plants that require little or no periodic re-establishment.

recharge area: An area in which water is absorbed and added to the *groundwater* reservoir.

reclamation: The entire process of *backfilling, grading,* topsoiling, planting, *revegetating,* and other work to restore land affected by surface mining.

reclamation bond: Insurance agreement by a mining company on a mining permit area basis; equal to the presumed expense of reclaiming the site in the event the company abandons its responsibilities for *reclamation.* Upon a finding by regulatory authorities that the land meets the criteria for successful reclamation, the bond is released and company's responsibility ends.

revegetation: Plants or growth that replace the original ground cover following land disturbance.

rights: As in mineral-rights or surface-rights, the term refers to the leasing or purchasing of a given entity, under contractual agreement.

rill: A small *gully.*

ripping: The act of breaking compacted soils or rock with a long, tractor-drawn, angled steel tooth into pieces small enough to be moved by scraper or bulldozer.

rip-rap: Broken rock, cobbles or boulders placed on earth surfaces (such as the bank of a stream or the lining of drainage channels) for slowing and protecting against the erosive action of running water.

runoff: Precipitation that flows over the land.

saltwater intrusion: The displacement of fresh surface water or *groundwater* by the advance of seawater, sometimes caused by *drawdown.*

scarification: Loosening or stirring the surface soil without turning it over.

seam: A stratum of a bed of coal.

sedimentation: The settling of solids to the bottom of a body of water, or the depositing of these solids on lands, after water flow ceases or recedes.

sediment pond: A reservoir for the confinement and retention of sand, gravel, rock, or other debris that settles out of captured runoff. (Also termed as *holding pond.*)

seed: One pound of seed of a single plant can contain 70,000 to 200,000 or more seeds. There are two measures of the amount of seed applied in seeding operations:

> *bulk:* Seed that contains a certain percentage of non-living seed that cannot be expected to germinate and grow.

> *pure live:* Seed made up of live, viable seeds, each of which has the potential to germinate and grow.

shrub: A low, woody plant, usually several-stemmed.

sodic: Laden with sodium.

spoil: *Overburden* or non-coal material removed in gaining access to the coal or mineral material in *surface mining.*

spoil ridge: The area created by the deposited *spoil* or *overburden* material prior to *backfilling.*

square mile: The equivalent of 640 *acres.*

stockpiling: Saving, and placing in storage piles, *topsoil* and/or *subsoil* for replacement on regraded land. This material generally provides the best medium in which plants can grow.

subsoil: The layer of organic material that may exist beneath the *topsoil* and above the *overburden.* It may be salvaged separately to be replaced on graded lands prior to the replacement of topsoil.

surface mining: Mining method in which the overlying materials are removed to expose the mineral for extraction. It is synonymous with strip mining. (See *deep mining* and *auger mining.*)

Surface Mining Control and Reclamation Act of 1977 (SMCRA): This act, Public Law 95-87, provides for cooperation between the Secretary of the Interior and the states regarding the regulation of surface coal mining operations, the acquisition and *reclamation* of abandoned mines, and other related purposes.

surface water: Waters falling upon, arising from, or naturally spreading over land; produced by rainfall, melting snow or springs.

tacking: The process of binding *mulch* fibers together by adding a sprayed chemical compound, or pushing into the soil.

ten-year/24-hour event: A precipitation event or *runoff* from melting snow that is equal to the maximum amount of precipitation within a 10-year interval that would accumulate in 24 hours.

terrace: Sloping ground cut into one of a succession of levels and inclines for cultivation or to control surface *runoff* and minimize soil *erosion.*

tipple: Facility where coal is loaded for transport off the mine.

topography: The configuration of the surface of the land.

topsoil: Soil material presumed to be fertile. It is replaced on regraded land, as it generally provides the best medium in which plants can grow.

Total Suspended Solids (TSS): The measure of non-dissolved solids contained in water, usually measured in milligrams per liter.

tree-spade transplants: Trees transplanted with a machine that lifts the tree, soil, and roots at one time.

tubelings: Containerized plant seedlings planted and nurtured in reinforced paper tubes. The tubeling is ready for transplanting when the root system develops.

warm-season plants: Plants that have their major growth in the warm months of the year, primarily summer and early autumn.

watershed: The upper surface of free *groundwater* in a zone of saturation; also called groundwater table.

Methodology

INFORM's study of the practices of 13 surface coal mining companies at 15 mine sites in the western United States proceeded in six basic stages over two and one-half years from September 1977 to April 1980. It included a one-year basic literature search and field work conducted in the summer of 1979, as well as eight months of writing, rechecking and editing. At many stages, INFORM sought comment and outside review from project advisers and other experts, although the organization and the project authors are solely responsible for the content.

Literature Search

From September 1977 to August 1978 INFORM conducted background research on all phases of surface coal mining, land reclamation, the coal industry and environmental issues related to coal mining. We reviewed annual and Form 10-K Reports (to the Securities and Exchange Commission) from 1976 and 1977 from more than 25 of the largest coal producers in the country, as well as the companies' information releases on their surface mining activities and land reclamation practices. Also reviewed were industry journals, including *Coal Age* and the *Mining Congress Journal*, and the proceedings from symposia on mining and reclamation sponsored by the National Coal Association and Bituminous Coal Research Incorporated from 1973 through 1977.

INFORM surveyed literature published by government, academic and technical experts outside the industry on many aspects of mining and reclamation. These included: *Coal Surface Mining and Reclamation*, prepared by the Council on Environmental Quality; *Economic Engineering Analysis of U.S. Surface Coal Mines and Effective Land Reclamation* by the consulting firm of Skelly &

Loy; the National Academy of Sciences study of the *Reclamation Potential of Western Coal Lands*; *Environmental Protection in Surface Mining of Coal* by the Environmental Protection Agency, and hearings on the Senate and House bills that eventually became the Surface Mining Control and Reclamation Act of 1977 (SMCRA). Materials from various state reclamation programs were also reviewed.

INFORM reviewed the many technical papers describing various aspects of land reclamation and environmental problems, lengthy legal treatises, and general articles on specific company practices or on reclamation. From these it became clear that nowhere was there a full and understandable picture of what an effective reclamation program should consist of, for the non-technical corporate official, regulatory or congressional staff, or concerned member of the public. Yet all the literature and concern in the field cried out for such an explanation, and with it, an evaluation of specific company performance at individual mine sites in the United States.

INFORM established an advisory board for the project, and advisors were consulted on directions the research should take and asked to make substantive reviews of INFORM questionnaires and later chapter drafts.

Scope of the Project

In August 1978 INFORM researchers visited 13 mine sites and reclamation sites in Indiana, Illinois and western Kentucky as part of a tour sponsored by the Southern Illinois University-Carbondale Coal Extraction and Utilization Research Center. On the tour INFORM researchers established contact with academic, regulatory and industry representatives and enlisted their help in further defining the project.

In September-October 1978 INFORM's project research, which had until then broadly explored surface mining and reclamation activities and problems nationwide, focused on mining and reclamation activities in the West. Two primary considerations dictated this decision: the rapid growth of mining in the West owing to large concentrations of readily accessible, low-sulfur coal reserves, and the difficult environmental factors--especially the low precipitation rates--faced by coal companies reclaiming lands in the West.

Some fundamental environmental questions confront the nation as it assesses the need to increase coal production in the West. How do we meet the basic issues of land reclamation, which seeks to restore to its pre-mining conditions land that is approximately equal in productivity and stability? What will be the effect of rapid industrial growth in this least populated area of the nation? What about air and water quality in a region known for its limited water resources? These and other issues have brought the question of expansion in the West into focus across the country--in executive boardrooms and state and federal law-making bodies, as well as in the homes of average citizens.

INFORM focused on the six western states that now produce the most coal and are currently projected by coal companies and government to continue to do so: Arizona, Colorado, Montana, New Mexico, North Dakota, and Wyoming. The next task was to select the sample of mines to be investigated. About 45 surface coal mines exist in these six states, of which about 27 were producing (or planning to produce) at least 1 million tons of coal per year. More than 120 million tons of coal were produced in these six states in 1977, and projections are far more than 560 million tons to be produced by 1987.[1] Mines that began operations after 1974 were excluded, as they could not be reasonably assumed to be far enough along with reclamation plans or programs to illustrate performance quality. From the resulting list of mines, a sampling of those that had been in operation for a relatively long period and of those that had opened more recently was chosen on the basis of 1) the amount of land disturbed at present and to be mined in the future according to the operating company's expansion plans in the West, and 2) the amount of coal the mines were providing to the national energy picture. Another factor was whether the mines selected would fairly represent performance and reclamation problems, as well as providing a geographical cross-section of the six western states where surface coal mining was having its major impact.

A questionnaire was sent to the operators of the 15 mine sites chosen for the INFORM sample. It was reviewed by environmental and industry representatives and members of the project advisory board. In January 1979 INFORM first contacted company officials, requesting information and a visit to their mine sites. On the whole the industry cooperated willingly. Certain companies, including Peabody, Utah International, CONSOL, AMAX, Knife River, Western Energy and Arch Mineral did so immediately. Others--NERCO, Peter Kiewit, NACCO, Energy Fuels, Westmoreland--became willing after initial concerns were resolved. The only company that refused throughout to work with INFORM was the Gulf Oil subsidiary, Pittsburg & Midway Coal. Paradoxically, although Pittsburg & Midway, unlike other coal companies, has run full-page advertisements in leading newspapers and magazines proclaiming that its reclamation practices create land that is as good as or better than that which existed prior to mining, it steadfastly refused to either communicate with INFORM or allow a visit to its McKinley Mine, the subject of its advertisements.

Field Work and Research

The field work stage was concentrated from February through August 1979. INFORM contacted and interviewed environmental regulatory authorities in the six states, officials at the newly established Office of Surface Mining's Region V office in Denver, and officials of the Region VI, VIII and IX offices of the Environmental Protection Agency. We consulted regional and local environmental organizations--including the Dakota Resource Council, Northern Plains Resource Council, Public Lands Institute, Powder River Basin Resource Council, and Colorado Open Space Council--familiar with the 15 sites, and state agricultural and highway departments knowledgeable about region-specific or state-specific information related to land reclamation and agriculture. Also contacted were officials at the U.S. Geological Survey, Soil Conservation Service and Bureau of Land Management.

INFORM researchers spent two weeks in April 1979 in West Germany touring the mining and reclamation operations of that country's largest coal producer, Rheinische Braunkohlenwerke A.G. The tour, sponsored by the Washington, D.C.-based German Marshall Fund, enabled INFORM to investigate the highly successful West German reclamation activities and to ascertain whether any of Rheinbraun's lessons could be applied to surface coal mining in the American West.

Although West Germany has shown that land reclamation can be highly successful, many factors differ significantly from those found in the West. This made direct comparison of their methods to those of western U.S. coal operations inapplicable:

- The terrain in North Rhine Westphalia (where Rheinbraun's operations are located) differs markedly from that in the American West. The Rhineland is largely level with thick, rich topsoil, few toxic materials, and a rainfall adequate for intensive farming and forests. By contrast, most of the western United States is arid or semi-arid with thin topsoil and a dry and windy climate.

- Germany has had reclamation legislation since 1865, so planning has traditionally been better there. The United States passed its first comprehensive federal legislation in 1977.

- In West Germany, production staff and reclamation staff are equals, in contrast to the U.S. system, where reclamation officials usually play a secondary role to production managers.

- Differences in land use between the Rhineland and the American West are considerable. As much of the land was farmland before mining began, Rheinbraun relies heavily on fertilizers and other supplements to reach and sustain these productive levels again. U.S. coal companies, on the other hand, are required by law to return western land--much of which was grazing land--to productivity levels that can be sustained without the use of these amendments. (For more information see INFORM News, October 1979, or contact IN-FORM.)

Over the summer of 1979, the INFORM staff began final research visits to 14 of the 15 mine sites profiled in this study. (As previously mentioned, INFORM was not allowed access to Pittsburg & Midway's McKinley Mine in New Mexico.) Researchers studied all aspects of mining and reclamation practices being employed and noted particularly those that were especially innovative, or environmentally unsound. We met with environmental groups, local landowners, Native American groups and state and federal regulators knowledgeable about each site, and attended a two-day conference in Lander, Wyoming, on the federal surface mining regulations. This meeting provided even greater contact with many of the key environmental and regulatory experts in the West.

Compilation of Data

In the two months following INFORM's return from the West, researchers compiled all data on each mine site into a "fact sheet." In addition, reclamation performance guidelines, against which company practices would be evaluated, were prepared (see Guidelines). The guidelines were designed to provide a lay person with a summary of those planning steps and practices that--from INFORM's research--emerged as essential for a good basic reclamation program. These practices fell into five categories: earthmoving, water resource protection, seeding and revegetation, wildlife concerns, and final goals for vegetative success.

From the end of August through the beginning of October 1979, fact sheets and drafts of INFORM's guidelines were sent to all companies operating the 15 mine sites in INFORM's sample, enabling them to comment on information that we had compiled from all sources. These would be used in profile chapters describing and evaluating their reclamation practices. This also afforded them a final opportunity to add new facts or elaborate on recent specific concerns about their mines. Industry representatives were asked to review INFORM's drafted guidelines, which also were sent to 21 technical experts for review and comment. Of the 34, ten corporate officials and 14 outside experts responded. They are listed at the end of the Guidelines chapter.

Writing

The report was written from September 1979 through April 1980, incorporating reviewer comments on the Guidelines chapter and company comments on the fact sheets. Writing, follow-up telephone interviews, and visits to some mines, as well as continuing review of new literature in the field, proceeded simultaneously. Drafts of each mine site chapter went to individuals with specific knowledge of the site for comment before content was considered final. Experts in the field reviewed the final product before publication.

Notes

Key to Abbreviations:

BLM	U.S. Bureau of Land Management
DEQ	(Wyoming) Department of Environmental Quality
DSL	(Montana) Department of State Lands
EPA	U.S. Environmental Protection Agency
EPC	(Navajo) Environmental Protection Commission
MLRB	(Colorado) Mined Land Reclamation Board
NACCO	North American Coal Company
NERCO	Northern Energy Resources Company
NOAA	National Oceanic and Atmospheric Administration
NPDES	National Pollutant Discharge Elimination System
NPRC	Northern Plains Resource Council
OSM	U.S. Office of Surface Mining
PSC	(North Dakota) Public Service Commission
SCS	U.S. Soil Conservation Service
USGS	U.S. Geological Survey

INTRODUCTION

1. National Coal Association, *Coal Facts 1978-1979* (Washington, D.C.: National Coal Association, 1979), p. 61.
2. *Ibid.*
3. U.S. Department of the Interior, OSM Reclamation and Enforcement, *Final Environmental Statement OSM-EIS-1, Permanent Regulatory Program Implementing Section 501(b) of the Surface Mining Control and Reclamation Act of 1977* (Washington, D.C.: U.S. Government Printing Office, January 1979), p. BIII-3.
4. *Ibid.*, p. BII-5.
5. National Coal Association, *Coal Facts 1978-1979*, p. 81.
6. National Coal Association, *Coal Facts 1974-1975* (Washington, D.C.: National Coal Association, 1975), p. 85.
7. *1979 Keystone Coal Industry Manual* (New York: Mining Information Services, McGraw-Hill, Inc., 1979), p. 665.
8. U.S. Congress, House, "Surface Mining Control and Reclamation Act of 1977," Public Law 95-87, 95th Congress, August 3, 1977, Title I, Section 101(c).

TECHNIQUES OF SURFACE MINING RECLAMATION

1.   Letter from Michael D. Loy, partner, Skelly and Loy Engineering Consultants, to John DiStefano, Research Assistant, INFORM, n.d.
2.   Public Law 91-190, The National Environmental Policy Act of 1969, (Washington, D.C.: 91st Congress), January 1, 1970.
3.   U.S. Congress, Senate, Committee on Energy and Natural Resources, *State Surface Mining Laws: A Survey, A Comparison with the Proposed Federal Legislation, and Background Information*, 95th Congress, First Session, June 1977, Committee Print, p. 136.
4.   Letter from Collier Goodlett, Inspection Specialist, OSM, U.S. Department of the Interior, to Daniel Wiener, Research Associate, INFORM, November 8, 1979.
5.   *Ibid.*
6.   U.S. Department of the Interior, OSM Reclamation and Enforcement, *Surface Coal Mining and Reclamation Operations--Permanent Regulatory Program, Federal Register*, Book 3, Section 745.11(f)(2), p. 15340.
7.   U.S. Department of the Interior, OSM, "News Release: Office of Surface Mining Calls for Comments on Possible Consolidation of Hydrologic Permits," September 28, 1979.
8.   INFORM interview with Lewis McNay, Water Specialist, Technical Services and Research, OSM, U.S. Department of the Interior, October 27, 1979.
9.   Letter from Collier Goodlett, Inspection Specialist, OSM, U.S. Department of the Interior, to Daniel Wiener, Research Associate, INFORM, November 8, 1979.
10. *Ibid.*
11. *Ibid.*
12. U.S. Department of the Interior, Bureau of Mines, Franklin H. Perse, David W. Lockard, and Alec E. Lindquist, *Coal Surface Mining Reclamation Costs in the Western United States*, Information Circular 8737, 1977, p. 9.
13. National Academy of Sciences, *Rehabilitation Potential of Western Coal Lands* (Cambridge, Mass.: Ballinger Publishing Co., 1974), p. 123.
14. U.S. Department of Commerce, Skelly and Loy Engineering Consultants, *U.S. Bureau of Mines Contract Report SO241049: Economic Engineering Analysis of U.S. Surface Coal Mines and Effective Land Reclamation*, February 1975, p. 3-120.
15. INFORM interview with Collier Goodlett, Inspection Specialist, OSM, U.S. Department of the Interior, November 8, 1979; Letter from Collier Goodlett, Inspection Specialist, OSM, U.S. Department of the Interior, to Daniel Wiener, Research Associate, INFORM, November 8, 1979.
16. INFORM interview with Collier Goodlett, Inspection Specialist, OSM, U.S. Department of the Interior, January 21, 1980.
17. U.S. Department of the Interior, OSM Reclamation and Enforcement, *Surface Coal Mining and Reclamation Operations--Permanent Regulatory Program, Federal Register*, Book 3, Section 701.5, p. 15318.
18. *Ibid.*
19. *Ibid.*
20. U.S. Department of the Interior, Bureau of Mines, Perse, Lockard, Lindquist, *Coal Surface Mining Reclamation Costs in the Western United States*, Information Circular 8741, p. 10.
21. U.S. Department of the Interior, OSM Reclamation and Enforcement, *Surface Coal Mining and Reclamation Operations--Permanent Regulatory Program, Federal Register*, Book 3, Section 701.5, p. 15319.
22. *Ibid.*, Section 816.116(b)(1), p. 15413.
23. *Ibid.*, Section 816.117(b)(4).

GUIDELINES

1.   Molly Ivins, "The Desert Creeps Up on the Rangelands of the West," *The New York Times*, July 29, 1979, p. 1.
2.   *1979 Keystone Coal Industry Manual* (New York: Mining Informational Ser-

vices, McGraw-Hill, 1979).

3. National Coal Association, *Coal Facts* (Washington, D.C.: National Coal Association, 1979), p. 75.

4. Francis X. Murray, ed., *Where We Agree: Report of the National Coal Policy Project*, Volume 2 (Boulder, Colo.: Westview Press, 1978), p. 96.

5. Robert Stobaugh and Daniel Yergin, eds., *Energy Future: Report of the National Coal Policy Project at the Harvard Business School* (New York: Random House, 1979), p. 87.

6. INFORM interview with Ed Grandis, Researcher, Environmental Policy Center, October 11, 1979.

7. U.S. Department of the Interior, OSM Reclamation and Enforcement, *Permanent Regulatory Program Implementing Section 501 (b) of the Surface Mining Control and Reclamation Act of 1977--Draft Environmental Statement* (Washington, D.C.: U.S. Government Printing Office, 1979), p. BII-5.

8. U.S. Department of the Interior, USGS, and Montana DSL, *Northern Powder River Basin Coal, Montana--Draft Environmental Statement* (Washington, D.C.: U.S. Government Printing Office, 1979), p. 1V-28.

9. Argonne National Laboratory, Land Reclamation Program, Becky B. Green, Edwin D. Pentecost, and John D. Taylor, *Jim Bridger Project Progress Report for 1977*, ANL/LRP--TM--1, July 1978, p. 1.

10. Colorado Department of Natural Resources, *Rules and Regulations of the Colorado Mined Land Reclamation Board*, 1977, Section 6.5 (d), p. 36.

11. Montana DSL, *Adopted Rules and Regulations Pursuant to Title 50, Chapter 10, R.C.M. 1947--Montana Administrative Code*, 1977, Section 26-2.10(10)-S10350(1)(5), p. 26-48.37.

12. U.S. Department of the Interior, USGS, and Montana DSL, *Northern Powder River Basin Coal, Montana--Draft Environmental Statement*, p. 1V-36.

13. Written communication from Gary Deveraux, Manager of Mined Land Reclamation, NERCO, to INFORM, October 22, 1979.

14. Montana DSL, *Title 50, Chapter 10, R.C.M. 1947--Montana Administrative Code*, Section 26-2.10(10)-S10310(1)(g)(i), p. 26-48.30.

15. U.S. Department of the Interior, OSM Reclamation and Enforcement, *Surface Coal Mining and Reclamation Operations--Permanent Regulatory Program, Federal Register*, Vol. 44, No. 50, Book 3, March 13, 1979, Section 816.101 (a)(3), p. 15411.

16. U.S. Department of the Interior, OSM Reclamation and Enforcement, *Surface Coal Mining and Reclamation Operations--Permanent Regulatory Program, Federal Register*, Book 2, Sec. 816.101(a)(3)(d), p. 15226.

17. New Mexico Coal Surfacemining Commission "Regulations of the State of New Mexico Coal Surfacemining Commission Pursuant to New Mexico Coal Surfacemining Act--Chapter 68, Laws 1972, Section 5(B)," p. 5.

18. Montana DSL, *Title 50, Chapter 10, R.C.M. 1947--Montana Administrative Code*, Section 11(1), p. 14.

19. Wyoming DEQ, *Land Quality Rules and Regulations, 1975*, Chapter II, Section 2(a)(1), p. 4.

20. U.S. Department of the Interior, OSM Reclamation and Enforcement, *Surface Coal Mining and Reclamation Operations--Permanent Regulatory Program, Federal Register*, Book 3, Section 816.102(a), p. 15411.

21. Written communication from Bruce Boyens, Region II Official, OSM, U.S. Department of the Interior, to Daniel Wiener, Research Associate, INFORM, October 19, 1979.

22. National Academy of Sciences, *Rehabilitation Potential of Western Coal Lands* (Cambridge, Mass.: Ballinger Publishing Co., 1974), p. 53.

23. EPA, Mining Pollution Control Branch, Elmore C. Grim and Ronald D. Hill, *Environmental Protection in Surface Mining of Coal*, EPA--670/2--74--093 (Washington, D.C.: U.S. Government Printing Office, 1974), p. 3.

24. McCormack, Donald E., "Soil Reconstruction: For the Best Soil After Mining" (Paper delivered at the Second Research and Applied Technology Symposium on Mined-Land Reclamation, Louisville, Kentucky, October 22-24, 1974), pp. 150, 156.

25. U.S. Department of the Interior, OSM Reclamation and Enforcement, *Surface Coal Mining and Reclamation Operations--Permanent Regulatory Program, Federal Register*, Book 2, Section 816.22, p. 15139.

26. *Ibid.*, Book 3, Section 816.22(b), p. 15397.

27. Written communication from Gary Deveraux, Manager of Mined Land Reclama-
     tion, NERCO, to INFORM, October 26, 1979.
28. EPA, Mining Pollution Control Branch, Grim and Hill, *Environmental Protec-
     tion in Surface Mining of Coal*, p. 170.
29. National Academy of Sciences, *Rehabilitation Potential of Western Coal Lands*,
     p. 54.
30. U.S. Department of the Interior, USGS, and Montana DSL, *Northern Powder
     River Basin Coal, Montana--Draft Environmental Statement*, p. 1V-35
31. Murray, *Report of the National Coal Policy Project*, Volume 2, p. 127.
32. Wyoming DEQ, *Land Quality Rules and Regulations, 1975*, Chapter II, Section
     4(a)(3), p. 6.
33. U.S. Department of the Interior, OSM Reclamation and Enforcement, *Surface
     Coal Mining and Reclamation Operations--Permanent Regulatory Program,
     Federal Register*, Book 3, Section 816.23(a), p. 15397.
34. Wyoming DEQ, *Land Quality Rules and Regulations, 1975*, Chapter II, Section
     4(a), p. 6.
35. Colorado Department of Natural Resources, *Rules and Regulations of the Col-
     orado Mined Land Reclamation Board*, 1977, Section 6.4(g), p. 35.
36. U.S. Department of the Interior, OSM Reclamation and Enforcement, *Surface
     Coal Mining and Reclamation Operations--Permanent Regulatory Program,
     Federal Register*, Book 3, Section 816.24(b)-816.24(b)(1)
37. EPA, Mining Pollution Control Branch, Grim and Hill, *Environmental Protec-
     tion in Surface Mining of Coal*, p. 102.
38. Wyoming DEQ, *Land Quality Rules and Regulations, 1975*, Chapter II, Section
     4(a)(2), p. 6; Colorado Department of Natural Resources, *Rules and Regula-
     tions of the Colorado MLRB*, 1977, Section 6.4(a), p. 35; Montana DSL, *Title
     50, Chapter 10, R.C.M. 1947--Montana Administrative Code*, Section 26-
     2.10(10)-S10340(2), p. 26-48.36.
39. U.S. Department of the Interior, OSM Reclamation and Enforcement, *Surface
     Coal Mining and Reclamation Operations--Permanent Regulatory Program,
     Federal Register*, Book 3, Section 816.23(b)-816.23(b)(1)(ii), p. 15397.
40. EPA, Mining Pollution Control Branch, Grim and Hill, *Environmental Protec-
     tion in Surface Mining of Coal*, pp. 197-199.
41. Written communication from Margaret MacDonald, staff member, Northern
     Plains Resource Council, to Daniel Wiener, Research Associate, INFORM,
     October 24, 1979.
42. Murray, *Report of the National Coal Policy Project*, Volume 2, p. 118.
43. Montana DSL, *Title 50, Chapter 10, R.C.M. 1947--Montana Administrative
     Code*, Section 10(3)-10(3)(a), p. 13.
44. U.S. Department of the Interior, OSM Reclamation and Enforcement, *Surface
     Coal Mining and Reclamation Operations--Permanent Regulatory Program,
     Federal Register*, Book 3, Section 816.103(a)(1)-816.103(a)(2), p. 18412.
45. Old West Regional Commission, Perry H. Rahn, *Potential of Coal Strip-Mine
     Spoils as Aquifers in the Powder River Basin*, Old West Project Number:
     10470025, 1976, p. 5.
46. *Ibid.*, p. 54.
47. Written communication from Russell Boulding, Reclamation Consultant, Na-
     tional Coal Policy Project, to Daniel Wiener, Research Associate, INFORM,
     October 22, 1979.
48. EPA, Mining Pollution Control Branch, Grim and Hill, *Environmental Protec-
     tion in Surface Mining of Coal*, pp. 101-103.
49. INFORM interview with Lewis McNay, Water Specialist, Technical Services
     and Research, OSM, U.S. Department of the Interior, November 13, 1979.
50. Montana DSL, *Title 50, Chapter 10, R.C.M. 1947--Montana Administrative
     Code*, Section 10(3)-10(3)(c), p. 13.
51. U.S. Department of the Interior, OSM Reclamation and Enforcement. *Surface
     Coal Mining and Reclamation Operations--Permanent Regulatory Program,
     Federal Register*, Book 3, Section 816.42(a)(1), p. 15398.
52. EPA, Mining Pollution Control Branch, Grim and Hill, *Environmental Protec-
     tion in Surface Mining of Coal*, p. 149.
53. Murray, *Report of the National Coal Policy Project*, Volume 2, p. 151.
54. Wyoming DEQ, *Land Quality Rules and Regulations, 1975*, Chapter III, Sec-
     tions 1(b), 1(b)(3)-1(b)(4), p. 10.
55. U.S. Department of the Interior, OSM Reclamation and Enforcement, *Surface*

Coal Mining and Reclamation Operations--Permanent Regulatory Program, Federal Register, Book 3, Section 816.43(b)-816.43(c), pp. 15398-15399.

56. Ibid.·, Section 816.47, p. 15401.

57. INFORM interview with Lewis McNay, Water Specialist, Technical Services and Research, OSM, U.S. Department of the Interior, October 2, 1979.

58. Written communication from Robert Coats, Staff Member, Environmental Defense Fund, to Daniel Wiener, Research Associate, INFORM, November 1, 1979.

59. National Academy of Sciences, Rehabilitation Potential of Western Coal Lands, p. 5.

60. Murray, Report of the National Coal Policy Project, Volume 2, p. 146.

61. Ibid., p. 160.

62. Geraghty and Miller, Inc., Consulting Groundwater Geologists and Hydrologists, Ground-Water Contamination: Fundamentals and Monitoring (Syosset, N.Y.: n.d.), pp. 3, 5.

63. Wyoming DEQ, Land Quality Rules and Regulations, 1975, Chapter III, Sections 3(b), 4(a), pp. 11-12.

64. U.S. Department of the Interior, OSM Reclamation and Enforcement, Surface Coal Mining and Reclamation Operations--Permanent Regulatory Program, Federal Register, Book 3, Section 816.52(a)(1), p. 15402.

65. Ibid., Sections 816.52(a)(2), 816.52(b)(1), 816.52(b)(1)(i), p. 15402.

66. Ibid., Section 816.52(b)(2), p. 15403

67. National Academy of Sciences, Rehabilitation Potential of Western Coal Lands, pp. 44-45.

68. U.S. Department of the Interior, OSM Reclamation and Enforcement, Surface Coal Mining and Reclamation Operations--Permanent Regulatory Program, Federal Register, Book 3, Sections 822.11(a), 822.12(a)-822.12(a)(iii), p. 15450.

69. Murray, Report of the National Coal Policy Project, Volume 2, pp. 146-147.

70. U.S. Department of the Interior, OSM Reclamation and Enforcement, Surface Coal Mining and Reclamation Operations--Permanent Regulatory Program, Federal Register, Book 3, Section 771.2, p. 15350.

71. Murray, Report of the National Coal Policy Project, Volume 2, p. 151.

72. Ibid., p. 129.

73. U.S. Department of the Interior, OSM Reclamation and Enforcement, Surface Coal Mining and Reclamation Operations--Permanent Regulatory Program, Federal Register, Book 2, p. 15231.

74. Ibid., Book 3, Section 816.106, p. 15413

75. Ibid., Book 2, p. 15231.

76. EPA, Mining Pollution Control Branch, Grim and Hill, Environmental Protection in Surface Mining of Coal, p. 154.

77. Environmental Policy Institute, John C. Doyle, Jr., State Strip Mining Laws: An Inventory and Analysis of Key Statutory Provisions in 28 Coal Producing States (Washington, D.C.: U.S. Government Printing Office, 1977), p. 3.

78. Wyoming DEQ, Land Quality Rules and Regulations, 1975, Chapter II, Section 4(a)(3), p. 6.

79. Montana DSL, Title 50, Chapter 10, R.C.M. 1947--Montana Administrative Code, Section 10(1), p. 13.

80. U.S. Department of the Interior, OSM Reclamation and Enforcement, Surface Coal Mining and Reclamation Operations--Permanent Regulatory Program, Federal Register, Book 3, Sections 816.101, 816.113, pp. 15411-15413.

81. EPA, Mining Pollution Control Branch, Grim and Hill, Environmental Protection in Surface Mining of Coal, p. 149.

82. Wayne C. Cook, Robert M. Hyde, and Phillip L. Sims, Revegetation Guidelines for Surface Mined Areas: Guidelines for Revegetation and Stabilization of Surface Mined Areas in the Western States, Range Science Series, No. 16, (Ft. Collins, Colo.: Colorado State University, 1974), p. 3.

83. Ibid., p. 14.

84. Vogel, Willis G., "All Season Seeding of Herbaceous Vegetation for Cover on Appalachian Strip-Mine Spoils" (Paper delivered at the Second Research and Applied Technology Symposium on Mined-Land Reclamation, Louisville, Kentucky, October 22-24, 1974), p. 175.

85. Montana DSL, Title 50, Chapter 10, R.C.M. 1947--Montana Administrative Code, Section 26-2.10(10)-S10350(1), p. 26-48.37.

86. U.S. Department of the Interior, OSM Reclamation and Enforcement, *Surface Coal Mining and Reclamation Operations--Permanent Regulatory Program, Federal Register*, Book 3, Section 816.113, p. 15413.

87. *Ibid.*

88. Cook, Hyde, and Sims, *Revegetation Guidelines for Surface Mined Areas*, p. 11.

89. Written communication from Russell Boulding, Reclamation Consultant, National Coal Policy Project, to Daniel Wiener, Research Associate, INFORM, October 22, 1979.

90. Written communication from Margaret McDonald, staff member, Northern Plains Resource Council, to Daniel Wiener, Research Associate, INFORM, October 24, 1979.

91. Cook, Hyde, and Sims, *Revegetation Guidelines for Surface Mined Areas*, p. 13.

92. U.S. Department of Agriculture, Forest Service, Surface Environment and Mining Program, Ray W. Brown and Robert S. Johnston, *Revegetation of an Alpine Mine Disturbance: Beartooth Plateau, Montana*, Research Note INT-206, 1976, p. 7.

93. U.S. Department of the Interior, OSM Reclamation and Enforcement, *Surface Coal Mining and Reclamation Operations--Permanent Regulatory Program, Federal Register*, Book 3, Section 816.111(b)(1), p. 15413.

94. *Ibid.*, Section 816.112(a)-816.112(b).

95. Cook, Hyde, and Sims, *Revegetation Guidelines for Surface Mined Areas*, pp. 9, 1.

96. EPA, *Processes, Procedures, and Methods to Control Pollution from Mining Activities*, EPA--430/9--73--011 (Washington, D.C.: U.S. Government Printing Office, 1973), p. 152.

97. Frank W. Schaller and Paul Sutten, eds., *Reclamation of Drastically Disturbed Lands* (Madison, Wis.: American Society of Agronomy, 1978), p. 401.

98. Wyoming DEQ, *Land Quality Rules and Regulations, 1975*, Chapter II, Section 5(c), p. 9.

99. Montana DSL, *Title 50, Chapter 10, R.C.M. 1947--Montana Administrative Code*, Section 26-2.10-S10350(6), p. 26-48.37.

100. John F. Vallentine, *Range Development and Improvements* (Provo, Utah: Brigham Young University Press, 1971), p. 301.

101. Cook, Hyde, and Sims, *Revegetation Guidelines for Surface Mined Areas*. p. 44.

102. EPA, *Processes, Procedures, and Methods to Control Pollution from Mining Activities*, p. 152.

103. B. T. Walquist, R. L. Dressler, and W. Sowards, "Mined Land Revegetation Without Supplemental Irrigation in the Arid Southwest" (Paper delivered at the Third Symposium on Surface Mining and Reclamation, Louisville, Kentucky, October 21-23, 1975), Volume I, p. 34.

104. Montana DSL, *Title 50, Chapter 10, R.C.M. 1947--Montana Administrative Code*, Section 26-2.10(10)-S10340(4), p. 24-48.36.

105. U.S. Department of the Interior, OSM Reclamation and Enforcement, *Surface Coal Mining and Reclamation Operations--Permanent Regulatory Program, Federal Register*, Book 3, Section 816.114(a), p. 15413.

106. Cook, Hyde, and Sims, *Revegetation Guidelines for Surface Mined Areas*. pp. 3, 5.

107. Vallentine, *Range Development and Improvements*, p. 274.

108. *Ibid.*, p. 264.

109. EPA, Mining Pollution Control Branch, Grim and Hill, *Environmental Protection in Surface Mining of Coal*, p. 191.

110. Schaller and Sutton, *Reclamation of Drastically Disturbed Lands*, p. 436.

111. Montana DSL, *Title 50, Chapter 10, R.C.M. 1947--Montana Administrative Code*, Section 26-2.10(10)-S10350(2), p. 26-48.37.

112. Schaller and Sutton, *Reclamation of Drastically Disturbed Lands*, p. 434; EPA, *Processes, Procedures, and Methods to Control Pollution from Mining Activities*, p. 151; Vogel, "All Season Seeding of Herbaceous Vegetation for Cover on Appalachian Strip-Mine Spoils," p. 153.

113. U.S. Department of the Interior, OSM Reclamation and Enforcement, *Surface Coal Mining and Reclamation Operations--Permanent Regulatory Program, Federal Register*, Book 3, Section 816.114, p. 15413.

114. EPA, Mining Pollution Control Branch, Grim and Hill, *Environmental Protection in Surface Mining of Coal*, p. 189.
115. Schaller and Sutton, *Reclamation of Drastically Disturbed Lands*, p. 434.
116. INFORM interview with Barbara West, Program Specialist, Division of State and Federal Programs, Region V OSM, U.S. Department of the Interior, October 10, 1979.
117. U.S. Department of the Interior, OSM Reclamation and Enforcement, *Surface Coal Mining and Reclamation Operations--Permanent Regulatory Program*, *Federal Register*, Book 3, Section 816.114(b), p. 15413.
118. *Ibid.*, Book 2, p. 15235.
119. Schaller and Sutton, *Reclamation of Drastically Disturbed Lands*, p. 516.
120. *Ibid.*, p. 515.
121. Cook, Hyde, and Sims, *Revegetation Guidelines for Surface Mined Areas*, p. 15.
122. Vallentine, *Range Development and Improvements*, p. 351.
123. U.S. Department of the Interior, OSM Reclamation and Enforcement, *Surface Coal Mining and Reclamation Operations--Permanent Regulatory Program*, *Federal Register*, Book 3, Section 816.25, p. 15397.
124. Frischknecht, Neil C., "Use of Shrubs for Mined Land Reclamation and Wildlife Habitat," reprinted from 1978 *Proceedings Workshop on Reclamation for Wildlife Habitat*.
125. Robert R. Humphrey, *Range Ecology* (New York: Ronald Press Company, 1962), p. 111.
126. *Ibid.*, p. 115.
127. U.S. Department of the Interior, OSM Reclamation and Enforcement, *Surface Coal Mining and Reclamation Operations--Permanent Regulatory Program*, *Federal Register*, Book 3, Section 816.111(a), p. 15413.
128. State of Montana Department of Highways Planning and Research Bureau, Ingvard B. Jensen and Richard L. Hodder, *Tubelings, Condensation Traps, Mature Tree Transplanting and Root Sprigging Techniques for Tree and Shrub Establishment in Semiarid Areas*, Vol. 2, 1977, p. 27.
129. *Ibid.*, pp. 42, 91.
130. Edward J. DePuit, Joe G. Coenenberg, and Waite H. Willmuth, *Research on Revegetation of Surface Mined Lands at Colstrip, Montana: Progress Report, 1975-1977*, Research Report 127 (Bozeman, Mt.: Montana State University, 1978), p. 31.
131. National Academy of Sciences, *Rehabilitation Potential of Western Coal Lands*, p. 57.
132. Murray, *Report of the National Coal Policy Project*, Volume 2, p. 123.
133. National Academy of Sciences, *Rehabilitation Potential of Western Coal Lands*, p. 12.
134. Colorado Department of Natural Resources, *Rules and Regulations of the Colorado Mined Land Reclamation Board*, 1977, Sections 6.3(b)-6.3(c), p. 35.
135. Wyoming DEQ, *Land Quality Rules and Regulations*, 1975, Chapter II, Section 1(b)(3), p. 3.
136. U.S. Department of the Interior, OSM Reclamation and Enforcement, *Surface Coal Mining and Reclamation Operations--Permanent Regulatory Program*, *Federal Register*, Book 3, Sections 816.97(d)(9)-816.97(d)(9)(i)(c), p. 15411.
137. Cook, Hyde, and Sims, *Revegetation Guidelines for Surface Mined Areas*, pp. 29-30.
138. Colorado Department of Natural Resources, *Rules and Regulations of the Colorado Mined Land Reclamation Board*, 1977, Section 6.3(b), p. 35.
139. U.S. Department of the Interior, OSM Reclamation and Enforcement, *Surface Coal Mining and Reclamation Operations--Permanent Regulatory Program*, *Federal Register*, Book 3, Section 816.97(d) - 816.97(d)(4), p. 15411.
140. *Ibid.*, Book 2, p. 15232.
141. Murray, *Report of the National Coal Policy Project*, Volume 2, p. 126.
142. Vallentine, *Range Development and Improvements*, p. 279.
143. Murray, *Report of the National Coal Policy Project*, Volume 2, p. 126.
144. *Ibid.*, p. 123.
145. U.S. Department of the Interior, USGS, and Montana DSL, *Northern Powder River Basin Coal, Montana--Draft Environmental Statement*, p. IV-33.

146. U.S. Department of the Interior, OSM Reclamation and Enforcement, *Surface Coal Mining and Reclamation Operations--Permanent Regulatory Program, Federal Register*, Book 3, Section 816.116(b)(3), p. 15414.
147. *Ibid.*, Section 816.115, p. 15413.

BLACK MESA

1. Professor John L. Thames and Tika R. Verma, Research Associate, University of Arizona, "Reclamation on the Black Mesa of Arizona," *Mining Congress Journal*, September 1977, p. 42-43.
2. Letter from William L. Benjamin, Project Officer, Bureau of Indian Affairs, U.S. Department of the Interior, to Ron Lanoue, Research Associate, INFORM, March 7, 1979.
3. U.S. Department of Commerce, NOAA, *Monthly Averages of Temperature and Precipitation for State Climatic Divisions 1941-70: Arizona*, "Monthly and Annual Divisional Averages Precipitation (Inches)" (Asheville, N.C.: National Climatic Center, July 1973), p. 4.
4. U.S. Bureau of Indian Affairs, 1964 Soil and Range Inventory--Arizona (Reprinted with soil classification 1975); Thames and Verma, "Reclamation on the Black Mesa of Arizona," p. 43.
5. Thames and Verma, "Reclamation on the Black Mesa of Arizona," p. 43.
6. INFORM interview with Leonard Robbins, Inspector, Navajo EPC, June 11, 1979.
7. *1979 Keystone Coal Industry Manual* (New York: Mining Informational Services, McGraw-Hill, 1979).
8. INFORM interviews with Gary Melvin, Environmental Quality Manager, Flagstaff Office, Peabody Coal Company, and Wayne Hilgedick, Manager of Reclamation, Black Mesa Mine, February 20, 1980.
9. INFORM interview with Gary Melvin, Environmental Quality Manager, Flagstaff Office, Peabody Coal Company, February 2, 1980.
10. Peabody Coal Company, Monthly Grading Report--Black Mesa 250, January 1980.
11. INFORM interview with Hall F. Susie, District Mining Supervisor, Arizona USGS, U.S. Department of the Interior, March 22, 1979.
12. Peabody Coal Company, Monthly Grading Report--Black Mesa 250, January 1980.
13. U.S. Department of the Interior, OSM, Region V On-Site Inspection: Peabody Coal Company--Black Mesa Mine, January 24, 1979.
14. Written communication from Peabody Coal Company to INFORM, January 3, 1979.
15. U.S. Department of the Interior, OSM, Region V On-Site Inspection: Peabody Coal Company--Black Mesa Mine, January 24, 1979.
16. INFORM interviews with Gary Melvin, Environmental Quality Manager, Flagstaff Office, Peabody Coal Company, and Wayne Hilgedick, Manager of Reclamation, Black Mesa Mine, February 20, 1980.
17. *Idem.*
18. *Idem.*
19. U.S. Department of the Interior, OSM, Region V On-Site Inspection: Peabody Coal Company--Black Mesa Mine, August 17, 1978.
20. Written communication from Peabody Coal Company to INFORM, n.d.
21. INFORM interviews with Alten F. Grandt, Director of Reclamation, Western Division, Peabody Coal Company, Gary Melvin, Environmental Quality Manager, Flagstaff Office, Peabody Coal Company; and Walter Begay, Reclamation Supervisor, Black Mesa Mine, June 12, 1979.
22. U.S. Department of the Interior, USGS, Inspection Report: Peabody Coal Company (Black Mesa Mine), April 14, 1978.
23. INFORM interview with Gary Melvin, Environmental Quality Manager, Flagstaff Office, Peabody Coal Company, February 20, 1980.
24. INFORM interviews with Gary Melvin, Environmental Quality Manager, Flagstaff Office, Peabody Coal Company, and Wayne Hilgedick, Manager of Reclamation, Black Mesa Mine, February 20, 1980.

25. Written communication from Peabody Coal Company to INFORM, January 3, 1979.

26. INFORM interviews with Gary Melvin, Environmental Quality Manager, Flagstaff Office, Peabody Coal Company, and Wayne Hilgedick, Manager of Reclamation, Black Mesa Mine, February 20, 1980.

27. Idem.; Written communication from Peabody Coal Company to INFORM, January 3, 1979.

28. Written communication from Peabody Coal Company to INFORM, January 3, 1979.

29. INFORM interviews with Gary Melvin, Environmental Quality Manager, Flagstaff Office, Peabody Coal Company, and Wayne Hilgedick, Manager of Reclamation, Black Mesa Mine, February 20, 1980.

30. Idem.

31. Written communication from Peabody Coal Company to INFORM, January 3, 1979.

32. INFORM interview with Alten F. Grandt, Director of Reclamation, Western Division, Peabody Coal Company, November 29, 1979.

33. Letter from William Benjamin, Project Officer, Bureau of Indian Affairs, U.S. Department of the Interior, to Ron Lanoue, Research Associate, INFORM, March 7, 1979.

34. INFORM interview with Wayne Hilgedick, Manager of Reclamation, Black Mesa Mine, February 20, 1980.

35. INFORM interviews with Gary Melvin, Environmental Quality Manager, Flagstaff Office, Peabody Coal Company, and Wayne Hilgedick, Manager of Reclamation, Black Mesa Mine, February 20, 1980.

36. INFORM interview with Wayne Hilgedick, Manager of Reclamation, Black Mesa Mine, February 20, 1980.

37. U.S. Department of the Interior, OSM, Region V On-Site Inspection: Peabody Coal Company--Black Mesa Mine, August 17, 1978.

38. Written communication from Peabody Coal Company to INFORM, January 3, 1979.

39. U.S. Department of the Interior, USGS, Inspection Report: Peabody Coal Company (Black Mesa Mine), November 29, 1978.

40. Written communication from Peabody Coal Company to INFORM, January 3, 1979.

41. U.S. Department of the Interior, OSM, Region V On-Site Inspection: Peabody Coal Company--Black Mesa Mine, January 24, 1979.

42. Written communication from Peabody Coal Company to INFORM, January 3, 1979.

43. INFORM interviews with Gary Melvin, Environmental Quality Manager, Flagstaff Office, Peabody Coal Company, and Wayne Hilgedick, Manager of Reclamation, Black Mesa Mine, February 20, 1980.

44. INFORM interview with Wayne Hilgedick, Manager of Reclamation, Black Mesa Mine, February 20, 1980.

45. INFORM interview with one of five Navajo Indians not wishing to be identified, February 21, 1980.

46. INFORM interview with Wayne Hilgedick, Manager of Reclamation, Black Mesa Mine, February 20, 1980.

47. Idem.; INFORM interview with one of five Navajo Indians not wishing to be identified, February 21, 1980.

48. INFORM interviews with Gary Melvin, Environmental Quality Manager, Flagstaff Office, Peabody Coal Company, and Wayne Hilgedick, Manager of Reclamation, Black Mesa Mine, February 20, 1980.

49. INFORM interviews with Alten F. Grandt, Director of Reclamation, Western Division, Peabody Coal Company, Gary Melvin, Environmental Quality Manager, Flagstaff Office, Peabody Coal Company, and Walter Begay, Reclamation Supervisor, Black Mesa Mine, June 12, 1979.

50. Tika R. Verma, Research Associate, and Professor John L. Thames, University of Arizona, "Rehabilitation of Land Disturbed by Surface Mining of Coal in Arizona," (1975), p. 7.

51. Professor John L. Thames and Tika R. Verma, Research Associate, University of Arizona, "Coal Mine Reclamation on the Black Mesa and The Four Corners Area of Northeastern Arizona," (1975).

52. U.S. Department of the Interior, OSM, Region V On-Site Inspection: Pea-

body Coal Company--Black Mesa Mine, August 17, 1978.

53. U.S. Department of the Interior, OSM, Region V On-Site Inspection: Pea-
     body Coal Company--Black Mesa Mine, January 24, 1979; U.S. Department
     of the Interior, OSM Reclamation and Enforcement, Notice of Violation 79-V-
     1-1, January 24, 1979; U.S. Department of the Interior, OSM, Region V
     On-Site Inspection: Peabody Coal Company--Black Mesa Mine, June 21,
     1979.

54. Written communication from Peabody Coal Company to INFORM, n.d.; Letter
     from James R. Jones, Vice President, Environmental Affairs, Peabody Coal
     Company, to Region IX Officials, Environmental Protection Agency, March
     28, 1978; Letter from Meade Stirland, Manager, Monitoring Section, Bureau
     of Water Quality Control, Arizona Department of Health Services, to Stan
     Liebowitz, Permits Branch E-4-1, Region IX, EPA, September 13, 1978.

55. Letter from Clyde B. Eller, Director, Enforcement Division, Region IX,
     EPA, to Ron Lanoue, Research Associate, INFORM, December 26, 1978.

56. INFORM interviews with Gary Melvin, Environmental Quality Manager, Flag-
     staff Office, Peabody Coal Company, and Wayne Hilgedick, Manager of Rec-
     lamation, Black Mesa Mine, February 20, 1980.

57. INFORM interview with Gary Melvin, Environmental Quality Manager, Flag-
     staff Office, Peabody Coal Company, February 20, 1980.

58. Written communication from Peabody Coal Company to INFORM, January 3,
     1979; Peabody Coal Company, Black Mesa Mine, "Mine Plan, Appendix 10,
     Exhibit G--Hydrological Monitoring Program," 1979.

59. Ibid.

60. INFORM interviews with Gary Melvin, Environmental Quality Manager, Flag-
     staff Office, Peabody Coal Company, and Wayne Hilgedick, Manager of Rec-
     lamation, Black Mesa Mine, February 20, 1980; Peabody Coal Company,
     Black Mesa Mine, "Mine Plan, Appendix 10, Exhibit G--Hydrological Moni-
     toring Program," 1979.

61. U.S. Department of the Interior, USGS, E.H.McGavock and Gary W. Lev-
     ings, "Groundwater in the Navajo Sandstone in the Black Mesa Area, Ari-
     zona,"Guidebook of Monument Valley and Vicinity: Arizona and Utah (New
     Mexico Geological Society, 24th Field Conference, October 4-6, 1973), pp.
     153-155.

62. INFORM interview with Edward Evans, Director of Land and Water Resources
     Management, Williams Brothers Engineering Co., Tulsa, Oklahoma, February
     27, 1980.

63. Written communication from Peabody Coal Company to INFORM, January 3,
     1979.

64. INFORM interview with James Goodman, Professor of Geography, University
     of Oklahoma, March 26, 1980.

65. U.S. Department of the Interior, OSM, Region V On-Site Inspection: Pea-
     body Coal Company--Black Mesa Mine, August 17, 1978.

66. Ibid.; U.S. Department of the Interior, OSM, Region V On-Site Inspection:
     Peabody Coal Company--Black Mesa Mine, January 24, 1979.

67. U.S. Department of the Interior, OSM, Region V On-Site Inspection: Pea-
     body Coal Company--Black Mesa Mine, June 21, 1979.

68. Tika R. Verma, Research Associate, University of Arizona, "Strip Mining
     and Hydrologic Environment on Black Mesa" (1975); Thames and Verma,
     "Coal Mine Reclamation on the Black Mesa and The Four Corners Area of
     Northeastern Arizona," (1975).

69. U.S. Department of the Interior, OSM, Region V On-Site Inspection: Pea-
     body Coal Company--Black Mesa Mine, August 17, 1978.

70. INFORM interview with Hall F. Susie, District Mining Supervisor, Arizona
     USGS, U.S. Department of the Interior, March 22, 1979.

71. Written communication from Peabody Coal Company to INFORM, n.d.

72. INFORM interview with Wayne Hilgedick, Manager of Reclamation, Black
     Mesa Mine, February 22, 1980.

73. Idem.

74. INFORM interviews with Gary Melvin, Environmental Quality Manager, Flag-
     staff Office, Peabody Coal Company, and Wayne Hilgedick, Manager of Rec-
     lamation, Black Mesa Mine, February 20, 1980.

75. INFORM interview with Wayne Hilgedick, Manager of Reclamation, Black
     Mesa Mine, February 22, 1980.

76. Memorandum from Leonard Robbins, Inspector, Navajo EPC, to Daniel Wiener, Research Associate, INFORM, March 23, 1979.

77. INFORM interview with Leonard Robbins, Inspector, Navajo EPC, June 11, 1979.

78. INFORM interviews with Gary Melvin, Environmental Quality Manager, Flagstaff Office, Peabody Coal Company, and Wayne Hilgedick, Manager of Reclamation, Black Mesa Mine, February 20, 1980.

79. *Idem.*

80. *Idem.*

81. *Idem.*

82. INFORM interview with Frank Farnsworth, Director of Engineering, Arizona Division, Peabody Coal Company, February 20, 1980.

83. INFORM interview with Gary Melvin, Environmental Quality Manager, Flagstaff Office, Peabody Coal Company, February 20, 1980.

84. INFORM interviews with Frank Farnsworth, Director of Engineering, Arizona Division, Peabody Coal Company, and Gary Melvin, Environmental Quality Manager, Flagstaff Office, Peabody Coal Company, February 20, 1980.

85. INFORM interview with one of five Navajo Indians not wishing to be identified, February 21, 1980.

86. INFORM interview with Wayne Hilgedick, Manager of Reclamation, Black Mesa Mine, February 20, 1980.

87. INFORM interviews with Gary Melvin, Environmental Quality Manager, Flagstaff Office, Peabody Coal Company, and Wayne Hilgedick, Manager of Reclamation, Black Mesa Mine, February 20, 1980.

88. INFORM interview with Gary Melvin, Environmental Quality Manager, Flagstaff Office, Peabody Coal Company, February 20, 1980.

89. U.S. Department of the Interior, OSM, Region V On-Site Inspection: Peabody Coal Company--Black Mesa Mine, August 17, 1978.

90. INFORM interviews with Alten F. Grandt, Director of Reclamation, Western Division, Peabody Coal Company, Gary Melvin, Environmental Quality Manager, Flagstaff Office, Peabody Coal Company, and Walter Begay, Reclamation Supervisor, Black Mesa Mine, June 12, 1979.

91. *Idem.*; February 20, 1980.

92. INFORM interviews with Gary Melvin, Environmental Quality Manager, Flagstaff Office, Peabody Coal Company, and Wayne Hilgedick, Manager of Reclamation, Black Mesa Mine, February 20, 1980.

93. INFORM interviews with Gary Melvin, Environmental Quality Manager, Flagstaff Office, Peabody Coal Company, and Frank Farnsworth, Director of Engineering, Arizona Division, Peabody Coal Company, February 20, 1980.

94. U.S. Department of the Interior, USGS, Inspection Report: Peabody Coal Company (Black Mesa Mine), March 13, 1978.

95. Memorandum from Hall F. Susie, District Mining Supervisor, Arizona USGS, U.S. Department of the Interior, to Ron Lanoue, Research Associate, INFORM, December 5, 1978.

96. INFORM interview with Hall F. Susie, District Mining Supervisor, Arizona USGS, U.S. Department of the Interior, February 20, 1980.

97. INFORM interviews with Gary Melvin, Environmental Quality Manager, Flagstaff Office, Peabody Coal Company, and Wayne Hilgedick, Manager of Reclamation, Black Mesa Mine, February 20, 1980.

98. INFORM interview with Andrew Benallie, Chairman, Black Mesa Review Board, March 21, 1980.

99. *Idem.*, March 12, 1979.

100. *Idem.*, March 12, 1979.

101. INFORM interviews with Gary Melvin, Environmental Quality Manager, Flagstaff Office, Peabody Coal Company, and Wayne Hilgedick, Manager of Reclamation, Black Mesa Mine, February 20, 1980.

102. *Idem.*

103. INFORM interviews with Gary Melvin, Environmental Quality Manager, Flagstaff Office, Peabody Coal Company, and Wayne Hilgedick, Manager of Reclamation, Black Mesa Mine, February 20, 1980.

104. INFORM interview with one of five Navajo Indians not wishing to be identified, February 21, 1980.

105. Written communication from Peabody Coal Company to INFORM, n.d.

106. *Ibid.*
107. U.S. Bureau of Indian Affairs, 1964 Soil and Range Inventory--Arizona (Reprinted with soil classification 1975); INFORM interviews with Gary Melvin, Environmental Quality Manager, Flagstaff Office, Peabody Coal Company, and Wayne Hilgedick, Manager of Reclamation, Black Mesa Mine, February 20, 1980.
108. Written communication from Peabody Coal Company to INFORM, January 3, 1979.
109. Written communication from Peabody Coal Company to INFORM, n.d.
110. A. Kent Evans, E.W. Uhleman, and P.A. Eby, *Atlas of Western Surface Mined Lands: Coal, Uranium and Phosphate* (Washington, D.C.: U.S. Department of the Interior, Fish and Wildlife Service, Biological Services Program, January 1978), p. 255.
111. Written communication from Peabody Coal Company to INFORM, n.d.
112. Letter from Thomas E. Ehmett, Reclamation Specialist, Region V, OSM, U.S. Department of the Interior, to Daniel Wiener, Research Associate, INFORM, March 8, 1979.
113. Written communication from Peabody Coal Company to INFORM, n.d.
114. INFORM interviews with Alten F. Grandt, Director of Reclamation, Western Division, Peabody Coal Company; Gary Melvin, Environmental Quality Manager, Flagstaff Office, Peabody Coal Company; and Walter Begay, Reclamation Supervisor, Black Mesa Mine, June 12, 1979.
115. U.S. Department of the Interior, OSM, Region V On-Site Inspection: Peabody Coal Company--Black Mesa Mine, August 17, 1978.
116. U.S. Department of the Interior, USGS, Inspection Report: Peabody Coal Company (Black Mesa Mine), September 29, 1978.
117. U.S. Department of the Interior, USGS, Inspection Report: Peabody Coal Company (Black Mesa Mine), November 29, 1978.
118. U.S. Department of the Interior, OSM, Region V On-Site Inspection: Peabody Coal Company--Black Mesa Mine, June 21, 1979.
119. INFORM interviews with Gary Melvin, Environmental Quality Manager, Flagstaff Office, Peabody Coal Company, and Wayne Hilgedick, Manager of Reclamation, Black Mesa Mine, February 20, 1980.
120. *Idem.*
121. *Idem.*
122. Peabody Coal Company, Black Mesa Mine, "Mine Plan, Appendix 6, Exhibit C--Fish and Wildlife Resources Information," 1979.
123. INFORM interview with Janet Bingham, teacher, Rock Point Community School, Rock Point, Arizona, March 24, 1979.
124. INFORM interviews with Alten F. Grandt, Director of Reclamation, Western Division, Peabody Coal Company; Gary Melvin, Environmental Quality Manager, Flagstaff Office, Peabody Coal Company; and Walter Begay, Reclamation Supervisor, Black Mesa Mine, June 12, 1979.
125. Written communication from Peabody Coal Company to INFORM, n.d.; INFORM interview with one of five Navajo Indians not wishing to be identified.
126. Leo C. Pachner, "An Experiment in Tilapia Cage Culture," *Farm Pond Harvest*, Winter 1976, p. 12.
127. Professor Boyd Kynard, "Black Mesa Fish Culture Project," *Peabody Magazine*, July 4, 1976.
128. *Ibid.*; Professor John L. Thames and Tika R. Verma, Research Associate, University of Arizona, "Multiple Uses of Surface-Mined Land in the Southwest" (Paper delivered at the Fourth Symposium on Surface Mining and Reclamation, Louisville, Kentucky, October 19-21, 1976), p. 234.

McKINLEY

1. U.S. Department of the Interior, BLM, *Draft Environmental Statement, Star Lake-Bisti Regional Coal--Regional Analysis* (Washington, D.C.: U.S. Government Printing Office, September 29, 1978), p. II-23.
2. INFORM interview with Leonard Robbins, Inspector, Navajo EPC, June 11, 1979.

3. U.S. Department of the Interior, OSM, Region V On-Site Inspection: Pittsburg & Midway Coal Co.--McKinley Mine, May 16, 1978.

4. U.S. Department of the Interior, BLM, *Draft Environmental Statement, Star Lake-Bisti Regional Coal--Regional Analysis*, p. IV-27.

5. INFORM interview with Leonard Robbins, Inspector, Navajo EPC, June 11, 1979.

6. *1979 Keystone Coal Industry Manual* (New York: Mining Informational Services, McGraw-Hill, 1979).

7. Earth Science Associates, Karl Vonder Linden, Mining Engineer and Engineering Geologist, "Strip Mine Inspection Report: Pittsburg & Midway Coal Mining Company--McKinley Mine," October 25, 1977-April 17, 1978.

8. U.S. Department of the Interior, USGS, Inspection Report: Pittsburg & Midway Coal Mining Company--McKinley Mine, April 21, 1977.

9. INFORM interview with Jack Reynolds, Inspector, New Mexico Coal Surface-mining Commission, May 17, 1979.

10. U.S. Department of the Interior, USGS, Inspection Report: Pittsburg & Midway Coal Mining Company--McKinley Mine, April 21, 1977.

11. Earth Science Associates, Vonder Linden, "Strip Mine Inspection Report: Pittsburg & Midway Coal Mining Company--McKinley Mine," April 18, 1977.

12. U.S. Department of the Interior, OSM, Region V On-Site Inspection: Pittsburg & Midway Coal Co.--McKinley Mine, Box M, May 22, 1979.

13. U.S. Department of the Interior, OSM, Region V On-Site Inspection: Pittsburg & Midway Coal Co.--McKinley Mine, May 16, 1978; Earth Science Associates, Vonder Linden, "Strip Mine Inspection Report: Pittsburg & Midway Coal Mining Company--McKinley Mine," April 17, 1978.

14. U.S. Department of the Interior, OSM, Region V On-Site Inspection: Pittsburg & Midway Coal Co.--McKinley Mine, Box M, May 22, 1979.

15. Earth Science Associates, Vonder Linden, "Strip Mine Inspection Report: Pittsburg & Midway Coal Mining Company--McKinley Mine," April 17, 1978.

16. U.S. Department of the Interior, OSM, Region V On-Site Inspection: Pittsburg & Midway Coal Co.--McKinley Mine, April 16, 1978.

17. Earth Science Associates, Vonder Linden, "Strip Mine Inspection Report: Pittsburg & Midway Coal Mining Company--McKinley Mine," October 24, 1977.

18. INFORM interview with Jack Reynolds, Inspector, New Mexico Coal Surface-mining Commission, May 17, 1979

19. U.S. Department of the Interior USGS, Inspection Report: Pittsburg & Midway Coal Mining Company--McKinley Mine, October 20, 1978.

20. INFORM interview with Harris Arthur, Bureau of Indian Affairs, U.S. Department of the Interior, Window Rock, Arizona, March 20, 1979.

21. Computer printout from H. Robert Moore, Assistant to the Director for Coal Management, BLM, U.S. Department of the Interior, November 20, 1978.

22. INFORM interview with Jack Reynolds, Inspector, New Mexico Coal Surface-mining Commission, May 17, 1979.

23. Letter from E. L. Noble, Soil Scientist, Intermountain Forest and Range Experiment Station, U.S. Department of Agriculture Forest Service, to Henry McCabe, Pittsburg & Midway Coal Mining Company, November 11, 1977; U.S. Department of the Interior, USGS, Inspection Report: Pittsburg & Midway Coal Mining Company--McKinley Mine, February 22, 1978.

24. U.S. Department of the Interior, USGS, Inspection Report: Pittsburg & Midway Coal Mining Company--McKinley Mine, April 24, 1978.

25. Earth Science Associates, Vonder Linden, "Strip Mine Inspection Report: Pittsburg & Midway Coal Mining Company--McKinley Mine," April 17, 1978.

26. INFORM interview with Al Taradash, Attorney, People's Legal Services (DNA), Window Rock, Arizona, June 11, 1979.

27. U.S. Department of the Interior, OSM, Region V On-Site Inspection: Pittsburg & Midway Coal Mining Company--McKinley Mine, Box M, May 22, 1979.

28. INFORM interview with Tom Tippeconic, Specialist in Indian Affairs, OSM, U.S. Department of the Interior, April 14, 1980.

29. Pittsburg & Midway Coal Mining Company, "Annual Reclamation Report Permit #3, Appendix C: Reclamation Summary," March 1, 1977-February 28, 1978.

30. Earth Science Associates, Vonder Linden, "Strip Mine Inspection Report: Pittsburg & Midway Coal Mining Company--McKinley Mine," April 17, 1978.

31. U.S. Department of the Interior, BLM, *Draft Environmental Statement, Star Lake-Bisti Regional Coal--Regional Analysis,* p. IV-27.
32. INFORM interview with Tom Tippeconic, Specialist in Indian Affairs, OSM, U.S. Department of the Interior, April 14, 1980.
33. INFORM interview with Leonard Robbins, Inspector, Navajo EPC, June 11, 1979.
34. Pittsburg & Midway Coal Mining Company, "Annual Reclamation Report Permit #3," April 27, 1978.
35. *Ibid.*
36. *Ibid.*
37. Letter from David G. Scholl, Range Scientist, Rocky Mountain Forest and Range Experiment Station, U.S. Department of Agriculture Forest Service, to Henry McCabe, Pittsburg & Midway Coal Company, December 13, 1977.
38. U.S. Department of the Interior, OSM, Region V On-Site Inspection: Pittsburg & Midway Coal Mining Company--McKinley Mine, May 16, 1978; Pittsburg & Midway Coal Mining Company, "Annual Reclamation Report Permit #3," April 27, 1978.
39. U.S. Department of the Interior, OSM, Region V On-Site Inspection: Pittsburg & Midway Coal Mining Company--McKinley Mine, April 18, 1977; Earth Science Associates, Vonder Linden, "Strip Mine Inspection Report: Pittsburg & Midway Coal Mining Company--McKinley Mine," April 17, 1978; Pittsburg & Midway Coal Mining Company, "Annual Reclamation Report Permit #3," April 27, 1978.
40. INFORM interview with Al Taradash, Attorney, People's Legal Services (DNA), Window Rock, Arizona, June 11, 1979.
41. Earth Science Associates, Vonder Linden, "Strip Mine Inspection Report: Pittsburg & Midway Coal Mining Company--McKinley Mine," April 17, 1978; U.S. Department of the Interior, OSM, Region V On-Site Inspection: Pittsburg & Midway Coal Mining Company--McKinley Mine, Box M, May 22, 1979; INFORM interview with Leonard Robbins, Inspector, Navajo EPC, June 11, 1979.
42. U.S. Department of the Interior, OSM, Region V On-Site Inspection: Pittsburg & Midway Coal Mining Company--McKinley Mine, Box M, May 22, 1979; Memorandum from John Burleson, Assistant Area Mining Supervisor, SRMA, Carlsbad, to Area Mining Supervisor, SMRA, Albuquerque, New Mexico.
43. Walter L. Gould, "Vegetative Cover and Herbage Production on Native Range and Seeded Mine Spoils at the McKinley Mine in October, 1977, A Report to Pittsburg & Midway Company" (New Mexico State University, 1977), p. 1.
44. A. Kent Evans, E.W. Uhleman, and P.A. Eby, *Atlas of Western Surface-Mined Lands: Coal, Uranium and Phosphate* (Washington, D.C.: U.S. Department of the Interior, Fish and Wildlife Service, Biological Services Program, January 1978), p. 184.
45. Gulf Oil Corporation, Advertisement, "We need the land as much as we need the coal. We found a way to have both," *The Wall Street Journal,* September 19, 1978, p. 23.
46. U.S. Department of the Interior, BLM, *Draft Environmental Statement, Star Loke-Bisti Regional Coal--Regional Analysis,* p.V-5.
47. INFORM interview with William Beck, Bureau of Indian Affairs, Window Rock, Arizona, June 13, 1979.

NAVAJO

1. U.S. Department of the Interior, Bureau of Reclamation, *Proposed Modification to the Four Corners Powerplant and Navajo Mine, New Mexico--Final Environmental Statement,* Vol. I (Washington, D.C.: U.S. Government Printing Office, 1976), pp. 1.32-1.33.
2. *Ibid.*
3. INFORM interviews with Sterling Grogan, Senior Environmental Engineer, Wes Karma, Assistant Mine Manager, and Ray Benally, Range Specialist, Navajo Mine, and Roger Nelson, Environmental Specialist, San Francisco Office, Utah

International, Inc., June 14, 1979.

4. Written communication from J. LeRoy Balzer, Director, Environmental Quality, Utah International, Inc., to INFORM, October 8, 1979.

5. U.S. Department of the Interior, Bureau of Reclamation, *Navajo Mine, New Mexico--Final Environmental Statement*, pp. 2.55-2.58.

6. U.S. Department of the Interior, OSM, Region V On-Site Inspection: Utah International, Inc.--Navajo Mine, August 18, 1979; Written communication from J. Leroy Balzer, Director, Environmental Quality, Utah International, Inc., to INFORM, May 22, 1979.

7. *Ibid.;* U.S. Department of the Interior, Bureau of Reclamation, *Navajo Mine, New Mexico--Final Environmental Statement*, p. 2.83.

8. U.S. Department of the Interior, Bureau of Reclamation, *Navajo Mine, New Mexico--Final Environmental Statement*, p. 2.124.

9. Written communication from J. LeRoy Balzer, Director, Environmental Quality, Utah International, Inc., to INFORM, May 22, 1979.

10. U.S. Department of the Interior, Bureau of Reclamation, *Navajo Mine, New Mexico--Final Environmental Statement*, p. 1.51.

11. INFORM interview with Orlando Estrada, Acting Senior Environmental Engineer, Navajo Mine, March 26, 1980.

12. U.S. Department of the Interior, OSM, Region V On-Site Inspection: Utah International, Inc.--Navajo Mine, August 18, 1979.

13. U.S. Department of the Interior, OSM, Region V On-Site Inspection: Utah International, Inc.--Navajo Mine, February 27, 1978; INFORM interview with Orlando Estrada, Acting Senior Environmental Engineer, Navajo Mine, November 6, 1979.

14. INFORM interview with Jack Reynolds, Inspector, New Mexico Coal Surface-mining Commission, May 17, 1979.

15. Utah International, Inc., "Annual Report of the Navajo Mine to the New Mexico Coal Surfacemining Commission," April 28, 1978.

16. INFORM interview with Orlando Estrada, Acting Senior Environmental Engineer, Navajo Mine, November 6, 1979.

17. INFORM interviews with Sterling Grogan, Senior Environmental Engineer, Wes Karma, Assistant Mine Manager, and Ray Benally, Range Specialist, Navajo Mine, and Roger Nelson, Environmental Specialist, San Francisco Office, Utah International, Inc., June 14, 1979.

18. Written communication from J. LeRoy Balzer, Director, Environmental Quality, Utah International, Inc., to INFORM, October 8, 1979.

19. U.S. Department of the Interior, USGS, Field Inspection Reports: Utah International, Inc.--Navajo Mine, February 27, 1978 and October 26, 1978.

20. Earth Science Associates, Karl Vonder Linden, Mining Engineer and Engineering Geologist, "Strip Mine Inspection Report: Utah International, Inc.--Navajo Mine," October 20, 1976-April 19, 1977.

21. INFORM interview with Richard Hughes, Attorney, People's Legal Services (DNA), Window Rock, Arizona, June 11, 1979.

22. U.S. Department of the Interior, USGS, Field Inspection Report: Utah International, Inc.--Navajo Mine, February 27, 1978; Written communication from J. LeRoy Balzer, Director, Environmental Quality, Utah International, Inc. to INFORM, October 8, 1979.

23. INFORM interview with Orlando Estrada, Acting Senior Environmental Engineer, Navajo Mine, November 6, 1979.

24. *Idem.*

25. U.S. Department of the Interior, USGS, Field Inspection Report: Utah International, Inc.--Navajo Mine, February 27, 1978.

26. U.S. Department of the Interior, OSM, Region V On-Site Inspection: Utah International Inc.--Navajo Mine, August 18, 1978; Earth Science Associates, Vonder Linden, "Strip Mine Inspection Report: Utah International, Inc.--Navajo Mine," July 20, 1977-October 25, 1977, and October 26, 1977-April 18, 1978.

27. U.S. Department of the Interior, OSM, Region V On-Site Inspection: Utah International, Inc.--Navajo Mine, August 18, 1978.

28. *Ibid.*

29. U.S. Department of the Interior, USGS, Inspection Report: Utah International, Inc.,--Navajo Mine, October 26, 1978; INFORM interview with Harris Arthur, Bureau of Indian Affairs, U.S. Department of the Interior, Window

Rock, Arizona, March 20, 1979; INFORM interview with Lee Bia, Inspector, Navajo EPC, June 11, 1979.

30. Written communication from J. LeRoy Balzer, Director, Environmental Quality, Utah International, Inc. to INFORM, October 8, 1979.

31. INFORM interviews with Sterling Grogan, Senior Environmental Engineer, Wes Karma, Assistant Mine Manager, and Ray Benally, Range Specialist, Navajo Mine, and Roger Nelson, Environmental Specialist, San Francisco Office, Utah International, Inc., June 14, 1979.

32. U.S. Department of the Interior, USGS, Inspection Report: Utah International, Inc. Navajo Mine, February 27, 1978.

33. INFORM interview with Jack Reynolds, Inspector, New Mexico Coal Surface-mining Commission, May 17, 1979.

34. Written communication from J. LeRoy Balzer, Director, Environmental Quality, Utah International, Inc. to INFORM, October 8, 1979.

35. EPA, "Authorization to Discharge Under the National Pollutant Discharge Elimination System, Permit No. NM0028193," March 28, 1977.

36. Letter from John T. Atkins, Manager, Navajo Mine, Region VI, EPA, August 9, 1977.

37. INFORM interview with Paul Robinson, Research Associate, Southwest Research and Information Center, Albuquerque, New Mexico, April 17, 1980.

38. U.S. Department of the Interior, USGS, Inspection Report: Utah International, Inc.--Navajo Mine, October 26, 1978; Written communication from J. LeRoy Balzer, Director, Environmental Quality, Utah International, October 8, 1979.

39. INFORM interviews with Sterling Grogan, Senior Environmental Engineer, Wes Karma, Assistant Mine Manager, and Ray Benally, Range Specialist, Navajo Mine, and Roger Nelson, Environmental Specialist, San Francisco Office, Utah International, Inc., June 14, 1979.

40. Written communication from J. LeRoy Balzer, Director, Environmental Quality, Utah International, Inc. to INFORM, October 8, 1979.

41. U.S. Department of the Interior, OSM Reclamation and Enforcement, Notice of Violation No. 79-V-1-16, May 24, 1979.

42. INFORM interviews with Sterling Grogan, Senior Environmental Engineer, Wes Karma, Assistant Mine Manager, and Ray Benally, Range Specialist, Navajo Mine, and Roger Nelson, Environmental Specialist, San Francisco Office, Utah International, Inc., June 14, 1979.

43. U.S. Department of the Interior, OSM, Region V On-Site Inspection: Utah International, Inc.--Navajo Mine, August 18, 1978.

44. INFORM interview with Jack Reynolds, Inspector, New Mexico Coal Surface-mining Commission, May 17, 1979.

45. INFORM interviews with Sterling Grogan, Senior Environmental Engineer, Wes Karma, Assistant Mine Manager, and Ray Benally, Range Specialist, Navajo Mine, and Roger Nelson, Environmental Specialist, San Francisco Office, Utah International, Inc., June 14, 1979.

46. U.S. Department of the Interior, USGS, Inspection Report: Utah International, Inc.--Navajo Mine, February 27, 1978.

47. U.S. Department of the Interior, Bureau of Reclamation, *Navajo Mine, New Mexico--Final Environmental Statement,* p. 1.51; Utah International, Inc., "Annual Report of the Navajo Mine to the New Mexico Coal Surfacemining Commission," April 28, 1978; Written communication from Utah International, Inc. to INFORM, n.d.

48. INFORM interview with Orlando Estrada, Acting Senior Environmental Engineer, Navajo Mine, March 11, 1980.

49. INFORM interview with Orlando Estrada, Acting Senior Environmental Engineer, Navajo Mine, March 26, 1980.

50. INFORM interviews with Sterling Grogan, Senior Environmental Engineer, Wes Karma, Assistant Mine Manager, and Ray Benally, Range Specialist, Navajo Mine, and Roger Nelson, Environmental Specialist, San Francisco Office, Utah International, Inc., June 14, 1979.

51. *Idem.*

52. Utah International, Inc., "Annual Report of the Navajo Mine to the New Mexico Coal Surfacemining Commission," April 28, 1978.

53. INFORM interviews with Sterling Grogan, Senior Environmental Engineer, Wes Karma, Assistant Mine Manager, and Ray Benally, Range Specialist,

Navajo Mine, and Roger Nelson, Environmental Specialist, San Francisco Office, Utah International, Inc., June 14, 1979.

54. *Idem.*

55. INFORM interview with Orlando Estrada, Acting Senior Environmental Engineer, Navajo Mine, November 5, 1979.

56. Earth Science Associates, Vonder Linden, "Strip Mine Inspection Report: Utah International, Inc.--Navajo Mine," October 26, 1977-April 18, 1978.

57. INFORM interviews with Sterling Grogan, Senior Environmental Engineer, Wes Karma, Assistant Mine Manager, and Ray Benally, Range Specialist, Navajo Mine, and Roger Nelson, Environmental Specialist, San Francisco Office, Utah International, Inc., June 14, 1979.

58. *Idem.*

59. U.S. Department of the Interior, Bureau of Reclamation, *Navajo Mine, New Mexico--Final Environmental Statement,* p. 1.57.

60. Utah International, Inc., "Annual Report of the Navajo Mine to the New Mexico Coal Surfacemining Commission," April 28, 1978.

61. INFORM interview with Orlando Estrada, Acting Senior Environmental Engineer, Navajo Mine, November 5, 1979.

62. Utah International, Inc., "Annual Report of the Navajo Mine to the New Mexico Coal Surfacemining Commission," April 28, 1978.

63. *Ibid.;* INFORM interviews with Sterling Grogan, Senior Environmental Engineer, Wes Karma, Assistant Mine Manager, and Ray Benally, Range Specialist, Navajo Mine, and Roger Nelson, Environmental Specialist, San Francisco Office, Utah International, Inc., June 14, 1979.

64. INFORM interview with Orlando Estrada, Acting Senior Environmental Engineer, Navajo Mine, November 5, 1979.

65. INFORM interviews with Sterling Grogan, Senior Environmental Engineer, Wes Karma, Assistant Mine Manager, and Ray Benally, Range Specialist, Navajo Mine, and Roger Nelson, Environmental Specialist, San Francisco Office, Utah International, Inc., June 14, 1979.

66. *Idem.;* INFORM interview with Orlando Estrada, Acting Senior Environmental Engineer, Navajo Mine, November 5, 1979.

67. Utah International, Inc., "Annual Report of the Navajo Mine to the New Mexico Coal Surfacemining Commission," April 28, 1978.

68. INFORM interviews with Sterling Grogan, Senior Environmental Engineer, Wes Karma, Assistant Mine Manager, and Ray Benally, Range Specialist, Navajo Mine, and Roger Nelson, Environmental Specialist, San Francisco Office, Utah International, Inc., June 14, 1979.

69. *Idem.;* U.S. Department of the Interior, OSM, Region V On-Site Inspection: Utah International, Inc.--Navajo Mine, August 18, 1978.

70. INFORM interviews with Sterling Grogan, Senior Environmental Engineer, Wes Karma, Assistant Mine Manager, and Ray Benally, Range Specialist, Navajo Mine, and Roger Nelson, Environmental Specialist, San Francisco Office, Utah International, Inc., June 14, 1979; Utah International, Inc., "Annual Report of the Navajo Mine to the New Mexico Coal Surfacemining Commission," April 28, 1978; J. LeRoy Balzer, Director, Environmental Quality, Utah International, Inc., "A Venture Into Reclamation," *Mining Congress Journal,* Vol. 61, No. I, (Washington, D.C.: American Mining Congress, 1975), January 1975.

71. U.S. Department of the Interior, USGS, Inspection Report: Utah International, Inc.--Navajo Mine, July 27, 1978.

72. U.S. Department of the Interior, Bureau of Reclamation, *Navajo Mine, New Mexico--Final Environmental Statement,* pp. 2.125-2.127.

73. INFORM interviews with Sterling Grogan, Senior Environmental Engineer, Wes Karma, Assistant Mine Manager, and Ray Benally, Range Specialist, Navajo Mine, and Roger Nelson, Environmental Specialist, San Francisco Office, Utah International, Inc., June 14, 1979; U.S. Department of the Interior, USGS, Inspection Report: Utah International, Inc.--Navajo Mine, July 25, 1977; INFORM interview with Orlando Estrada, Acting Senior Environmental Engineer, Navajo Mine, November 5, 1979.

74. INFORM interview with Don Crane, Director, Region V, OSM, August 10, 1979.

75. U.S. Department of the Interior, Bureau of Reclamation, *Navajo Mine, New Mexico--Final Environmental Statement,* pp. 2.125-2.127; Written communication from J. LeRoy Balzer, Director, Environmental Quality, Utah Inter-

national, Inc. to INFORM, October 8, 1979.

76. Written communications from J. LeRoy Balzer, Director, Environmental Quality, Utah International, Inc. to INFORM, October 8, 1979 and May 22, 1979; INFORM interviews with Sterling Grogan, Senior Environmental Engineer, Wes Karma, Assistant Mine Manager, and Ray Benally, Range Specialist, Navajo Mine, and Roger Nelson, Environmental Specialist, San Francisco Office, Utah International, Inc., June 14, 1979.

77. Earth Science Associates, Vonder Linden, "Strip Mine Inspection Report: Utah International, Inc.--Navajo Mine," October 20, 1976-April 19, 1977, July 20, 1977--October 25, 1977 and October 26, 1977-April 18, 1978.

78. Earth Science Associates, Vonder Linden, "Strip Mine Inspection Report: Utah International, Inc.--Navajo Mine," October 26, 1977-April 18, 1978.

79. U.S. Department of the Interior, USGS, Inspection Report: Utah International, Inc.--Navajo Mine, April 21, 1977.

80. A. Kent Evans, E.W. Uhleman, and P.A. Eby, *Atlas of Western Surface-Mined Lands: Coal, Uranium and Phosphate* (Washington, D.C.: U.S. Department of the Interior, Fish and Wildlife Service, Biological Services Program, January 1978), p. 27.

81. U.S. Department of the Interior, OSM, Region V On-Site Inspection: Utah International, Inc.--Navajo Mine, July 18, 1978.

82. INFORM interview with Jack Reynolds, Inspector, New Mexico Coal Surface-mining Commission, May 17, 1979; Written communication from J. LeRoy Balzer, Director, Environmental Quality, Utah International, Inc. to INFORM, October 8, 1979.

83. INFORM interview with Jack Reynolds, Inspector, New Mexico Surfacemining Commission, May 17, 1979.

84. INFORM interviews with Sterling Grogan, Senior Environmental Engineer, Wes Karma, Assistant Mine Manager, and Ray Benally, Range Specialist, Navajo Mine, and Roger Nelson, Environmental Specialist, San Francisco Office, Utah International, Inc., June 14, 1979.

85. INFORM interview with Don Crane, Director, Region V, OSM, August 10, 1979.

86. Written communication from J. LeRoy Balzer, Director, Environmental Quality, Utah International, Inc. to INFORM, October 8, 1979.

87. INFORM interviews with Sterling Grogan, Senior Environmental Engineer, Wes Karma, Assistant Mine Manager, and Ray Benally, Range Specialist, Navajo Mine, and Roger Nelson, Environmental Specialist, San Francisco Office, Utah International, Inc., June 14, 1979.

88. Written communication from J. LeRoy Balzer, Director, Environmental Quality, Utah International, Inc. to INFORM, October 8, 1979.

89. INFORM interviews with Sterling Gorgan, Senior Environmental Engineer, Wes Karma, Assistant Mine Manager, and Ray Benally, Range Specialist, Navajo Mine, and Roger Nelson, Environmental Specialist, San Francisco Office, Utah International, Inc., June 14, 1979.

90. INFORM interview with Harris Arthur, Bureau of Indian Affairs, U.S. Department of the Interior, Window Rock, Arizona, March 20, 1979.

91. Written communication from J. LeRoy Balzer, Director, Environmental Quality, Utah International, Inc. to INFORM, October 8, 1979.

92. *Ibid.*

93. INFORM interview with Brant Caulkins, Director, Sierra Club, Albuquerque, New Mexico, April 17, 1980.

94. U.S. Department of the Interior, Bureau of Reclamation, *Navajo Mine, New Mexico--Final Environmental Statement,* pp. 2.136-2.142, 3.19; INFORM interviews with Sterling Grogan, Senior Environmental Engineer, Wes Karma, Assistant Mine Manager, and Ray Benally, Range Specialist, Navajo Mine, and Roger Nelson, Environmental Specialist, San Francisco Office, Utah International, Inc., June 14, 1979.

95. *Ibid.*

96. *Ibid.;* Written communication from J. LeRoy Balzer, Director, Environmental Quality, Utah International, Inc., to INFORM, October 8, 1979.

ENERGY FUELS

1. INFORM interviews with Kent A. Crofts, Range Scientist, Energy Fuels Corporation, October 15, 1979 and November 26, 1979.
2. *Idem.*
3. U.S. Department of the Interior, *Northwest Colorado Coal Final Environmental Statement--Site Specific Analysis* (Washington, D.C.: U.S. Government Printing Office, 1976), p. EII-1.
4. Written communication from Energy Fuels Corporation to INFORM, October 5, 1979.
5. Energy Fuels Corporation, Engineering Staff, *Modification to the USGS Mine Plan for Energy Fuels Corporation Mines No. 1 and 2*, 1977, pp. 10-14.
6. INFORM interview with Kent A. Crofts, Range Scientist, Energy Fuels Corporation, November 26, 1979.
7. INFORM interviews with officials from the Colorado MLRB, July 30, 1979.
8. U.S. Department of Commerce, NOAA, *Monthly Averages of Temperature and Precipitation for State Climatic Divisions 1941-70: Colorado*, "Monthly and Annual Divisional Averages Precipitation (Inches)" (Asheville, N.C.: National Climatic Center, July 1973), p. 3.
9. Written communication from Energy Fuels Corporation to INFORM, October 5, 1979.
10. INFORM interviews with Kent A. Crofts, Range Scientist, Energy Fuels Corporation, Gary W. Myers, Chief Geologist, Energy Fuels Corporation, Jeff Saunders, Geologist, Energy Fuels Corporation, and James A. Larson, President, Energy Fuels Corporation, June 20, 1979.
11. Energy Fuels Corporation, Engineering Staff, *Modification to the USGS Mine Plan for Energy Fuels Corporation Mines No. 1 and 2*, 1977, pp. 34-38.
12. *Ibid.*, p. 38.
13. INFORM interviews with Kent A. Crofts, Range Scientist, Energy Fuels Corporation, Gary W. Myers, Chief Geologist, Energy Fuels Corporation, Jeff Saunders, Geologist, Energy Fuels Corporation, and James A. Larson, President, Energy Fuels Corporation, June 20, 1979.
14. INFORM interviews with officials from the Colorado MLRB, July 30, 1979.
15. Written communication from Energy Fuels Corporation to INFORM, October 5, 1979.
16. INFORM interview with Kent A. Crofts, Range Scientist, Energy Fuels Corporation, November 26, 1979.
17. Colorado MLRB, Coal Mine Inspection Report: Energy Fuels Corporation, July 11, 1979.
18. Colorado MLRB, Coal Mine Inspection Report: Energy Fuels Corporation, December 20, 1977; INFORM interviews with officials from the MLRB, July 30, 1979 and January 9, 1980.
19. Energy Fuels Corporation, *Annual Reclamation Report*, January 27, 1978.
20. INFORM interview with Kent A. Crofts, Range Scientist, Energy Fuels Corporation, November 26, 1979.
21. INFORM interviews with Kent A. Crofts, Range Scientist, Energy Fuels Corporation, Gary W. Myers, Chief Geologist, Energy Fuels Corporation, Jeff Saunders, Geologist, Energy Fuels Corporation, and James A. Larson, President, Energy Fuels Corporation, June 20, 1979.
22. Letter from Bjarne Flaatten, adjacent land owner, to Colorado MLRB, February 15, 1977.
23. U.S. Department of the Interior, *Northwest Colorado Coal Final Environmental Statement*, p. EIII-2.
24. INFORM interview with Kent A. Crofts, Range Scientist, Energy Fuels Corporation, October 15, 1979.
25. Written communication from Energy Fuels Corporation to INFORM, October 5, 1979.
26. U.S. Department of the Interior, USGS, Inspection Report: Energy Fuels Corporation, July 20, 1976; Written communication from Energy Fuels Corporation to INFORM, October 5, 1979.
27. U.S. Department of the Interior, OSM, Region V On-Site Inspection: Energy Fuels Corporation, November 1-3, 1978; INFORM interviews with officials from the Colorado MLRB, January 9, 1980.

28. INFORM interviews with Kent A. Crofts, Range Scientist, Energy Fuels Corporation, Gary W. Myers, Chief Geologist, Energy Fuels Corporation, Jeff Saunders, Geologist, Energy Fuels Corporation, and James A. Larson, President, Energy Fuels Corporation, June 20, 1979.

29. *Idem.*; INFORM interviews with officials from the Colorado MLRB, July 30, 1979.

30. Written communication from Energy Fuels Corporation to INFORM, October 5, 1979.

31. INFORM interviews with Kent A. Crofts, Range Scientist, Energy Fuels Corporation, Gary W. Myers, Chief Geologist, Energy Fuels Corporation, Jeff Saunders, Geologist, Energy Fuels Corporation, and James A. Larson, President, Energy Fuels Corporation, June 20, 1979.

32. Colorado MLRB, Coal Mine Inspection Report: Energy Fuels Corporation, July 11, 1978.

33. U.S. Department of the Interior, OSM, Region V On-Site Inspection: Energy Fuels Corporation, November 1-3, 1978.

34. INFORM interviews with officials from the Colorado MLRB, July 30, 1979; Written communication from Energy Fuels Corporation to INFORM, October 5, 1979.

35. Written communication from Energy Fuels Corporation to INFORM, October 5, 1979.

36. INFORM interviews with officials from the Colorado MLRB, July 30, 1979.

37. Energy Fuels Corporation, Maps, *Energy Mines 1 and 2 Reclamation*, Revision 4, February 20, 1979, and *Energy Mine No. 3 Reclamation*, Revision 4, February 20, 1979.

38. Energy Fuels Corporation, NPDES Discharge Monitoring Report, reporting period from January 1977 to January 1979.

39. Energy Fuels Corporation, NPDES Discharge Monitoring Report, reporting period from April 1, 1979 to July 1, 1979.

40. National Oceanic Administration Association, Atlas 2, Vol. 3, for Colorado, reviewed by Daniel Wiener, Research Associate, INFORM, at the Colorado MLRB office, March 9, 1979.

41. U.S. Department of the Interior, OSM, Region V On-Site Inspection: Energy Fuels Corporation, November 1-3, 1978.

42. INFORM interview with Kent A. Crofts, Range Scientist, Energy Fuels Corporation, November 26, 1979; Written communication from Energy Fuels Corporation to INFORM, October 5, 1979.

43. INFORM interviews with Kent A. Crofts, Range Scientist, Energy Fuels Corporation, Gary W. Myers, Chief Geologist, Energy Fuels Corporation, Jeff Saunders, Geologist, Energy Fuels Corporation, and James A. Larson, President, Energy Fuels Corporation, June 20, 1979.

44. Written communication from Energy Fuels Corporation to INFORM, October 5, 1979.

45. INFORM interviews with officials from the Colorado MLRB, July 30, 1979.

46. INFORM interviews with Kent A Crofts, Range Scientist, Energy Fuels Corporation, Gary W. Myers, Chief Geologist, Energy Fuels Corporation, Jeff Saunders, Geologist, Energy Fuels Corporation, and James A. Larson, President, Energy Fuels Corporation, June 20, 1979; Written communication from Energy Fuels Corporation to INFORM, October 5, 1979.

47. Written communication from Energy Fuels Corporation to INFORM, October 5, 1979.

48. U.S. Department of the Interior, *Northwest Colorado Coal Final Environmental Statement*, pp. EIII-1-3.

49. Written communication from Energy Fuels Corporation to INFORM, October 5, 1979.

50. *Ibid.*

51. *Ibid.*

52. *Ibid.*

53. INFORM interviews with Kent A. Crofts, Range Scientist, Energy Fuels Corporation, Gary W. Myers, Chief Geologist, Energy Fuels Corporation, Jeff Saunders, Geologist, Energy Fuels Corporation, and James A. Larson, President, Energy Fuels Corporation, June 20, 1979.

54. *Idem.*
55. Colorado MLRB, Coal Mine Inspection Report: Energy Fuels Corporation, March 31, 1978.
56. U.S. Department of the Interior, OSM, Region V On-Site Inspection: Energy Fuels Corporation, November 1-3, 1978.
57. Colorado MLRB, Coal Mine Inspection Report: Energy Fuels Corporation, July 11, 1978.
58. INFORM interviews with Kent A. Crofts, Range Scientist, Energy Fuels Corporation, Gary W. Myers, Chief Geologist, Energy Fuels Corporation, Jeff Saunders, Geologist, Energy Fuels Corporation, and James A. Larson, President, Energy Fuels Corporation, June 20, 1979.
59. *Idem.*
60. U.S. Department of the Interior, OSM, Region V On-Site Inspection: Energy Fuels Corporation, November 1-3, 1978.
61. Energy Fuels Corporation, Maps, *Energy Mines 1 and 2 Reclamation*, Revision 4, and *Energy Mine No. 3 Reclamation*, Revision 4, February 20, 1979; INFORM interview with Kent A. Crofts, Range Scientist, Energy Fuels Corporation, November 26, 1979.
62. INFORM interview with Kent A. Crofts, Range Scientist, Energy Fuels Corporation, November 26, 1979.
63. *Idem.*
64. Written communication from Energy Fuels Corporation to INFORM, October 5, 1979.
65. INFORM interviews with Kent A. Crofts, Range Scientist, Energy Fuels Corporation, Gary W. Myers, Chief Geologist, Energy Fuels Corporation, Jeff Saunders, Geologist, Energy Fuels Corporation, and James A. Larson, President, Energy Fuels Corporation, June 20, 1979.
66. Colorado MLRB, Coal Mine Inspection Report: Energy Fuels Corporation, August 8, 1978.
67. INFORM interviews with officials from the Colorado MLRB, July 30, 1979.
68. Written communication from Energy Fuels Corporation to INFORM, October 5, 1979; U.S. Department of the Interior, OSM, Region V On-Site Inspection: Energy Fuels Corporation, November 1-3, 1978.
69. Energy Fuels Corporation, Maps, *Energy Mines 1 and 2 Reclamation*, Revision 4, and *Energy Mine No. 3 Reclamation*, Revision 4, February 20, 1979; Howard M. Kaplan, "'A Time to Plant,'" *The Denver Post*, October 3, 1976.
70. Energy Fuels Corporation, Maps, *Energy Mines 1 and 2 Reclamation*, Revision 4, and *Energy Mine No. 3 Reclamation*, Revision 4, February 20, 1979.
71. INFORM interviews with Kent A. Crofts, Range Scientist, Energy Fuels Corporation, Gary W. Myers, Chief Geologist, Energy Fuels Corporation, Jeff Saunders, Geologist, Energy Fuels Corporation, and James A. Larson, President, Energy Fuels Corporation, June 20, 1979.
72. *Idem.*
73. INFORM interviews with officials from the Colorado MLRB, July 30, 1979.
74. INFORM interviews with Kent A. Crofts, Range Scientist, Energy Fuels Corporation, Gary W. Myers, Chief Geologist, Energy Fuels Corporation, Jeff Saunders, Geologist, Energy Fuels Corporation, and James A. Larson, President, Energy Fuels Corporation, June 20, 1979; Written communication from Energy Fuels Corporation to INFORM, October 5, 1979.
75. Written communication from Energy Fuels Corporation to INFORM, October 5, 1979.
76. INFORM interview with Kent A. Crofts, Range Scientist, Energy Fuels Corporation, November 26, 1979.
77. Energy Fuels Corporation, Maps, *Energy Mines 1 and 2 Reclamation*, Revision 4, and *Energy Mine No. 3 Reclamation*, Revision 4, February 20, 1979.
78. INFORM interviews with officials from the Colorado MLRB, July 30, 1979.
79. INFORM interviews with officials from the Colorado MLRB, March 9, 1979.
80. INFORM interviews with Kent A. Crofts, Range Scientist, Energy Fuels Corporation, Gary W. Myers, Chief Geologist, Energy Fuels Corporation, Jeff Saunders, Geologist, Energy Fuels Corporation, and James A. Larson, President, Energy Fuels Corporation, June 20, 1979.
81. INFORM interview with Kent A. Crofts, Range Scientist, Energy Fuels Corporation, November 26, 1979; INFORM interviews with Kent A. Crofts, Range Scientist, Energy Fuels Corporation, Gary W. Myers, Chief Geologist, Energy Fuels Corporation, Jeff Saunders, Geologist, Energy Fuels Corporation,

and James A. Larson, President, Energy Fuels Corporation, June 20, 1979.

82. INFORM interviews with Kent A. Crofts, Range Scientist, Energy Fuels Corporation, Gary W. Myers, Chief Geologist, Energy Fuels Corporation, Jeff
Saunders, Geologist, Energy Fuels Corporation, and James A. Larson, President, Energy Fuels Corporation, June 20, 1979.

83. Energy Fuels Corporation, Engineering Staff, *Modification to the USGS Mine
Plan for Energy Fuels Corporation Mine No. 1 and 2*, 1977, p. 58.

84. *Ibid.*

85. Amendment to Energy Fuels Corporation State Mining Permit, January 1976,
viewed at Mined Land Reclamation offices in Denver, March 1979.

86. Written communication from Energy Fuels Corporation to INFORM, October 5,
1979.

87. INFORM interviews with Kent A. Crofts, Range Scientist, Energy Fuels Corporation, Gary W. Myers, Chief Geologist, Energy Fuels Corporation, Jeff
Saunders, Geologist, Energy Fuels Corporation, and James A. Larson, President, Energy Fuels Corporation, June 20, 1979.

88. Written communication from Energy Fuels Corporation to INFORM, October 5,
1979.

89. *Ibid.*

90. U.S. Department of the Interior, *Northwest Colorado Coal Final Environmental Statement*, p. EV-3.

SENECA II

1.   INFORM interview with Alten F. Grandt, Director of Reclamation, Western
Division, Peabody Coal Company, September 28, 1979.

2.   A. Kent Evans, E. W. Uhleman, and P. A. Eby, *Atlas of Western Surface-
Mined Lands: Coal, Uranium and Phosphate* (Washington, D.C.: U.S. Department of the Interior, Fish and Wildlife Service, Biological Services Program, January 1978), p. 59.

3.   U.S. Department of the Interior, *Northwest Colorado Coal Final Environmental Statement--Site Specific Analysis* (Washington, D.C.: U.S. Government
Printing Office, 1976), p. PII-18; Written communication from James R. Jones,
Vice President, Environmental Affairs, Peabody Coal Company, to INFORM,
n.d.

4.   Written communication from James R. Jones, Vice President, Environmental
Affairs, Peabody Coal Company, to INFORM, n.d.

5.   U.S. Department of Commerce, NOAA, *Monthly Averages of Temperature and
Precipitation for State Climatic Divisions 1941-70: Colorado*, "Monthly and
Annual Divisional Averages Precipitation (Inches)" (Asheville, N.C.: Nation-
al Climatic Center, July 1973), p. 3; Peabody Coal Company, "Mining and
Reclamation Plan, Pursuant to the Rules and Regulations of Title 30, Chapter
II, Coal Mining Operating Regulations--Submission by Seneca Coals, Ltd. for
Seneca II-W Mine," submitted September-December 1977, p. G-7.

6.   Russell T. Moore, David A. Koehler, and Craig L. Kling, "Surface Mine Rehabilitation in Northwestern Colorado" (Papers presented before the Third
Symposium on Surface Mining and Reclamation, Louisville, Kentucky, October
21-23, 1975), Volume II, p. 183.

7.   U.S. Department of the Interior, *Final Environmental Statement*, p. K-8.

8.   U.S. Department of Commerce, NOAA, *Monthly Averages of Temperature and
Precipitation: Colorado*, p. 3.

9.   INFORM interviews with Alten F. Grandt, Director of Reclamation, Western
Division, Peabody Coal Company, James R. Jones, Vice President, Environ-
mental Affairs, Peabody Coal Company, and Don Zulian, Superintendent,
Seneca II Mine, June 21, 1979.

10. *Idem.*

11. INFORM interview with Thomas E. Ehmett, Reclamation Specialist, Region V,
OSM, U.S. Department of the Interior, March 8, 1979.

12. Peabody Coal Company, "Monthly Reclamation Grading Report: Seneca 260,"
May 1979.

13. Written communication from Roy Karo, Reclamation Supervisor, Seneca II
Mine, to INFORM, February 16, 1979.

14. Colorado Department of Natural Resources, MLRB, Coal Mine Inspection Report: Peabody--Seneca II, December 20, 1977.
15. *Ibid.*
16. Colorado MLRB, Coal Mine Inspection Report: Peabody--Seneca II, July 11, 1978.
17. INFORM interviews with Alten F. Grandt, Director of Reclamation, Western Division, Peabody Coal Company, James R. Jones, Vice President, Environmental Affairs, Peabody Coal Company, and Don Zulian, Superintendent, Seneca II Mine, June 21, 1979.
18. INFORM interviews with officials from the Colorado MLRB, July 30, 1979.
19. *Idem.*
20. INFORM interviews with Alten F. Grandt, Director of Reclamation, Western Division, Peabody Coal Company, James R. Jones, Vice President, Environmental Affairs, Peabody Coal Company, and Don Zulian, Superintendent, Seneca II Mine, June 21, 1979.
21. INFORM interviews with officials from the Colorado MLRB, July 30, 1979.
22. Colorado MLRB, Coal Mine Inspection Report: Peabody--Seneca II, October 6, 1978.
23. INFORM interviews with officials from the Colorado MLRB, July 30, 1979; Letter from Hamlet J. Barry III, Director, Colorado MLRB, to Daniel Wiener, Research Associate, INFORM, December 13, 1979.
24. INFORM interviews with officials from the Colorado MLRB, March 9, 1979.
25. *Idem.*, July 30, 1979.
26. INFORM interviews with Alten F. Grandt, Director of Reclamation, Western Division, Peabody Coal Company, James R. Jones, Vice President, Environmental Affairs, Peabody Coal Company, and Don Zulian, Superintendent, Seneca II Mine, June 21, 1979.
27. Peabody Coal Company, "Mining and Reclamation Plan, Seneca II-W Mine," Exhibit E: Reclamation Plan, December 13, 1977, p. E-11.
28. INFORM interviews with officials from the Colorado MLRB, March 9, 1979; Letter from Hamlet J. Barry III, Director, Colorado MLRB, to Daniel Wiener, Research Associate, INFORM, December 13, 1979.
29. INFORM interviews with officials from the Colorado MLRB, March 9, 1979.
30. INFORM interviews with Alten F. Grandt, Director of Reclamation, Western Division, Peabody Coal Company, James R. Jones, Vice President, Environmental Affairs, Peabody Coal Company, and Don Zulian, Superintendent, Seneca II Mine, June 21, 1979.
31. INFORM interview with Alten F. Grandt, Director of Reclamation, Western Division, Peabody Coal Company, September 28, 1979
32. INFORM interview with officials from the Colorado MLRB, July 30, 1979.
33. U.S. Department of the Interior, OSM, Region V On-Site Inspection, Peabody Coal Company--Seneca Mine, November 2, 1978.
34. INFORM interview with Alten F. Grandt, Director of Reclamation, Western Division, Peabody Coal Company, September 28, 1979.
35. Colorado Department of Health, "Authorization to Discharge Under the National Pollutant Discharge Elimination System, Permit No. CO-0000221," April 1, 1976.
36. Peabody Coal Company, "Mining and Reclamation Plan--Seneca II-W Mine," September-December 1977.
37. Colorado Department of Health, "Authorization to Discharge Under the National Pollutant Discharge Elimination System, Permit No. CO-0000221," April 1, 1976.
38. Peabody Coal Company, NPDES Discharge Monitoring Report--Seneca Mine, reporting period from September 1 to November 31, 1977.
39. U.S. Department of the Interior, OSM, Region V On-Site Inspection, Peabody Coal Company--Seneca Mine, November 2, 1978.
40. *Ibid.*
41. U.S. Department of the Interior, OSM, Region V On-Site Inspection, Peabody Coal Company--Seneca Mine, reporting period from December 1 to December 31, 1977.
42. INFORM interviews with officials from the Colorado MLRB, July 30, 1979.
43. Peabody Coal Company, NPDES Discharge Monitoring Report--Seneca Mine, reporting period from July 1, 1978 to January 28, 1979.
44. Peabody Coal Company, "Mining and Reclamation Plan--Seneca II-W Mine,"

September-December 1977.

45. INFORM interviews with Alten F. Grandt, Director of Reclamation, Western Division, Peabody Coal Company, James R. Jones, Vice President, Environmental Affairs, Peabody Coal Company, and Don Zulian, Superintendent, Seneca II Mine, June 21, 1979.

46. Peabody Coal Company, "Mining and Reclamation Plan--Seneca II-W Mine," September-December 1977.

47. INFORM interviews with Alten F. Grandt, Director of Reclamation, Western Division, Peabody Coal Company, James R. Jones, Vice President, Environmental Affairs, Peabody Coal Company, and Don Zulian, Superintendent, Seneca II Mine, June 21, 1979.

48. INFORM interview with Alten F. Grandt, Director of Reclamation, Western Division, Peabody Coal Company, September 28, 1979.

49. *Idem.*

50. *Idem.*

51. Peabody Coal Company, "Mining and Reclamation Plan--Seneca II-W Mine," September-December 1977.

52. INFORM interviews with Alten F. Grandt, Director of Reclamation, Western Division, Peabody Coal Company, James R. Jones, Vice President, Environmental Affairs, Peabody Coal Company, and Don Zulian, Superintendent, Seneca II Mine, June 21, 1979.

53. Written communication from George F. Brown, Acting Conservation Manager, USGS, U.S. Department of the Interior, to Ron Lanoue, Research Associate, INFORM, March 21, 1979; INFORM interviews with Colorado MLRB Officials, March 9, 1979; Computer printout from H. Robert Moore, Assistant to the Director for Coal Management, BLM, U.S. Department of the Interior, November 28, 1978; Peabody Coal Company, "Mining and Reclamation Plan--Seneca II-W Mine,"September-December 1977, p. G-5; Written communication from James R. Jones, Vice President, Environmental Affairs, Peabody Coal Company, to INFORM, n.d.

54. Colorado MLRB, Coal Mine Inspection Report: Peabody--Seneca II, December 20, 1977.

55. INFORM interview with Alten F. Grandt, Director of Reclamation, Western Division, Peabody Coal Company, September 28, 1979.

56. INFORM interviews with Alten F. Grandt, Director of Reclamation, Western Division, Peabody Coal Company, James R. Jones, Vice President, Environmental Affairs, Peabody Coal Company, and Don Zulian, Superintendent, Seneca II Mine, June 21, 1979.

57. Colorado MLRB, Coal Mine Inspection Report: Peabody--Seneca II, July 11, 1978.

58. INFORM interviews with Alten F. Grandt, Director of Reclamation, Western Division, Peabody Coal Company, James R. Jones, Vice President, Environmental Affairs, Peabody Coal Company, and Don Zulian, Superintendent, Seneca II Mine, June 21, 1979.

59. INFORM interview with Alten F. Grandt, Director of Reclamation, Western Division, Peabody Coal Company, September 28, 1979.

60. Colorado MLRB, Coal Mine Inspection Report: Peabody--Seneca II, December 20, 1977.

61. Colorado MLRB, Coal Mine Inspection Report: Peabody--Seneca II, July 11, 1978.

62. INFORM interview with Alten F. Grandt, Director of Reclamation, Western Division, Peabody Coal Company, September 28, 1979.

63. Written communication from James R. Jones, Vice President, Environmental Affairs, Peabody Coal Company, to INFORM, n.d.

64. INFORM interviews with Alten F. Grandt, Director of Reclamation, Western Division, Peabody Coal Company, James R. Jones, Vice President, Environmental Affairs, Peabody Coal Company, and Don Zulian, Superintendent, Seneca II Mine, June 21, 1979.

65. *Idem.*

66. Written communication from James R. Jones, Vice President, Environmental Affairs, Peabody Coal Company, to INFORM, n.d.

67. INFORM interviews with Alten F. Grandt, Director of Reclamation, Western Division, Peabody Coal Company, James R. Jones, Vice President, Environmental Affairs, Peabody Coal Company, and Don Zulian, Superintendent,

Seneca II Mine, June 21, 1979.

68. *Idem.*, INFORM interview with Alten F. Grandt, Director of Reclamation, Western Division, Peabody Coal Company, September 28, 1979.

69. INFORM interviews with Alten F. Grandt, Director of Reclamation, Western Division, Peabody Coal Company, James R. Jones, Vice President, Environmental Affairs, Peabody Coal Company, and Don Zulian, Superintendent, Seneca II Mine, June 21, 1979; Written communication from James R. Jones, Vice President, Environmental Affairs, Peabody Coal Company, to INFORM, n.d.

70. INFORM interviews with Alten F. Grandt, Director of Reclamation, Western Division, Peabody Coal Company, James R. Jones, Vice President, Environmental Affairs, Peabody Coal Company, and Don Zulian, Superintendent, Seneca II Mine, June 21, 1979.

71. Written communication from James R. Jones, Vice President, Environmental Affairs, Western Division, Peabody Coal Company, n.d.

72. INFORM interview with Alten F. Grandt, Director of Reclamation, Western Division, Peabody Coal Company, September 28, 1979.

73. *Idem.*

74. *Idem.*

75. Letter from Hamlet J. Barry III, Director, Colorado MLRB, to Daniel Wiener, Research Associate, INFORM, December 13, 1979.

76. Written communication from James R. Jones, Vice President, Environmental Affairs, Peabody Coal Company, to INFORM, n.d.

77. INFORM interview with Alten F. Grandt, Director of Reclamation, Western Division, Peabody Coal Company, September 28, 1979.

78. *Idem.*

79. Written communication from James R. Jones, Vice President, Environmental Affairs, Peabody Coal Company, to INFORM, n.d.; INFORM interviews with officials from the Colorado MLRB, July 30, 1979.

80. Written communication from James R. Jones, Vice President, Environmental Affairs, Peabody Coal Company, to INFORM, n.d.

81. INFORM interview with Alten F. Grandt, Director of Reclamation, Western Division, Peabody Coal Company, September 28, 1979.

82. *Idem.*

83. INFORM interviews with officials from the Colorado MLRB, July 30, 1979; INFORM interviews with Alten F. Grandt, Director of Reclamation, Western Division, Peabody Coal Company, James R. Jones, Vice President, Environmental Affairs, Peabody Coal Company, and Don Zulian, Superintendent, Seneca II Mine, June 21, 1979.

84. INFORM interview with Alten F. Grandt, Director of Reclamation, Western Division, Peabody Coal Company, September 28, 1979.

85. *Idem.*

86. INFORM interviews with officials from the Colorado MLRB, July 30, 1979.

87. Colorado MLRB, Coal Mine Inspection Report: Peabody--Seneca II, July 11, 1978.

88. INFORM interviews with Alten F. Grandt, Director of Reclamation, Western Division, Peabody Coal Company, James R. Jones, Vice President, Environmental Affairs, Peabody Coal Company, and Don Zulian, Superintendent, Seneca II Mine, June 21, 1979.

89. Memorandum from R. T. Moore, Ecology Consultants, Inc., to J. L. Beckner, Peabody Coal Company, September 11, 1974.

90. W. A. Berg, "Revegetation of Land Disturbed by Surface Mining in Colorado," *Practices and Problems of Land Reclamation in Western North America* (University of North Dakota Press: 1975), pp. 79-89.

91. INFORM interviews with Alten F. Grandt, Director of Reclamation, Western Division, Peabody Coal Company, James R. Jones, Vice President, Environmental Affairs, Peabody Coal Company, and Don Zulian, Superintendent, Seneca II Mine, June 21, 1979.

92. *Idem.*

93. Evans, Uhleman, and Eby, *Atlas of Western Surface Mined Lands*, p. 59; INFORM interviews with Alten F. Grandt, Director of Environmental Affairs, Western Division, Peabody Coal Company, James R. Jones, Vice President, Environmental Affairs, Peabody Coal Company, and Don Zulian, Superintendent, Seneca II Mine, June 21, 1979.

94. *Ibid.*
95. INFORM interviews with Alten F. Grandt, Director of Environmental Affairs, Western Division, Peabody Coal Company, James R. Jones, Vice President, Environmental Affairs, Peabody Coal Company, and Don Zulian, Superintendent, Seneca II Mine, June 21, 1979.
96. *Idem.*
97. *Idem.*
98. INFORM interview with Wallace McRae, Montana rancher and former board member, NPRC, November 12, 1979.
99. INFORM interviews with Alten F. Grandt, Director of Reclamation, Western Division, Peabody Coal Company, James R. Jones, Vice President, Environmental Affairs, Peabody Coal Company, and Don Zulian, Superintendent, Seneca II Mine, June 21, 1979.
100. U.S. Department of the Interior, *Northwest Colorado Coal Final Environmental Statement*, p. PII-22.
101. Memorandum from R. T. Moore, Ecology Consultants, Inc., to J. L. Beckner, Peabody Coal Company, September 11, 1974.
102. Moore, Koehler, and Kling, "Surface Mine Rehabilitation in Northwestern Colorado," p. 183.
103. Letter from E. F. Sedgeley, Resource Conservationist, U.S. Soil Conservation Service, to Daniel Wiener, Research Associate, INFORM, March 23, 1979.
104. INFORM interview with Alten F. Grandt, Director of Reclamation, Western Division, Peabody Coal Company, September 28, 1979.
105. Moore, Koehler, and Kling, "Surface Mine Rehabilitation in Northwestern Colorado," p. 183.
106. INFORM interview with Alten F. Grandt, Director of Reclamation, Western Division, Peabody Coal Company, September 28, 1979.
107. INFORM interviews with Alten F. Grandt, Director of Reclamation, Western Division, Peabody Coal Company, James R. Jones, Vice President, Environmental Affairs, Peabody Coal Company, and Don Zulian, Superintendent, Seneca II Mine, June 21, 1979.
108. *Idem.*
109. INFORM interview with Larry Damrau, Reclamation Specialist, Region V, OSM, U.S. Department of the Interior, December 11, 1979.
110. INFORM interview with Alten F. Grandt, Director of Reclamation, Western Division, Peabody Coal Company, September 28, 1979.
111. Written communication from James R. Jones, Vice President, Environmental Affairs, Peabody Coal Company, to INFORM, n.d.

JIM BRIDGER

1. Bridger Coal Company, "Mining and Reclamation Plan Pursuant to the Rules and Regulations of Title 30, Chapter II, Coal Mining Operating Regulations," Federal Mine Plan, submitted September 29, 1978, p. 5-1.
2. *Ibid.*, p. 5-3.
3. *Ibid.*, p. D-4-1.
4. *Ibid.*, p. D-4-2.
5. *Ibid.*; INFORM interview with Richard M. Kail, Director of Reclamation, Jim Bridger Mine, November 10, 1979.
6. Bridger Coal Company, "Mining and Reclamation Plan," p. D-1-1.
7. U.S. Department of the Interior, OSM, Region V On-Site Inspection: Bridger Coal Company--Jim Bridger Mine, March 21, 1979.
8. Written communication from George F. Brown, Acting Conservation Manager, USGS, U.S. Department of the Interior, to Ron Lanoue, Research Associate, INFORM, March 26, 1979; Bridger Coal Company, "Mining and Reclamation Plan," p. D-1-1.
9. INFORM interviews with Richard M. Kail, Director of Reclamation, and Tom Schreeg, Reclamation Specialist, Jim Bridger Mine, June 26, 1979.
10. Wyoming DEQ, Land Quality Division, District II, Quarterly Inspection Report: Bridger Coal Company, July 27, 1977.
11. Wyoming DEQ, Land Quality Division, District II, Quarterly Inspection Report: Bridger Coal Company, August 25, 1978.

12. INFORM interviews with Richard M. Kail, Director of Reclamation, and Tom Schreeg, Reclamation Specialist, Jim Bridger Mine, June 26, 1979.

13. U.S. Department of the Interior, OSM, Region V On-Site Inspection: Bridger Coal Company--Jim Bridger Mine, March 21, 1979.

14. INFORM interview with Richard M. Kail, Director of Reclamation, Jim Bridger Mine, November 30, 1979.

15. U.S. Department of the Interior, OSM, Region V On-Site Inspection: Bridger Coal Company--Jim Bridger Mine, March 21, 1979.

16. Written communication from Gary Deveraux, Director of Mined Land Reclamation, NERCO, to INFORM, October 26, 1979.

17. INFORM interviews with Richard M. Kail, Director of Reclamation, and Tom Schreeg, Reclamation Specialist, Jim Bridger Mine, June 26, 1979.

18. U.S. Department of the Interior, OSM, Region V On-Site Inspection: Bridger Coal Company--Jim Bridger Mine, March 21, 1979.

19. INFORM interviews with Richard M. Kail, Director of Reclamation, and Tom Schreeg, Reclamation Specialist, Jim Bridger Mine, June 26, 1979.

20. Bridger Coal Company, "Mining and Reclamation Plan," p. II-38.

21. INFORM interviews with Richard M. Kail, Director of Reclamation, and Tom Schreeg, Reclamation Specialist, Jim Bridger Mine, June 26, 1979.

22. Bridger Coal Company, "Mining and Reclamation Plan," p. III-9.

23. INFORM interviews with Richard M. Kail, Director of Reclamation, and Tom Schreeg, Reclamation Specialist, Jim Bridger Mine, June 26, 1979.

24. *Idem.*; INFORM interview with Richard M. Kail, Director of Reclamation, Jim Bridger Mine, November 30, 1979.

25. *Idem.*

26. Written communication from Gary Deveraux, Director of Mined land Reclamation, NERCO, to INFORM, October 26, 1979.

27. INFORM interviews with Richard M. Kail, Director of Reclamation, and Tom Schreeg, Reclamation Specialist, Jim Bridger Mine, June 26, 1979.

28. U.S. Department of the Interior, USGS, Inspection Report: Bridger Coal Company, March 2, 1978.

29. U.S. Department of the Interior, OSM, Region V On-Site Inspection: Bridger Coal Company--Jim Bridger Mine, March 21, 1979.

30. *Ibid.*

31. INFORM interview with Richard M. Kail, Director of Reclamation, Jim Bridger Mine, January 30, 1980.

32. Wyoming DEQ, Land Quality Division, District II, Quarterly Inspection Report: Bridger Coal Company, April 3, 1978.

33. INFORM interviews with Richard M. Kail, Director of Reclamation, and Tom Schreeg, Reclamation Specialist, Jim Bridger Mine, June 26, 1979.

34. *Idem.*

35. U.S. Department of the Interior, USGS, Inspection Report: Bridger Coal Company, March 2, 1978.

36. U.S. Department of the Interior, OSM, Region V On-Site Inspection: Bridger Coal Company--Jim Bridger Mine, March 21, 1979.

37. Bridger Coal Company, "Mining and Reclamation Plan," p. II-16.

38. *Ibid.*

39. U.S. Department of the Interior, USGS, Inspection Report: Bridger Coal Company, March 2, 1978.

40. U.S. Department of the Interior, OSM, Region V On-Site Inspection: Bridger Coal Company--Jim Bridger Mine, March 21, 1979.

41. Letter from Lance Vinson, Director, Enforcement Division, EPA, to Ron Lanoue, Research Associate, INFORM, December 21, 1978.

42. Letter from Roger E. Frenette, Acting Director, Environmental Division, to EPA, to Daniel Wiener, Research Associate, INFORM, November 7, 1979; in-INFORM interview with Richard M. Kail, Director of Reclamation, Jim Bridger Mine, January 30, 1980.

43. Bridger Coal Company, "Mining and Reclamation Plan," p. II-18.

44. U.S. Department of the Interior, OSM, Region V On-Site Inspection: Bridger Coal Company--Jim Bridger Mine, March 21, 1979.

45. U.S. Department of the Interior, OSM Reclamation and Enforcement, Notice of Violation No. 79-V-3-13, March 21, 1979.

46. Written communication from Richard M. Kail, Director of Reclamation, Jim Bridger Mine, to INFORM, June 26, 1979.

47. INFORM interview with Richard M. Kail, Director of Reclamation, Jim Bridger Mine, January 30, 1980.

48. Written communication from Gary Deveraux, Director of Mined Land Reclamation, NERCO, to INFORM, October 26, 1979.

49. Letter from Roger E. Frenette, Acting Director, Environmental Division, EPA, to Daniel Wiener, Research Associate, INFORM, November 7, 1979.

50. Bridger Coal Company, "Mining and Reclamation Plan," p. II-39.

51. *Ibid.*, p. D-6-3.

52. Written communication from Gary Deveraux, Director of Mined Land Reclamation, NERCO, to INFORM, October 26, 1979.

53. INFORM interviews with Richard M. Kail, Director of Reclamation, and Tom Schreeg, Reclamation Specialist, Jim Bridger Mine, June 26, 1979.

54. *Idem.*

55. Bridger Coal Company, "Mining and Reclamation Plan," p. D-1-1.

56. Written communication from Gary Deveraux, Director of Mined Land Reclamation, NERCO, to INFORM, October 26, 1979.

57. INFORM interviews with Richard M. Kail, Director of Reclamation, and Tom Schreeg, Reclamation Specialist, Jim Bridger Mine, June 26, 1979.

58. NERCO, "1978 Annual DEQ Report--Jim Bridger Mine Permit No. 338C," May 2, 1978.

59. Written communication from Gary Deveraux, Director of Mined Land Reclamation, NERCO, to INFORM, October 26, 1979; Written communication from Richard M. Kail, Director of Reclamation, Jim Bridger Mine, to INFORM, June 26, 1979.

60. INFORM interview with Richard M. Kail, Director of Reclamation, Jim Bridger Mine, November 30, 1979.

61. Bridger Coal Company, "Mining and Reclamation Plan," p. III-14.

62. INFORM interviews with Richard M. Kail, Director of Reclamation, and Tom Schreeg, Reclamation Specialist, Jim Bridger Mine, June 26, 1979.

63. U.S. Department of the Interior, USGS, Inspection Report: Bridger Coal Company, July 12, 1978.

64. Wyoming DEQ, Land Quality Division, District II, Quarterly Inspection Report: Bridger Coal Company, April 3, 1978.

65. U.S. Department of the Interior, USGS, Inspection Report: Bridger Coal Company, December 1, 1978.

66. Wyoming DEQ, Land Quality Division, District II, Quarterly Inspection Report: Bridger Coal Company, July 27, 1977.

67. INFORM interview with Richard M. Kail, Director of Reclamation, Jim Bridger Mine, January 30, 1980.

68. *Idem.*

69. INFORM interviews with Richard M. Kail, Director of Reclamation, and Tom Schreeg, Reclamation Specialist, Jim Bridger Mine, June 26, 1979.

70. NERCO, "Update and Modification of Permit to Mine No. 338C--Jim Bridger Mine," September 29, 1978, Appendix B: Supplemental Irrigation Test.

71. INFORM interviews with Richard M. Kail, Director of Reclamation, and Tom Schreeg, Reclamation Specialist, Jim Bridger Mine, June 26, 1979.

72. Written communication from Gary Deveraux, Director of Mined Land Reclamation, NERCO, to INFORM, October 26, 1979.

73. INFORM interview with Richard M. Kail, Director of Reclamation, Jim Bridger Mine, November 30, 1979.

74. INFORM interviews with Richard M. Kail, Director of Reclamation, and Tom Schreeg, Reclamation Specialist, Jim Bridger Mine, June 26, 1979.

75. Written communication from Howard L. Millsap, District Conservationist, SCS, U.S. Department of Agriculture, to Daniel Wiener, Research Associate, INFORM, March 22, 1979.

76. INFORM interviews with Richard M. Kail, Director of Reclamation, and Tom Schreeg, Reclamation Specialist, Jim Bridger Mine, June 26, 1979.

77. U.S. Department of the Interior, USGS, Inspection Reports: Bridger Coal Company, September 21, 1978, July 12, 1978, and October 3, 1977; Wyoming DEQ, Land Quality Division, District II, Quarterly Inspection Report: Bridger Coal Company, February 7, 1977.

78. INFORM interviews with Richard M. Kail, Director of Reclamation, and Tom Schreeg, Reclamation Specialist, Jim Bridger Mine, June 26, 1979; U.S. Department of the Interior, USGS, Inspection Report: Bridger Coal Company,

September 21, 1978.

79. Wyoming DEQ, Land Quality Division, District II, Quarterly Inspection Report: Bridger Coal Company, July 27, 1977.

80. *Ibid.*

81. INFORM interviews with Richard M. Kail, Director of Reclamation, and Tom Schreeg, Reclamation Specialist, Jim Bridger Mine, June 26, 1979.

82. U.S. Department of the Interior, USGS, Inspection Reports: Bridger Coal Company, September 21, 1978 and July 12, 1978; Wyoming DEQ, Land Quality Division, District II, Quarterly Inspection Report: Bridger Coal Company, January 7, 1977.

83. INFORM interviews with Richard M. Kail, Director of Reclamation, and Tom Schreeg, Reclamation Specialist, Jim Bridger Mine, June 26, 1979.

84. Wyoming DEQ, Land Quality Division, District II, Quarterly Inspection Report: Bridger Coal Company, October 3, 1977.

85. INFORM interview with Gary Deveraux, Director of Mined Land Reclamation, NERCO, October 18, 1979.

ROSEBUD

1. Rosebud Coal Sales Company, "Mining and Reclamation Plan Pursuant to the Rules and Regulations of Title 30, Chapter II, Part 211--Coal Mining Operating Regulations," submitted May 12, 1977, p. 49.

2. *Ibid.*

3. U.S. Department of Commerce, NOAA, *Monthly Averages of Temperature and Precipitation for State Climatic Divisions 1941-1970: Wyoming,* "Monthly and Annual Divisional Averages Precipitation (Inches)" (Asheville, N.C.: National Climatic Center, July 1973), p. 8.

4. Rosebud Coal Sales Company, "Mining and Reclamation Plan," submitted May 12, 1977, p. 48.

5. INFORM interviews with David Evans, Reclamation Supervisor, Rosebud Coal Sales Company, and Sam Scott, Reclamation Director, Sheridan Office, Peter Kiewit Sons' Company, June 25, 1979.

6. Rosebud Coal Sales Company, "Mining and Reclamation Plan," submitted May 12, 1977, p. 38.

7. Written communication from George F. Brown, Acting Conservation Manager, USGS, U.S. Department of the Interior, to Ron Lanoue, Research Associate, INFORM, March 31, 1979.

8. INFORM interview with Sam Scott, Reclamation Director, Sheridan Office, Peter Kiewit Sons' Company, January 8, 1980.

9. U.S. Department of the Interior, OSM, Region V On-Site Inspections: Rosebud Coal Sales Company--Rosebud Mine, August 1, 1978, June 26, 1979, and December 18, 1979.

10. Peter Kiewit Sons' Company, *Annual Reclamation Report,* July 1, 1977-June 30, 1978.

11. INFORM interviews with David Evans, Reclamation Supervisor, Rosebud Coal Sales Company, and Sam Scott, Reclamation Director, Sheridan Office, Peter Kiewit Sons' Company, June 25, 1979.

12. *Idem.*

13. U.S. Department of the Interior, OSM, Region V On-Site Inspection: Rosebud Coal Sales Company--Rosebud Mine, August 1, 1978.

14. INFORM interviews with David Evans, Reclamation Supervisor, Rosebud Coal Sales Company, and Sam Scott, Reclamation Director, Sheridan Office, Peter Kiewit Sons' Company, June 25, 1979.

15. U.S. Department of the Interior, OSM, Region V On-Site Inspection: Rosebud Coal Sales Company--Rosebud Mine, August 1, 1978.

16. Peter Kiewit Sons' Company, *Annual Reclamation Report,* July 1, 1977-June 30, 1978.

17. INFORM interviews with David Evans, Reclamation Supervisor, Rosebud Coal Sales Company, and Sam Scott, Reclamation Director, Sheridan Office, Peter Kiewit Sons' Company, June 25, 1979.

18. *Idem.*

19. U.S. Department of the Interior, USGS, Inspection Report: Rosebud Coal

Sales Company, September 2, 1976.
20. *Ibid.*
21. U.S. Department of the Interior, USGS, Inspection Report: Rosebud Coal Sales Company, March 1, 1977.
22. U.S. Department of the Interior, USGS, Inspection Report: Rosebud Coal Sales Company, August 17, 1977.
23. U.S. Department of the Interior, USGS, Inspection Report: Rosebud Coal Sales Company, November 21, 1977.
24. U.S. Department of the Interior, USGS, Inspection Report: Rosebud Coal Sales Company, April 1978.
25. U.S. Department of the Interior, USGS, Inspection Report: Rosebud Coal Sales Company, November 30, 1978; Letter from Jerry Ellis, Chief Mining Engineer, Rosebud Coal Sales Company, to J. P. Storrs, Area Mining Supervisor, USGS, U.S. Department of the Interior, May 12, 1978.
26. Wyoming DEQ, Water Quality Division, "Authorization to Discharge Under the National Pollutant Discharge Elimination System, Permit No. WY-0022853," November 23, 1977.
27. *Ibid.*, Part I, p. 2.
28. Rosebud Coal Sales Company, NPDES Discharge Monitoring Reports--Rosebud Mine, reporting periods from April 1, 1978 to July 1, 1978 and from April 1, 1979 to July 1, 1979.
29. Rosebud Coal Sales Company, NPDES Discharge Monitoring Reports--Rosebud Mine, reporting periods from April 1, 1978 to July 1, 1978 and from April 1, 1979 to July 1, 1979.
30. INFORM interviews with David Evans, Reclamation Supervisor, Rosebud Coal Sales Company, and Sam Scott, Reclamation Director, Sheridan Office, Peter Kiewit Sons' Company, June 25, 1979.
31. *Idem.*
32. Written communication from Sam Scott, Reclamation Director, Sheridan Office, Peter Kiewit Sons' Company, to INFORM, December 17, 1979.
33. *Ibid.*
34. INFORM interviews with David Evans, Reclamation Supervisor, Rosebud Coal Sales Company, and Sam Scott, Reclamation Director, Sheridan Office, Peter Kiewit Sons' Company, June 25, 1979.
35. Wyoming DEQ, Land Quality Division--District II, Quarterly Inspection Reports: Rosebud Coal Sales Company, April 3, 1978 and September 27, 1978.
36. INFORM interview with Sam Scott, Reclamation Director, Sheridan Office, Peter Kiewit Sons' Company, January 8, 1980.
37. Peter Kiewit Sons' Company, *Annual Reclamation Report*, July 1, 1977-June 30, 1978.
38. INFORM interviews with David Evans, Reclamation Supervisor, Rosebud Coal Sales Company, and Sam Scott, Reclamation Director, Sheridan Office, Peter Kiewit Sons' Company, June 25, 1979.
39. Written communication from Sam Scott, Reclamation Director, Sheridan Office, Peter Kiewit Sons' Company, to INFORM, December 17, 1979.
40. INFORM interviews with David Evans, Reclamation Supervisor, Rosebud Coal Sales Company, and Sam Scott, Reclamation Director, Sheridan Office, Peter Kiewit Sons' Company, June 25, 1979.
41. *Idem.*
42. *Idem.*
43. Peter Kiewit Sons' Company, *Annual Reclamation Report*, July 1, 1977-June 30, 1978.
44. INFORM interviews with David Evans, Reclamation Supervisor, Rosebud Coal Sales Company, and Sam Scott, Reclamation Director, Sheridan Office, Peter Kiewit Sons' Company, June 25, 1979.
45. U.S. Department of the Interior, OSM, Region V On-Site Inspection: Rosebud Coal Sales Company--Rosebud Mine, August 1, 1978.
46. INFORM interviews with David Evans, Reclamation Supervisor, Rosebud Coal Sales Company, and Sam Scott, Reclamation Director, Sheridan Office, Peter Kiewit Sons' Company, June 25, 1979.
47. *Idem.*
48. *Idem.*
49. *Idem.*
50. INFORM interview with Sam Scott, Reclamation Director, Sheridan Office,

Peter Kiewit Sons' Company, January 8, 1980.

51. Rosebud Coal Sales Company, "Mining and Reclamation Plan," submitted May 12, 1977, p. 80.

52. INFORM interviews with David Evans, Reclamation Supervisor, Rosebud Coal Sales Company, and Sam Scott, Reclamation Director, Sheridan Office, Peter Kiewit Sons' Company, June 25, 1979.

53. *Idem.*

54. INFORM interview with Sam Scott, Reclamation Director, Sheridan Office, Peter Kiewit Sons' Company, January 8, 1980.

55. U.S. Department of the Interior, USGS, Inspection Reports: Rosebud Coal Sales Company, October 21, 1977 and October 30, 1978; Wyoming DEQ, Land Quality Division--District II, Quarterly Inspection Report: Rosebud Coal Sales Company, July 13, 1978.

56. INFORM interviews with David Evans, Reclamation Supervisor, Rosebud Coal Sales Company, and Sam Scott, Reclamation Director, Sheridan Office, Peter Kiewit Sons' Company, June 25, 1979.

57. Written communication from Marvin C. Suhr, Soil Scientist, SCS, U.S. Department of Agriculture, to Daniel Wiener, Research Associate, INFORM, March 5, 1979.

58. Written communication from Sam Scott, Reclamation Director, Sheridan Office, Peter Kiewit Sons' Company, to INFORM, December 17, 1979.

59. INFORM interviews with David Evans, Reclamation Supervisor, Rosebud Coal Sales Company, and Sam Scott, Reclamation Director, Sheridan Office, Peter Kiewit Sons' Company, June 25, 1979.

60. *Idem.*; Written communication from Sam Scott, Reclamation Director, Sheridan Office, Peter Kiewit Sons' Company, to INFORM, December 17, 1979.

61. *Ibid.*

62. *Ibid.*

63. *Ibid.*

64. Written communication from Sam Scott, Reclamation Director, Sheridan Office, Peter Kiewit Sons' Company, to INFORM, December 17, 1979.

65. INFORM interview with Sam Scott, Reclamation Director, Sheridan Office, Peter Kiewit Sons' Company, January 8, 1980.

SEMINOE II

1. U.S. Department of the Interior, OSM, Region V On-Site Inspection: Arch Mineral Corporation--Seminoe II Mine, August 1, 1978.

2. Arch Mineral Corporation, "Supplementary and Modified Seminoe No. 2 Mine Plan--South Seminoe Area, Pursuant to the Rules and Regulations of Title 30, Chapter II, Part 211--Coal Mining Operating Regulations," submitted October 1976, p. 6.

3. *Ibid.*, p. 17.

4. U.S. Department of Commerce, NOAA, *Monthly Averages of Temperature and Precipitation for State Climatic Divisions 1941-1970: Wyoming*, "Monthly and Annual Divisional Averages Precipitation (Inches)" (Asheville, N.C.: National Climatic Center, July 1973), p. 8.

5. Arch Mineral Corporation, "Seminoe No. 2 Mine Plan," submitted October 1976, p. 51.

6. INFORM interview with Greg Bierei, Reclamation Director, Western Division, Arch Mineral Corporation, December 11, 1979.

7. Wyoming DEQ, Land Quality Division--District II, Joint Annual Quarterly Inspection Report: Arch Mineral Corporation--Seminoe II, September 1, 1978, p. 4.

8. INFORM interviews with Greg Bierei, Reclamation Director, Western Division, Arch Mineral Corporation, Benton Kelly, Permitting Director, Arch Mineral Corporation, and Steve Menlove, Manager of Engineering, Arch Mineral Corporation, June 22, 1979.

9. INFORM interview with Greg Bierei, Reclamation Director, Western Division, Arch Mineral Corporation, December 11, 1979.

10. U.S. Department of the Interior, OSM, Region V On-Site Inspection: Arch Mineral Corporation--Seminoe II Mine, November 14, 1978.

11. INFORM interview with Thomas E. Ehmett, Reclamation Specialist, Region V, OSM, U.S. Department of the Interior, March 8, 1979.

12. Arch Mineral Corporation, Seminoe Mine No. 2, "Re: Annual Reclamation Report, July 1, 1977–June 30, 1978," to Jim A. Kandolin, Wyoming DEQ, Land Quality Division, and J. Paul Storrs, Area Mining Supervisor, USGS, U.S. Department of the Interior, June 27, 1978.

13. INFORM interview with Greg Bierei, Reclamation Director, Western Division, Arch Mineral Corporation, December 11, 1979.

14. U.S. Department of the Interior, OSM, Region V On-Site Inspections: Arch Mineral Corporation--Seminoe II Mine, August 1, 1978, November 14, 1978, April 2, 1979, and September 10, 1979.

15. Written communication from Arch Mineral Corporation to INFORM, June 22, 1979, and September 10, 1979.

16. Don Jackson, "Arch Mineral Puts Three on the Line," Coal Age (New York: McGraw-Hill, 1976), September 1976.

17. Wyoming DEQ, Land Quality Division--District II, Quarterly Inspection Report: Arch Mineral Corporation--Seminoe II, April 5, 1978.

18. INFORM interviews with Greg Bierei, Reclamation Director, Western Division, Arch Mineral Corporation, Benton Kelly, Permitting Director, Arch Mineral Corporation, and Steve Menlove, Manager of Engineering, Arch Mineral Corporation.

19. U.S. Department of the Interior, OSM, Region V On-Site Inspection: Arch Mineral Corporation--Seminoe II Mine, August 1, 1978.

20. Written communication from Arch Mineral Corporation to INFORM, June 22, 1979.

21. Written communication from Sheridan A. Glen, Director of Environmental Affairs, Arch Mineral Corporation, to INFORM, November 15, 1979.

22. Wyoming DEQ, Land Quality Division--District II, Joint Annual Quarterly Inspection Report: Arch Mineral Corporation--Seminoe II, September 13, 1978.

23. INFORM interviews with Greg Bierei, Reclamation Director, Western Division, Arch Mineral Corporation, Benton Kelly, Permitting Director, Arch Mineral Corporation, and Steve Menlove, Manager of Engineering, Arch Mineral Corporation, June 22, 1979.

24. INFORM interview with Greg Bierei, Reclamation Director, Western Division, Arch Mineral Corporation--Seminoe II Mine, ·August 1, 1978.

25. U.S. Department of the Interior, OSM, Region V On-Site Inspection: Arch Mineral Corporation--Seminoe II Mine, August 1, 1978.

26. U.S. Department of the Interior, OSM, Region V On-Site Inspection: Arch Mineral Corporation, June 22, 1979.

27. INFORM interviews with Greg Bierei, Reclamation Director, Western Division, Arch Mineral Corporation, Benton Kelly, Permitting Director, Arch Mineral Corporation, and Steve Menlove, Manager of Engineering, Arch Mineral Corporation, June 22, 1979.

28. U.S. Department of the Interior, OSM, Region V On-Site Inspection: Arch Mineral Corporation--Seminoe II Mine, August 1, 1978.

29. Written communication from Sheridan A. Glen, Director of Environmental Affairs, Arch Mineral Corporation, to INFORM, November 15, 1979.

30. U.S. Department of the Interior, OSM, Region V On-Site Inspection: Arch Mineral Corporation--Seminoe II Mine, August 1, 1978.

31. U.S. Department of the Interior, OSM Reclamation and Enforcement, Notices of Violation No. 79-V-1-7, April 2, 1979.

32. Written communication from Sheridan A. Glen, Director of Environmental Affairs, Arch Mineral Corporation, to INFORM, November 15, 1979.

33. Wyoming DEQ, Land Quality Division--District II, Quarterly Inspection Report: Arch Mineral Corporation--Seminoe II, April 5, 1978; Wyoming DEQ, Land Quality Division--District II, Joint Annual Quarterly Inspection Report: Arch Mineral Corporation--Seminoe II, September 13, 1978.

34. INFORM interview with Greg Bierei, Reclamation Director, Western Division, Arch Mineral Corporation, December 11, 1979.

35. INFORM interviews with Greg Bierei, Reclamation Director, Western Division, Arch Mineral Corporation, Benton Kelly, Permitting Director, Arch Mineral Corporation, and Steve Menlove, Manager of Engineering, Arch Mineral Corporation, June 22, 1979.

36. Wyoming DEQ, Land Quality Division--District II, Quarterly Inspection Re-

port: Arch Mineral Corporation--Seminoe II, June 28, 1978.

37. INFORM interview with Greg Bierei, Reclamation Director, Western Division, Arch Mineral Corporation, December 11, 1979.

38. Arch Mineral Corporation, "Seminoe No. 2 Mine Plan," submitted October 1976, p. 44.

39. U.S. Department of the Interior, OSM Reclamation and Enforcement, Notices of Violation No. 79-V-1-7, April 2, 1979.

40. U.S. Department of the Interior, OSM, Region V On-Site Inspection: Arch Mineral Corporation--Seminoe II Mine, April 2, 1979.

41. INFORM interviews with Greg Bierei, Reclamation Director, Western Division, Arch Mineral Corporation, Benton Kelly, Permitting Director, Arch Mineral Corporation, and Steve Menlove, Manager of Engineering, Arch Mineral Corporation, June 22, 1979.

42. INFORM interview with Robert E. Walline, Chemical Engineer, Region VIII, EPA, December 19, 1979.

43. Arch Mineral Corporation, "Seminoe No. 2 Mine Plan," submitted October 1976, p. 45.

44. *Ibid.*, p. 67; Written communication from Arch Mineral Corporation to IN-FORM, June 22, 1979.

45. INFORM interviews with Greg Bierei, Reclamation Director, Western Division, Arch Mineral Corporation, Benton Kelly, Permitting Director, Arch Mineral Corporation, and Steve Menlove, Manager of Engineering, Arch Mineral Corporation, June 22, 1979.

46. Written communication from George F. Brown, Acting Conservation Manager, USGS, U.S. Department of the Interior, to Ron Lanoue, Research Associate, INFORM, March 21, 1979.

47. Computer printout from H. Robert Moore, Assistant to the Director for Coal Management, BLM, U.S. Department of the Interior, November 20, 1978.

48. Written communication from Sheridan A. Glen, Director of Environmental Affairs, Arch Mineral Corporation, to INFORM, November 15, 1979.

49. U.S. Department of the Interior, OSM, Region V On-Site Inspection: Arch Mineral Corporation--Seminoe II Mine, October 1, 1978.

50. INFORM interviews with Greg Bierei, Reclamation Director, Western Division, Arch Mineral Corporation, Benton Kelly, Permitting Director, Arch Mineral Corporation, and Steve Menlove, Manager of Engineering, Arch Mineral Corporation, June 22, 1979.

51. INFORM interview with Greg Bierei, Reclamation Director, Western Division, Arch Mineral Corporation, December 11, 1979.

52. Wyoming DEQ, Land Quality Division--District II, Quarterly Inspection Report: Arch Mineral Corporation--Seminoe II, January 20, 1979.

53. INFORM interview with Greg Bierei, Reclamation Director, Western Division, Arch Mineral Corporation, December 11, 1979.

54. Written communication from Arch Mineral Corporation to INFORM, June 22, 1979.

55. *Ibid.*

56. INFORM interviews with Greg Bierei, Reclamation Director, Western Division, Arch Mineral Corporation, Benton Kelly, Permitting Director, Arch Mineral Corporation, and Steve Menlove, Manager of Engineering, Arch Mineral Corporation, June 22, 1979.

57. *Idem.*

58. *Idem.*

59. INFORM interview with Greg Bierei, Reclamation Director, Western Division, Arch Mineral Corporation, December 11, 1979.

60. *Idem.*

61. *Idem.*

62. Written communication from Arch Mineral Corporation to INFORM, June 22, 1979.

63. *Ibid.*

64. Wyoming DEQ, Land Quality Division--District II, Quarterly Inspection Report: Arch Mineral Corporation--Seminoe II, January 20, 1977.

65. Mimi Cunningham, "Coal Mining: Mapping Aids Reclamation Effort," CH2M Hill Reports, Winter 1978, pp. 6-7.

66. Written communication from Arch Mineral Corporation to INFORM, June 22, 1979.

67. *Ibid.*
68. INFORM interviews with Greg Bierei, Reclamation Director, Western Division, Arch Mineral Corporation, Benton Kelly, Permitting Director, Arch Mineral Corporation, and Steve Menlove, Manager of Engineering, Arch Mineral Corporation, June 22, 1979.
69. *Idem.*
70. INFORM interview with Greg Bierei, Reclamation Director, Western Division, Arch Mineral Corporation, December 11, 1979.
71. Wyoming DEQ, Land Quality Division--District II, Joint Annual Quarterly Inspection Report: Arch Mineral Corporation--Seminoe II, September 13, 1978.
72. *Ibid.*
73. Wyoming DEQ, Land Quality Division--District II, Quarterly Inspection Report: Arch Mineral Corporation--Seminoe II, January 20, 1977.
74. Wyoming DEQ, Land Quality Division--District II, Joint Annual Quarterly Inspection Report: Arch Mineral Corporation--Seminoe II, September 13, 1978.
75. INFORM interviews with Greg Bierei, Reclamation Director, Western Division, Arch Mineral Corporation, Benton Kelly, Permitting Director, Arch Mineral Corporation, and Steve Menlove, Manager of Engineering, Arch Mineral Corporation, June 22, 1979.
76. *Idem.*
77. Written communication from Arch Mineral Corporation to INFORM, June 22, 1979.
78. INFORM interviews with Greg Bierei, Reclamation Director, Western Division, Arch Mineral Corporation, Benton Kelly, Permitting Director, Arch Mineral Corporation, and Steve Menlove, Manager of Engineering, Arch Mineral Corporation, June 22, 1979.
79. A. Kent Evans, E. W. Uhleman, and P. A. Eby, *Atlas of Western Surface-Mined Lands: Coal, Uranium and Phosphate* (Washington, D.C.: U.S. Department of the Interior, Fish and Wildlife Service, Biological Services Program, January 1978), p. 184.
80. U.S. Department of the Interior, *Draft Environmental Statement: Proposed Development of Coal Resources in Southcentral Wyoming* (Washington, D.C.: U.S. Government Printing Office, October 13, 1978), p. HS3-24.
81. Written communication from Arch Mineral Corporation to INFORM, June 22, 1979.
82. Montana State University, Cooperative Extension Service, "Livestock Range--Its Nature and Use, Part III: Plant Relationships," Bulletin 1015, January 1977.
83. Arch Mineral Corporation, Seminoe Mine No. 2, "Re: Annual Reclamation Report, July 1, 1977-June 30, 1978," to Jim A. Kandolin, Wyoming DEQ, Land Quality Division, and J. Paul Storrs, Area Mining Supervisor, USGS, U.S. Department of the Interior, June 27, 1978; INFORM interviews with Greg Bierei, Reclamation Director, Western Division, Arch Mineral Corporation, Benton Kelly, Permitting Director, Arch Mineral Corporation, and Steve Menlove, Manager of Engineering, Arch Mineral Corporation, June 22, 1979.
84. *Ibid.*

ABSALOKA

1. *Rand McNally Road Atlas*, 54th Edition (Chicago: Rand McNally, Inc., 1978), p. 122.
2. Montana DSL, *Final Environmental Impact Statement on Leasing of School Section 36 Lands to Westmoreland Resources, Inc.*, December 1977, p. 22.
3. U.S. Department of Commerce, NOAA, *Monthly Averages of Temperature and Precipitation for State Climatic Divisions 1941-70: Montana*, "Monthly and Annual Divisional Averages Precipitation (Inches)" (Asheville, N.C.: National Climatic Center, 1973), p. 7.
4. INFORM interview with David Simpson, Environmental Administrator, Westmoreland Resources, Inc., June 20, 1979.
5. Montana DSL, *Final Environmental Impact Statement on Leasing of School Section 36 Lands to Westmoreland Resources, Inc.*, p. 22.
6. INFORM interview with David Simpson, Environmental Administrator, West-

moreland Resources, Inc., June 20, 1979.

7. National Academy of Sciences, *Rehabilitation Potential of Western Coal Lands* (Cambridge, Mass.: Ballinger Publishing Co., 1974), p. 122.

8. U.S. Department of the Interior, Bureau of Indian Affairs, *Draft Environmental Statement 76-30, Crow Ceded Area Coal Lease, Tracts II and III, Westmoreland Resources*, Vol. I, p. II-80.

9. INFORM interview with Dennis Hemmer, Vegetation Specialist, Montana DSL, February 15, 1980.

10. U.S. Department of the Interior, Bureau of Indian Affairs, *Draft Environmental Statement 76-30, Crow Ceded Area Coal Lease, Tracts II and III, Westmoreland Resources*, Vol. I, p. II-79.

11. INFORM interview with Jack Schmidt, Former Inspector, Montana DSL, April 4, 1979.

12. INFORM interview with Dennis Hemmer, Vegetation Specialist, Montana DSL, February 15, 1980.

13. *Idem.*

14. Written communication from George F. Brown, Acting Conservation Manager, USGS, U.S. Department of the Interior, to Ron Lanoue, Research Associate, INFORM, April 9, 1979; U.S. Department of the Interior, USGS, Inspection Report: Westmoreland Resources, Inc.--Absaloka, May 1, 1978.

15. INFORM interview with David Simpson, Environmental Administrator, Westmoreland Resources, Inc., January 8, 1980.

16. U.S. Department of the Interior, USGS, Inspection Report: Westmoreland Resources, Inc.--Absaloka, May 1, 1978.

17. INFORM interviews with Montana DSL Inspectors Dennis Hemmer, Vegetation Specialist, Mike Bishop, Hydrogeologist, and Neil Harrington, Soil Scientist, April 2, 1979.

18. U.S. Department of the Interior, USGS, Inspection Report: Westmoreland Resources, Inc.--Absaloka, May 1, 1978.

19. INFORM interviews with Montana DSL Inspectors Dennis Hemmer, Vegetation Specialist, Mike Bishop, Hydrogeologist, and Neil Harrington, Soil Scientist, April 2, 1979.

20. INFORM interview with David Simpson, Environmental Administrator, Westmoreland Resources, Inc., June 20, 1979.

21. Montana DSL, Field Inspection Reports: Absaloka Mine, June 7, 1977, July 13, 1977, and December 27, 1977.

22. INFORM interview with David Simpson, Environmental Administrator, Westmoreland Resources, Inc., June 20, 1979.

23. INFORM interviews with Montana DSL Inspectors Dennis Hemmer, Vegetation Specialist, Mike Bishop, Hydrogeologist, and Neil Harrington, Soil Scientist, April 2, 1979.

24. Montana DSL, Field Inspection Reports: Absaloka Mine, June 7, 1977, July 13, 1977, and December 27, 1977.

25. INFORM interview with David Simpson, Environmental Administrator, Westmoreland Resources, Inc., January 8, 1980.

26. Montana DSL, Field Inspection Reports: Absaloka Mine, October 14, 1977, November 17, 1977, and May 4, 1978; INFORM interview with Dennis Hemmer, Vegetation Specialist, Montana DSL, July 10, 1979.

27. Written communication from David Simpson, Environmental Administrator, Westmoreland Resources, Inc., to INFORM, October 24, 1979.

28. U.S. Department of the Interior, OSM, Region V On-Site Inspection: Westmoreland Resources, Inc.--Absaloka Mine, July 19, 1979; U.S. Department of the Interior, OSM Reclamation and Enforcement, Notices of Violation No. 79-V-1-22.

29. U.S. Department of the Interior, OSM, Region V On-Site Inspection: Westmoreland Resources, Inc.--Absaloka Mine, October 9-10, 1979.

30. INFORM interview with David Simpson, Environmental Administrator, Westmoreland Resources, Inc., June 20, 1979.

31. Montana DSL, Field Inspection Report: Absaloka Mine, August 22, 1978.

32. INFORM interview with David Simpson, Environmental Administrator, Westmoreland Resources, Inc., January 8, 1980.

33. INFORM interview with David Simpson, Environmental Administrator, Westmoreland Resources, Inc., June 20, 1979.

34. U.S. Department of the Interior, OSM, Region V On-Site Inspection: West-

moreland Resources, Inc.--Absaloka Mine, July 19, 1979.

35. INFORM interview with David Simpson, Environmental Administrator, Westmoreland Resources, Inc., January 8, 1980.

36. Montana DSL, Field Inspection Report: Absaloka Mine, March 21, 1977.

37. Montana DSL, Field Inspection Report: Absaloka Mine, February 23, 1977.

38. Montana DSL, Field Inspection Report: Absaloka Mine, March 23, 1978.

39. Written communication from Daniel Simpson, Environmental Administrator, Westmoreland Resources, Inc., to INFORM, October 24, 1979.

40. Westmoreland Resources, Inc., to Absaloka Mine, "Mine Plan Exhibit B-1," 1979.

41. Written communication from David Simpson, Environmental Administrator, Westmoreland Resources, Inc., to INFORM, October 24, 1979.

42. Montana DSL, Field Inspection Reports: Absaloka Mine, February 23, 1977 and May 31, 1978; INFORM interview with Dennis Hemmer, Vegetation Specialist, Montana DSL, February 15, 1980.

43. Montana DSL, Field Inspection Report: Absaloka Mine, October 20, 1978.

44. INFORM interview with David Simpson, Environmental Administrator, Westmoreland Resources, Inc., June 20, 1979.

45. INFORM interview with Dennis Hemmer, Vegetation Specialist, Montana DSL, February 15, 1980.

46. Montana Bureau of Mines and Geology, Wayne A. Van Voast, Robert B. Hedges, and John J. McDermott, "Strip Coal Mining and Mined-Land Recmation in the Hydrologic System, Southeastern Montana" (Project Completion Report, sponsored in part by the Old West Regional Commission, Billings, Montana, December 1978).

47. Montana DSL, *Draft Environmental Impact Statement on the Proposed Mining and Reclamation Plan for Westmoreland Resources, Inc., Absaloka Mine, Big Horn County, Montana*, September 14, 1979, p. II-5.

48. INFORM interview with David Simpson, Environmental Administrator, Westmoreland Resources, Inc., January 8, 1980.

49. INFORM interview with David Simpson, Environmental Administrator, Westmoreland Resources, Inc., January 8, 1980; U.S. Department of the Interior, OSM, Region V On-Site Inspection: Westmoreland Resources, Inc.--Absaloka Mine, July 19, 1979.

50. U.S. Department of the Interior, OSM, Region V On-Site Inspection: Westmoreland Resources, Inc.--Absaloka Mine, September 27, 1978; INFORM interview with David Simpson, Environmental Administrator, Westmoreland Resources, Inc., January 8, 1980.

51. Montana DSL, Field Inspection Report: Absaloka Mine, April 7, 1978; U.S. Department of the Interior, OSM, Region V On-Site Inspection: Westmoreland Resources, Inc.--Absaloka Mine, September 27, 1978.

52. Montana DSL, Field Inspection Report: Absaloka Mine, November 17, 1977; U.S. Department of the Interior, USGS, Inspection Report: Westmoreland Resources, Inc.--Absaloka, October 5, 1978.

53. Montana DSL, Field Inspection Report: Absaloka Mine, October 20, 1978; U.S. Department of the Interior, OSM, Region V On-Site Inspection: Westmoreland Resources, Inc.--Absaloka Mine, September 27, 1978.

54. U.S. Department of the Interior, OSM, Region V On-Site Inspection: Westmoreland Resources, Inc.--Absaloka Mine, July 19, 1979.

55. U.S. Department of the Interior, OSM, Region V On-Site Inspection: Westmoreland Resources, Inc.--Absaloka Mine, September 27, 1978; Westmoreland Resources, Inc., NPDES Discharge Monitoring Reports--Absaloka Mine, reporting periods from January 1, 1979 through September 30, 1979.

56. Montana DSL, Field Inspection Report: Absaloka Mine, April 7, 1978; INFORM interviews with Montana DSL Inspectors Dennis Hemmer, Vegetation Specialist, Mike Bishop, Hydrogeologist, and Neil Harrington, Soil Scientist, April 2, 1979.

57. INFORM interview with Dennis Hemmer, Vegetation Specialist, Montana DSL, February 15, 1980.

58. U.S. Department of the Interior, OSM, Region V On-Site Inspection: Westmoreland Resources, Inc.--Absaloka Mine, October 9-10, 1979.

59. INFORM interview with David Simpson, Environmental Administrator, Westmoreland Resources, Inc., June 20, 1979.

60. INFORM interview with David Simpson, Environmental Administrator, West-

moreland Resources, Inc., January 8, 1980.

61. INFORM interviews with Montana DSL Inspectors Dennis Hemmer, Vegetation Specialist, Mike Bishop, Hydrogeologist, and Neil Harrington, Soil Scientist, April 2, 1979.

62. Montana Bureau of Mines and Geology, Van Voast, Hedges, and McDermott, "Strip Coal Mining and Mined-Land Reclamation in the Hydrologic System, Southeastern Montana," December 1978.

63. INFORM interview with David Simpson, Environmental Administrator, Westmoreland Resources, Inc., June 20, 1979.

64. INFORM interviews with Montana DSL Inspectors Dennis Hemmer, Vegetation Specialist, Mike Bishop, Hydrogeologist, and Neil Harrington, Soil Scientist, April 2, 1979.

65. Montana DSL, *Draft Environmental Impact Statement on the Proposed Mining and Reclamation Plan for Westmoreland Resources, Inc., Absaloka Mine*, September 14, 1979, p. III-5-7.

66. *Ibid.*, p. III-4.

67. Ben A. Franklin, "Millions of Dollars are at Stake in Environmental Dispute Over Strip Mine in Montana," *The New York Times*, Monday, August 27, 1979.

68. INFORM interview with David Simpson, Environmental Administrator, Westmoreland Resources, Inc., June 20, 1979.

69. Montana DSL, Field Inspection Reports: Absaloka Mine, February 9, 1977 and February 23, 1977.

70. Montana DSL, Field Inspection Report: Absaloka Mine, May 6, 1977.

71. Montana DSL, Field Inspection Report: Absaloka Mine, September 14, 1977.

72. Montana DSL, Field Inspection Report: Absaloka Mine, June 15, 1978.

73. Montana DSL, Field Inspection Report: Absaloka Mine, August 22, 1978.

74. U.S. Department of the Interior, OSM, Region V On-Site Inspection: Westmoreland Resources, Inc.--Absaloka Mine, September 27, 1978.

75. INFORM interviews with Montana DSL Inspectors Dennis Hemmer, Vegetation Specialist, Mike Bishop, Hydrogeologist, and Neil Harrington, Soil Scientist, April 2, 1979.

76. INFORM interview with David Simpson, Environmental Administrator, Westmoreland Resources, Inc., June 20, 1979.

77. Montana DSL, Field Inspection Reports: Absaloka Mine, March 9, 1977, March 7, 1978, February 9, 1977 and July 13, 1977.

78. Montana DSL, Field Inspection Report: Absaloka Mine, November 17, 1977.

79. Montana DSL, Field Inspection Reports: Absaloka Mine, July 13, 1977 and October 20, 1978; INFORM interviews with Montana DSL Inspectors Dennis Hemmer, Vegetation Specialist, Mike Bishop, Hydrogeologist, and Neil Harrington, Soil Scientist, April 2, 1979.

80. Written communication from David Simpson, Environmental Administrator, Westmoreland Resources, Inc., to INFORM, October 24, 1979.

81. Montana DSL, Field Inspection Report: Absaloka Mine, June 15, 1978.

82. INFORM interviews with Montana DSL Inspectors Dennis Hemmer, Vegetation Specialist, Mike Bishop, Hydrogeologist, and Neil Harrington, Soil Scientist, April 2, 1979.

83. Westmoreland Resources, Inc., Absaloka Mine, "Mine Plan Exhibit B-1," 1979.

84. INFORM interviews with Montana DSL Inspectors Dennis Hemmer, Vegetation Specialist, Mike Bishop, Hydrogeologist, and Neil Harrington, Soil Scientist, April 2, 1979.

85. *Idem.*

86. U.S. Department of the Interior, OSM, Region V On-Site Inspection, Westmoreland Resources, Inc.--Absaloka Mine, September 27, 1978.

87. INFORM interview with David Simpson, Environmental Administrator, Westmoreland Resources, Inc., January 8, 1980.

88. U.S. Department of the Interior, OSM, Region V On-Site Inspection: Westmoreland Resources, Inc.--Absaloka Mine, September 27, 1978.

89. INFORM interview with David Simpson, Environmental Administrator, Westmoreland Resources, Inc., June 20, 1979.

90. Written communication from David Simpson, Environmental Administrator, Westmoreland Resources, Inc., to INFORM, October 24, 1979.

91. INFORM interview with David Simpson, Environmental Administrator, West-

moreland Resources, Inc., June 20, 1979.

92. *Idem.*

93. INFORM interview with David Simpson, Environmental Administrator, Westmoreland Resources, Inc., January 8, 1980.

94. *Idem.*

95. *Idem.*

96. Written communication from David Simpson, Environmental Administrator, Westmoreland Resources, Inc., to INFORM, October 24, 1979.

97. *Ibid.*

98. INFORM interview with David Simpson, Environmental Administrator, Westmoreland Resources, Inc., January 8, 1980.

99. *Idem.*

100. Written communication from David Simpson, Environmental Administrator, Westmoreland Resources, Inc., to INFORM, October 24, 1979.

101. Westmoreland Resources, Inc., Absaloka Mine, "Mine Plan Exhibit B-1," 1979.

102. U.S. Department of the Interior, USGS, Inspection Report: Westmoreland Resources, Inc.--Absaloka, May 1, 1978.

103. U.S. Department of the Interior, USGS, Inspection Report: Westmoreland Resources, Inc.--Absaloka, October 5, 1978.

104. Montana DSL, Field Inspection Report: Absaloka Mine, May 31, 1978.

105. Montana DSL, Field Inspection Report: Absaloka Mine, June 15, 1978.

106. Written communication from David Simpson, Environmental Administrator, Westmoreland Resources, Inc., to INFORM, October 24, 1979.

107. INFORM interview with David Simpson, Environmental Administrator, Westmoreland Resources, Inc., June 20, 1979.

108. INFORM interview with Dennis Hemmer, Vegetation Specialist, Montana DSL, February 15, 1980.

109. U.S. Department of the Interior, USGS, Inspection Report: Westmoreland Resources, Inc.--Absaloka, May 1, 1978.

110. U.S. Department of the Interior, Bureau of Indian Affairs, *Draft Environmental Statement 76-30, Crow Ceded Area Coal Lease, Tract II, Westmoreland Resources*, Vol. I, p. II-153.

111. INFORM interview with David Simpson, Environmental Administrator, Westmoreland Resources, Inc., June 20, 1979; Letter from Daniel Simpson, Environmental Administrator, to Brace Hayden, Administrator, Reclamation Division, Montana DSL, July 20, 1977.

112. INFORM interview with David Simpson, Environmental Administrator, Westmoreland Resources, Inc., June 20, 1979.

113. *Idem.*

114. *Idem.*; U.S. Department of the Interior, USGS, Inspection Report: Westmoreland Resources, Inc.--Absaloka, October 5, 1978; A. Kent Evans, E. W. Uhleman, and P. A. Eby, *Atlas of Western Surface Mined Lands: Coal, Uranium and Phosphate* (Washington, D.C.: U.S. Department of the Interior, Fish and Wildlife Service, Biological Services Program, January 1978), p. 184.

115. Written communication from David Simpson, Environmental Administrator, Westmoreland Resources, Inc., June 20, 1979.

116. INFORM interview with Dennis Hemmer, Vegetation Specialist, Montana DSL, July 10, 1979.

117. INFORM interview with David Simpson, Environmental Administrator, Westmoreland Resources, Inc., June 20, 1979.

118. INFORM interviews with Montana DSL Inspectors Dennis Hemmer, Vegetation Specialist, Mike Bishop, Hydrogeologist, and Neil Harrington, Soil Scientist, April 2, 1979.

119. Montana DSL, Field Inspection Report: Absaloka Mine, December 27, 1977; U.S. Department of the Interior, USGS, Inspection Report: Westmoreland Resources, Inc.--Absaloka, May 1, 1978.

120. INFORM interview with David Simpson, Environmental Administrator, Westmoreland Resources, Inc., June 20, 1979.

121. *Idem.*

122. Written communication from David Simpson, Environmental Administrator, Westmoreland Resources, Inc., to INFORM, October 24, 1979.

BELLE AYR

1. U.S. Department of the Interior, *Final Environmental Statement--Belle Ayr South Mine* (Washington, D.C.: U.S. Government Printing Office, October 1975), p. 38.
2. *Ibid.*, p. 42.
3. Written communication from Jon M. Cassady, Director of Regulatory Affairs, AMAX, to INFORM, December 3, 1979; U.S. Department of Commerce, NOAA, *Monthly Averages of Temperature and Precipitation for State Climatic Divisions 1941-1970: Wyoming*, "Monthly and Annual Divisional Averages Precipitation (Inches)" (Asheville, N.C.: National Climatic Center, July 1973), p. 8.
4. Written communication from Jon M. Cassady, Director of Regulatory Affairs, AMAX, to INFORM, December 3, 1979.
5. National Academy of Sciences, *Rehabilitation Potential of Western Coal Lands*, Appendix A (Cambridge, Mass.: Ballinger Publishing Co., 1974), p. 124.
6. Written communication from AMAX to INFORM, May 29, 1979.
7. *Ibid.*
8. U.S. Department of the Interior, OSM, Region V On-Site Inspections: AMAX Coal Company--Belle Ayr Mine, July 24, 1978, November 1978, and November 5, 1979.
9. INFORM interview with Thomas E. Ehmett, Reclamation Specialist, Region V, OSM, U.S. Department of the Interior, March 8, 1979.
10. Written communication from AMAX Coal Company to INFORM, May 29, 1979.
11. INFORM interviews with Jon M. Cassady, Director of Regulatory Affairs, AMAX, David Ham, Regulatory Affairs Counsel, AMAX, Lyle Randen, Manager, Environmental Services, Western Division, AMAX, Jerry Shepperson, Reclamation Foreman, Belle Ayr Mine, and Randy Shinn, Reclamation Supervisor, Belle Ayr Mine, June 19, 1979.
12. *Idem.*
13. *Idem.*
14. Written communication from Jon M. Cassady, Director of Regulatory Affairs, AMAX, to INFORM, December 3, 1979; Wyoming DEQ, Land Quality Division--District IV, "Annual Inspection Report, AMAX Coal Company's Belle Ayr Mine," January 21, 1977.
15. INFORM interview with David Ham, Regulatory Affairs Counsel, Belle Ayr Mine, January 9, 1980.
16. *Idem.*
17. INFORM interviews with Jon M. Cassady, Director of Regulatory Affairs, AMAX, David Ham, Regulatory Affairs Counsel, AMAX, Lyle Randen, Manager, Environmental Services, Western Division, AMAX, Jerry Shepperson, Reclamation Foreman, Belle Ayr Mine, and Randy Shinn, Reclamation Supervisor, Belle Ayr Mine, June 19, 1979.
18. Wyoming DEQ, Land Quality Division--District IV, Trip Report, October 3, 1977.
19. Wyoming DEQ, Land Quality Division--District IV, "Annual Inspection Report, AMAX Coal Company's Belle Ayr Mine," January 21, 1977.
20. Written communication from Jon M. Cassady, Director of Regulatory Affairs, AMAX, to INFORM.
21. INFORM interviews with Jon M. Cassady, Director of Regulatory Affairs, AMAX, David Ham, Regulatory Affairs Counsel, AMAX, Lyle Randen, Manager, Environmental Services, Western Division, AMAX, Jerry Shepperson, Reclamation Foreman, Belle Ayr Mine, and Randy Shinn, Reclamation Supervisor, Belle Ayr Mine, June 19, 1979.
22. A. Kent Evans, E. W. Ehleman, and P. A. Eby, *Atlas of Western Surface-Mined Lands: Coal, Uranium and Phosphate* (Washington, D.C.: U.S. Department of the Interior, Fish and Wildlife Service, Office of Biological Services, January 1978), p. 156.
23. INFORM interviews with Jon M. Cassady, Director of Regulatory Affairs, AMAX, David Ham, Regulatory Affairs Counsel, AMAX, Lyle Randen, Manager, Environmental Services, Western Division, AMAX, Jerry Shepperson, Reclamation Foreman, Belle Ayr Mine, and Randy Shinn, Reclamation Supervisor, Belle Ayr Mine, June 19, 1979.
24. Written communication from George F. Brown, Acting Conservation Manager,

USGS, U.S. Department of the Interior, to Ron Lanoue, Research Associate, INFORM, April 9, 1979.

25. U.S. Department of the Interior, USGS, "Inspection Report: AMAX/Belle Ayr," September 29, 1978.

26. INFORM interview with David Ham, Regulatory Affairs Counsel, Belle Ayr Mine, January 9, 1980.

27. U.S. Department of the Interior, *Final Environmental Statement--Belle Ayr South Mine*, p. 70.

28. Wyoming DEQ, Land Quality Division--District IV, Second Quarter Inspection Report: AMAX Belle Ayr Mine, June 20, 1978.

29. *Ibid.*

30. U.S. Department of the Interior, USGS, "Inspection Report: AMAX/Belle Ayr," July 3, 1978.

31. Written communication from AMAX to INFORM, May 29, 1979.

32. U.S. Department of the Interior, *Final Environmental Statement--Belle Ayr South Mine*, p. 69.

33. Wyoming DEQ, Land Quality Division--District IV, "Annual Inspection Report, AMAX Coal Company's Belle Ayr Mine," February 10, 1978.

34. Letter from Jack Lautenschlager, General Manager, Western Division, AMAX, to Kathryn Rittmueller, Water Quality Specialist, Water Quality Division, Wyoming DEQ, March 28, 1979.

35. INFORM interviews with Jon M. Cassady, Director of Regulatory Affairs, AMAX, David Ham, Regulatory Affairs Counsel, AMAX, Lyle Randen, Manager, Environmental Services, Western Division, AMAX, Jerry Shepperson, Reclamation Foreman, Belle Ayr Mine, and Randy Shinn, Reclamation Supervisor, Belle Ayr Mine, June 19, 1979.

36. AMAX Coal Company, NPDES Discharge Monitoring Report--Belle Ayr Mine, reporting period from July 1, 1979 to July 31, 1979; Wyoming DEQ, Water Quality Division, "Authorization to Discharge Under the National Pollutant Discharge Elimination System, Permit No. WY-0003514," March 13, 1978.

37. INFORM interviews with Jon M. Cassady, Director of Regulatory Affairs, AMAX, David Ham, Regulatory Affairs Counsel, AMAX, Lyle Randen, Manager, Environmental Services, Western Division, AMAX, Jerry Shepperson, Reclamation Foreman, Belle Ayr Mine, and Randy Shinn, Reclamation Supervisor, Belle Ayr Mine, June 19, 1979.

38. Written communication from Jon M. Cassady, Director of Regulatory Affairs, AMAX, to INFORM, December 3, 1979.

39. INFORM interviews with Jon M. Cassady, Director of Regulatory Affairs, AMAX, David Ham, Regulatory Affairs Counsel, AMAX, Lyle Randen, Manager, Environmental Services, Western Division, AMAX, Jerry Shepperson, Reclamation Foreman, Belle Ayr Mine, and Randy Shinn, Reclamation Supervisor, Belle Ayr Mine, June 19, 1979.

40. Written communication from Jon M. Cassady, Director of Regulatory Affairs, AMAX, to INFORM, December 3, 1979.

41. U.S. Department of the Interior, *Final Environmental Statement--Belle Ayr South Mine*, Appendix--Comments from EPA.

42. Written communication from AMAX to INFORM, May 29, 1979; Written communication from George F. Brown, Acting Conservation Manager, USGS, U.S. Department of the Interior, to Ron Lanoue, Research Associate, INFORM, April 9, 1979.

43. Written communication from AMAX to INFORM, May 29, 1979.

44. INFORM interviews with Jon M. Cassady, Director of Regulatory Affairs, AMAX, David Ham, Regulatory Affairs Counsel, AMAX, Lyle Randen, Manager, Environmental Services, Western Division, AMAX, Jerry Shepperson, Reclamation Foreman, Belle Ayr Mine, and Randy Shinn, Reclamation Supervisor, Belle Ayr Mine, June 19, 1979.

45. INFORM interview with Dennis Morrow, District IV Engineer, Land Quality Division, Wyoming DEQ, June 18, 1979.

46. Wyoming DEQ, Land Quality Division--District IV, Second Quarter Inspection Report: AMAX Belle Ayr Mine, June 13, 1977.

47. *Ibid.*

48. INFORM interviews with Jon M. Cassady, Director of Regulatory Affairs, AMAX, David Ham, Regulatory Affairs Counsel, AMAX, Lyle Randen, Manager, Environmental Services, Western Division, AMAX, Jerry Shepperson,

Reclamation Foreman, Belle Ayr Mine, and Randy Shinn, Reclamation Supervisor, Belle Ayr Mine, June 19, 1979.

COLSTRIP

1. Montana DSL, *Final Environmental Impact Statement for the Proposed Expansion of Western Energy Company's Rosebud Mine into Area B*, July 26, 1976, p. 9.
2. INFORM interviews with Don Bailey, Nick Golden, and Wallace McRae, Rosebud Protective Association, April 4, 1980.
3. Montana DSL, *Final Environmental Impact Statement*, 1976, p. 47.
4. INFORM interviews with Montana DSL Inspectors Dennis Hemmer, Vegetation Specialist, Mike Bishop, Hydrogeologist, and Neil Harrington, Soil Scientist, April 2, 1979.
5. Letter from Dennis Hemmer, Inspector, Montana DSL, to Michael Grende, Supervisor, Lands and Permits, Western Energy Company, March 29, 1979.
6. Montana DSL, *Final Environmental Impact Statement*, pp. 9-12.
7. *Ibid.*; U.S. Department of the Interior, USGS, and Montana DSL, *Northern Powder River Basin Coal, Montana--Draft Environmental Statement* (Washington, D.C.: U.S. Government Printing Office, 1979), p. II-22.
8. INFORM interviews with Chris Cull, Senior Reclamation Engineer, and Earl Murray, Reclamation Engineer, Western Energy Company, June 23, 1979.
9. U.S. Department of the Interior, USGS, and Montana DSL, *Northern Powder River Basin--Draft Environmental Statement*.
10. Montana DSL, *Final Environmental Impact Statement*, 1976, p. 55; Written communication from Western Energy Company to INFORM, n.d.
11. INFORM interviews with Chris Cull, Senior Reclamation Engineer, and Earl Murray, Reclamation Engineer, Western Energy Company, June 23, 1979.
12. INFORM interviews with Montana DSL Inspectors Dennis Hemmer, Vegetation Specialist, Mike Bishop, Hydrogeologist, and Neil Harrington, Soil Scientist, April 2, 1979.
13. INFORM interview with Chris Cull, Senior Reclamation Engineer, Western Energy Company, February 28, 1980.
14. INFORM interviews with Earl Murray, Reclamation Engineer, Western Energy Company, P. Russel Brown, NPRC, and William Gillin and Wallace McRae, Rosebud Protective Association, April 4, 1980.
15. INFORM interviews with Chris Cull, Senior Reclamation Engineer, and Earl Murray, Reclamation Engineer, Western Energy Company, June 23, 1979.
16. U.S. Department of the Interior, OSM, Region V On-Site Inspection: Western Energy Company--Rosebud (Colstrip) Mine, June 5, 1978; Written communication from Western Energy Company to INFORM, n.d.
17. INFORM interviews with Montana DSL Inspectors Dennis Hemmer, Vegetation Specialist, Mike Bishop, Hydrogeologist, and Neil Harrington, Soil Scientist, April 2, 1979.
18. Montana DSL, Field Inspection Report: Rosebud (Colstrip) Mine, June 5, 1978; INFORM interviews with Chris Cull, Senior Reclamation Engineer, and Earl Murray, Reclamation Engineer, Western Energy Company, June 23, 1979.
19. INFORM interview with Dennis Hemmer, Inspector, Montana DSL, April 14, 1980.
20. Montana DSL, Field Inspection Report: Rosebud Mine, January 16, 1976, February 9, 1976, February 24, 1976, May 6, 1976, June 10, 1976, August 12, 1976, June 11, 1977, June 21, 1977, and September 7, 1977.
21. Memorandum from Neil Harrington and Charles van Hook, Inspectors, Montana DSL, to C.C. McCall and Richard Juntunen, Reclamation Specialists, Montana DSL, June 10, 1976.
22. INFORM interviews with Earl Murray, Reclamation Engineer, Western Energy Company, P. Russell Brown, NPRC, and William Gillin and Wallace McRae, Rosebud Protective Association, April 4, 1980.
23. U.S. Department of the Interior, USGS, Inspection Report: Western Energy Company--Rosebud (Colstrip) Mine, August 19, 1976; U.S. Department of the Interior, OSM, Region V On-Site Inspection: Western Energy Company--Rosebud Mine, July 18, 1979.

24. Montana DSL, Field Inspection Report: Rosebud Mine, August 4, 1977;
    U.S. Department of the Interior, USGS, Inspection Reports: Western
    Energy Company--Rosebud Mine, October 14, 1977 and March 8, 1978;
    U.S. Department of the Interior, OSM, Region V On-Site Inspection:
    Western Energy Company--Rosebud Mine, July 18, 1979.
25. INFORM interviews with Chris Cull, Senior Reclamation Engineer, and Earl
    Murray, Reclamation Engineer, Western Energy Company, June 23, 1979.
26. INFORM interviews with Earl Murray, Reclamation Engineer, Western Ener-
    gy Company, P. Russell Brown, NPRC, and William Gillin and Wallace Mc-
    Rae, Rosebud Protective Association, April 4, 1980.
27. *Idem.*
28. Montana DSL, Field Inspection Reports: Rosebud Mine, February 2, 1978
    and June 5, 1979; U.S. Department of the Interior, OSM, Region V On-
    Site Inspections: Western Energy Company--Rosebud Mine, June 5, 1978
    and July 18, 1979.
29. Montana DSL, Field Inspection Reports: Rosebud Mine, August 4, 1977
    and October 26, 1978; U.S. Department of the Interior, USGS, Inspection
    Report: Western Energy Company--Rosebud Mine, March 18, 1977.
30. Letter from Dennis Hemmer, Inspector, Montana DSL, to Michael Grende,
    Supervisor, Lands and Permits, Western Energy Company, March 29, 1979.
31. Montana DSL, *Draft Environmental Impact Statement, Western Energy Com-
    pany's Rosebud Mine, Area B Extension,* (Washington, D.C.: U.S. Govern-
    ment Printing Office, February 1980), p. II-1.
32. INFORM interviews with Don Bailey, Nick Golden, and Wallace McRae, Rose-
    bud Protective Association, April 4, 1980.
33. INFORM interviews with Chris Cull, Senior Reclamation Engineer, and Earl
    Murray, Reclamation Engineer, Western Energy Company, June 23, 1979.
34. Montana DSL, Field Inspection Reports: Rosebud Mine, February 2-3,
    1978, May 23-24, 1978, and July 12, 1978; U.S. Department of the Interior,
    OSM, Region V On-Site Inspection: Western Energy Company--Rosebud
    Mine, June 5, 1978.
35. Montana DSL, Field Inspection Report: Rosebud Mine, March 2, 1978; U.S.
    Department of the Interior, OSM, Region V On-Site Inspection: Western
    Energy Company--Rosebud Mine, June 5, 1978; Written communication from
    Western Energy Company to INFORM, n.d.
36. Montana DSL, Field Inspection Report: Rosebud Mine, March 2, 1978.
37. INFORM interview with Dennis Hemmer, Inspector, Montana DSL, March 17,
    1980.
38. Montana DSL, *Final Environmental Impact Statement,* Letter 26: from J.R.
    Lee to Alan Merson, Regional Administrator, Region VIII, U.S. Environ-
    mental Protection Agency, p. IX-88,89.
39. INFORM interview with Patty Kluver, local landowner, Colstrip, Montana,
    April 4, 1980.
40. *Idem.*; U.S. Department of the Interior, OSM, Region V On-Site Inspec-
    tion: Western Energy Company--Rosebud Mine, October 11, 1979.
41. U.S. Department of the Interior, OSM, Region V On-Site Inspection:
    Western Energy Company--Rosebud Mine, July 18, 1979.
42. U.S. Department of the Interior, BLM, "Technical Examination and Envi-
    ronmental Assessment--Western Energy Area A Coal Lease Application,"
    August 1979, p. II-19.
43. Written communication from Western Energy Company to INFORM, n.d.
44. *Ibid.*
45. Montana Bureau of Mines and Geology, Wayne A. Van Voast, Robert B.
    Hedges, and John J. McDermott, "Strip Coal Mining and Mined Land Rec-
    lamation in the Hydrologic System, Southeastern Montana: Project Comple-
    tion Report," December 1978, pp. 12-13.
46. *Ibid.*, p. 13.
47. *Ibid.*, pp. 69-70.
48. INFORM interview with Dennis Hemmer, Inspector, Montana DSL, March 17,
    1980.
49. INFORM interviews with Don Bailey, Nick Golden, and Wallace McRae, Rose-
    bud Protective Association, April 4, 1980.
50. Letter from Dennis Hemmer, Inspector, Montana DSL, to Michael Grende,
    Supervisor, Lands and Permits, Western Energy Company, March 29, 1979.

51. U.S. Department of the Interior, BLM, "Western Energy Area A Coal Lease Application," p. III-6.
52. INFORM interview with Dennis Hemmer, Inspector, Montana DSL, March 17, 1980; Montana DSL, Field Inspection Report: Rosebud Mine, June 5, 1978; Montana DSL, *Final Environmental Impact Statement*, p. 47.
53. Written communication from Western Energy Company to INFORM, n.d.; INFORM interviews with Chris Cull, Senior Reclamation Engineer, and Earl Murray, Reclamation Engineer, Western Energy Company, June 23, 1979.
54. Montana DSL, *Draft Environmental Impact Statement, Western Energy Company's Rosebud Mine, Area B Extension*, p. II-3.
55. *Ibid.*, p. III-3.
56. Montana DSL, Field Inspection Reports: Rosebud Mine, January 16, 1976, January 20, 1977, February 17, 1977, February 1, 1978, and May 23-24, 1978; Memorandum from Neil Harrington and Charles van Hook, Inspectors, Montana DSL, to C.C. McCall and Richard Juntunen, Reclamation Specialists, Montana DSL, June 10, 1976; U.S. Department of the Interior, OSM, Region V On-Site Inspection: Western Energy Company--Rosebud Mine, June 5, 1978.
57. Montana DSL, Field Inspection Reports: Rosebud Mine, February 9, 1976, February 24, 1976, May 6-8, 1976, July 19, 1976, February 17, 1977, March 31, 1977, August 4, 1977, September 8, 1977, November 8-9, 1977, March 29, 1978, May 23-24, 1978, June 5, 1978, and October 26, 1978; Memorandum from Neil Harrington and Greg Mills, Inspectors, Montana DSL, November 24, 1976; Memorandum from Neil Harrington and Charles van Hook, Inspectors, Montana DSL, to C.C. McCall and Richard Juntunen, Reclamation Specialists, Montana DSL, June 10, 1976; U.S. Department of the Interior, USGS, Inspection Report: Western Energy Company--Rosebud Mine, August 19, 1976.
58. Thomas Kotynski, "Reclamation Ills 'Serious' at 2 Colstrip Pits," *Great Falls Tribune*, July 21, 1977, p. 5.
59. Letter from Richard L. Juntunen, Reclamation Specialist, Montana DSL, to Michael Grende, Supervisor, Lands and Permits, Western Energy Company, May 2, 1977.
60. Montana DSL, Field Inspection Reports: Rosebud Mine, February 17, 1977, March 5, 1977, February 1, 1978, May 23, 1978, and July 12, 1978.
61. Montana DSL, Field Inspection Reports: Rosebud Mine, January 20, 1977, August 4, 1977, March 29, 1978, and August 8, 1978.
62. INFORM interviews with Chris Cull, Senior Reclamation Engineer, and Earl Murray, Reclamation Engineer, Western Energy Company, June 23, 1979.
63. INFORM interview with Chris Cull, Senior Reclamation Engineer, February 28, 1980.
64. INFORM interviews with Montana DSL Inspectors Dennis Hemmer, Vegetation Specialist, Mike Bishop, Hydrogeologist, and Neil Harrington, Soil Scientist, April 2, 1979.
65. Western Energy Company, "Revised Revegetation Plan," March 3, 1979.
66. Edward J. DePuit, Joe G. Coenenberg, and Waite H. Willmuth, *Research on Revegetation of Surface Mined Lands at Colstrip, Montana: Progress Report, 1975-1977*, Research Report 127 (Bozeman, MT: Montana State University, 1978), p. 42.
67. INFORM interviews with Chris Cull, Senior Reclamation Engineer, and Earl Murray, Reclamation Engineer, Western Energy Company, June 23, 1979.
68. *Idem.*; Western Energy Company, "Revised Revegetation Plan," March 3, 1979.
69. INFORM interviews with Earl Murray, Reclamation Engineer, Western Energy Company, P. Russell Brown, NPRC, and William Gillin and Wallace McRae, Rosebud Protective Association, April 4, 1980.
70. *Idem.*
71. *Idem.*; "Machine Changes Aid Seed Stand," *The Land Reclamation Report*, Vol. 2, No. 19 (Cleveland, Ohio: The Harvest Publishing Company, 1979), October 1, 1979, p. 1.
72. INFORM interviews with Earl Murray, Reclamation Engineer, Western Energy Company, P. Russell Brown, NPRC; and William Gillin and Wallace McRae, Rosebud Protective Association, April 4, 1980.

73. INFORM interviews with Chris Cull, Senior Reclamation Engineer, and Earl
     Murray, Reclamation Engineer, Western Energy Company, June 23, 1979.
74. INFORM interviews with Earl Murray, Reclamation Engineer, Western Energy
     Company, P. Russell Brown, NPRC and William Gillin and Wallace McRae,
     Rosebud Protective Association, April 4, 1980.
75. INFORM interviews with Chris Cull, Senior Reclamation Engineer, and Earl
     Murray, Reclamation Engineer, Western Energy Company, June 23, 1979.
76. INFORM interviews with Dennis Hemmer, Inspector, Montana DSL, July 10,
     1979 and March 17, 1980.
77. Montana DSL, *Final Environmental Impact Statement*, p. 60; Western Energy
     Company, "Annual Vegetation Description, Areas A, B, and E," 1978.
78. Written communication from Western Energy Company to INFORM, n.d.;
     Northern Plains Resource Council, "Fact Sheet on HB 406 [Bill to change
     revegetation standards for the Strip and Underground Mine Reclamation
     Act]," n.d.
79. U.S. Department of the Interior, USGS, and Montana DSL, *Northern Pow-
     der River Basin--Final Environmental Statement*, (Washington, D.C.: U.S.
     Government Printing Office, 1980), p. II-63.
80. INFORM interviews with Chris Cull, Senior Reclamation Engineer, and Earl
     Murray, Reclamation Engineer, Western Energy Company, June 23, 1979.
81. *Idem.*
82. A. Kent Evans, E.W. Uhleman, and P.A. Eby, *Atlas of Western Surface-
     Mined Lands: Coal, Uranium and Phosphate* (Washington, D.C.: U.S.
     Department of the Interior, Fish and Wildlife Service, Biological Services
     Program, January 1978), p. 74.
83. Written communication from Western Energy Company to INFORM, n.d.
84. DePuit, *et al.*, *Research on Revegetation of Surface Mined Lands at Col-
     strip*, pp. 64, 67-69, 81.
85. INFORM interview with William Gillin, Rosebud Protective Association,
     April 4, 1980.
86. INFORM interviews with Chris Cull, Senior Reclamation Engineer, and Earl
     Murray, Reclamation Engineer, Western Energy Company, June 23, 1979.
87. Meyn, *et al.*, *Plant Response and Forage Quality for Controlled Grazing
     on Coal Mine Spoils Pastures* (Bozeman, Mt.: Montana State University,
     1977), pp. 48-51.
88. INFORM interviews with Don Bailey, Nick Golden, and Wallace McRae,
     Rosebud Protective Association, April 4, 1980.
89. U.S. Department of the Interior, USGS, and Montana DSL, *Northern Pow-
     der River Basin--Final Environmental Statement*, pp. II-75, 76.
90. INFORM interviews with Dennis Hemmer, Inspector, Montana DSL, July 10,
     1979 and March 17, 1980.
91. INFORM interviews with Montana DSL Inspectors Dennis Hemmer, Vegeta-
     tion Specialist, Mike Bishop, Hydrogeologist, and Neil Harrington, Soil
     Scientist, April 2, 1980.
92. INFORM interviews with Chris Cull, Senior Reclamation Engineer, and
     Earl Murray, Reclamation Engineer, Western Energy Company, June 23,
     1979.
93. INFORM interview with Wallace McRae, Rosebud Protective Association,
     April 4, 1980.

DECKER

1. U.S. Department of the Interior, USGS, and Montana DSL, *Final Environmen-
    tal Impact Statement, Proposed Plan of Mining and Reclamation, East Decker,
    and North Extension Mines, Decker Coal Co., Big Horn County, Montana*
    (Washington, D.C.: U.S. Government Printing Office, 1976), pp. 93-98.
2. *Ibid.*
3. *Ibid.*, pp. 101-102.
4. INFORM interview with Sam Scott, Reclamation Director, Sheridan Office,
    Peter Kiewit Sons' Company, June 21, 1979; INFORM interviews with Dale
    Johnson, Reclamation Specialist, Peter Kiewit Sons' Company, and P. Russell
    Brown, NPRC, April 8, 1980.

5. U.S. Department of the Interior, USGS, and Montana DSL, *Northern Powder River Basin Coal, Montana--Draft Environmental Statement* (Washington, D.C.: U.S. Government Printing Office, 1979), p. II-22; U.S. Department of the Interior, USGS, and Montana DSL, *Final Environmental Impact Statement, Proposed Plan of Mining and Reclamation, East Decker and North Extension Mines*, p. 105.

6. U.S. Department of the Interior, USGS, and Montana DSL, *Final Environmental Impact Statement, Proposed Plan of Mining and Reclamation, East Decker and North Extension Mines*, Appendix E.

7. INFORM interview with Sam Scott, Reclamation Director, Sheridan Office, Peter Kiewit Sons' Company, June 21, 1979.

8. INFORM interview with Dennis Hemmer, Inspector, Montana DSL, April 10, 1980.

9. INFORM interviews with Montana DSL Inspectors Dennis Hemmer, Vegetation Specialist, Mike Bishop, Hydrogeologist, and Neil Harrington, Soil Scientist, April 2, 1979.

10. U.S. Department of the Interior, OSM, Region V On-Site Inspection: Decker Coal Company--Decker Mine, December 17, 1979.

11. INFORM interview with Sam Scott, Reclamation Director, Sheridan Office, Peter Kiewit Sons' Company, April 24, 1980.

12. Montana DSL, Field Inspection Report: Decker Coal Company, April 4, 1977; U.S. Department of the Interior, OSM, Region V On-Site Inspection: Decker Coal Company--Decker Mine, October 17, 1978; INFORM interviews with Dale Johnson, Reclamation Specialist, Peter Kiewit Sons' Company, and P. Russell Brown, NPRC, April 8, 1980.

13. INFORM interviews with Dale Johnson, Reclamation Specialist, Peter Kiewit Sons' Company, and P. Russell Brown, NPRC, April 8, 1980.

14. INFORM interview with Dennis Hemmer, Inspector, Montana DSL, April 23, 1980.

15. Montana DSL, Field Inspection Report: Decker Coal Company, March 14, 1979.

16. Montana DSL, Field Inspection Reports: Decker Coal Company, September 1, 1976, February 24, 1977, April 4, 1977, April 12, 1979, and November 28-29, 1979.

17. Written communication from Peter Kiewit Sons' Company to INFORM, December 26, 1979.

18. Montana DSL, Field Inspection Report: Decker Coal Company, August 4, 1977.

19. Montana DSL, Field Inspection Report: Decker Coal Company, March 27, 1979.

20. INFORM interviews with Dale Johnson, Reclamation Specialist, Peter Kiewit Sons' Company, and P. Russell Brown, NPRC, April 8, 1980.

21. *Idem.*

22. *Idem.*

23. Montana DSL, Field Inspection Reports: Decker Coal Company, November 7, 1977, November 16, 1977, June 30, 1978, January 16, 1980, and March 22, 1978; U.S. Department of the Interior, OSM, Region V On-Site Inspection: Decker Coal Company--Decker Mine, October 17, 1978.

24. Montana DSL, Field Inspection Reports: Decker Coal Company, January 10, 1979 and May 5, 1977; U.S. Department of the Interior, Region V On-Site Inspection: Decker Coal Company--Decker Mine, December 19, 1979.

25. INFORM interviews with Dale Johnson, Reclamation Specialist, Peter Kiewit Sons' Company, and P. Russell Brown, NPRC, April 8, 1980.

26. *Idem.*; Montana DSL, Field Inspection Reports: Decker Coal Company, May 2, 1978, April 12, 1979, and September 11, 1979.

27. U.S. Department of the Interior, OSM, Region V On-Site Inspection: Decker Coal Company--Decker Mine, December 17, 1979.

28. Montana DSL, Field Inspection Reports: Decker Coal Company, March 16, 1979, October 13, 1977, January 6, 1978, March 22, 1978, April 6, 1978, and August 21, 1978.

29. Montana DSL, Field Inspection Reports: Decker Coal Company, March 16, 1979 and May 29, 1979.

30. Letter from J.F. Power, Agricultural Research Service, Mandan, North Dakota Office, U.S. Department of Agriculture, to C.C. McCall, Reclamation Specialist, Montana DSL, May 8, 1975.

31. INFORM interviews with Dale Johnson, Reclamation Specialist, Peter Kiewit Sons' Company, and P. Russell Brown, NPRC, April 8, 1980.

32. Montana DSL, Field Inspection Reports: Decker Coal Company, June 8, 1977 and July 12, 1977.
33. Montana DSL, Field Inspection Report: Decker Coal Company, March 27, 1979.
34. Montana DSL, Field Inspection Report: Decker Coal Company, March 23, 1977.
35. INFORM interviews with Montana DSL Inspectors Dennis Hemmer, Vegetation Specialist, Mike Bishop, Hydrogeologist, and Neil Harrington, Soil Scientist, April 2, 1979.
36. *Idem.*
37. Montana Department of Health and Environmental Sciences, "Authorization to Discharge Under the Montana Pollutant Discharge Elimination System, Permit No. MT-0000892," June 27, 1977.
38. INFORM interviews with Montana DSL Inspectors Dennis Hemmer, Vegetation Specialist, Mike Bishop, Hydrogeologist, and Neil Harrington, Soil Scientist, April 2, 1979.
39. U.S. Department of the Interior, OSM, Region V On-Site Inspection: Decker Coal Company--Decker Mine, December 17, 1979.
40. Montana DSL, Field Inspection Reports: Decker Coal Company, February 8, 1977, March 23, 1977, April 4, 1977, July 12, 1977, August 4, 1977, February 22, 1979, March 8, 1979, April 6, 1978 and June 1, 1978.
41. Montana DSL, Field Inspection Report: Decker Coal Company, July 12, 1977.
42. Montana DSL, Field Inspection Report: Decker Coal Company, April 6, 1978.
43. Montana DSL, Field Inspection Report: Decker Coal Company, May 12, 1978.
44. Montana DSL, Field Inspection Report: Decker Coal Company, June 1, 1978.
45. U.S. Department of the Interior, BLM, *Decker North Extension Coal Lease Application--Final Technical Examination and Environmental Assessment* (Washington, D.C.: U.S. Government Printing Office, July 1979), p. V-2.
46. Decker Coal Company, NPDES Discharge Monitoring Report--Decker Mine, reporting periods from January 3, 1978 to April 3, 1978.
47. Decker Coal Company, NPDES Discharge Monitoring Reports--Decker Mine, reporting periods from January 1, 1979 to March 31, 1979 and April 1, 1979 to June 30, 1979.
48. Montana DSL, Field Inspection Report: Decker Coal Company, February 11, 1980.
49. U.S. Department of the Interior, BLM, *Decker North Extension Coal Lease Application--Final Technical Examination and Environmental Assessment*, pp. II-19, II-20.
50. *Ibid.*, p. III-6.
51. *Ibid.*, p. II-19.
52. *Ibid.*, pp. II-17, II-18.
53. U.S. Department of the Interior, BLM, *West Decker Coal Lease Application, Land Use Analysis--Draft Technical Examination and Environmental Assessment* (Washington, D.C.: U.S. Government Printing Office, August 1979), p. III-10.
54. *Ibid.*, p. III-13.
55. INFORM interview with Sam Scott, Reclamation Director, Sheridan Office, Peter Kiewit Sons' Company, June 21, 1979.
56. U.S. Department of the Interior, BLM, *West Decker Coal Lease Application, Land Use Analysis--Draft Technical Examination and Environmental Assessment*, p. II-23.
57. U.S. Department of the Interior, BLM, *Decker North Extension Coal Lease Application--Final Technical Examination and Environmental Assessment*, p. II-23.
58. *Ibid.*
59. Letter from Dennis Hemmer, Inspector, Montana DSL, to David Shelso, Decker Coal Company, November 28, 1979; Letter from Brace Hayden, Administrator, Reclamation Division, Montana DSL, to Jack Reed, Engineer, Decker Coal Company, January 30, 1979; Memorandum from Mike Bishop, Inspector, Montana DSL, to Brace Hayden, Administrator, Reclamation Division, and Leo Berry, Jr., Commissioner, Montana DSL, January 29, 1979, "Re: North Decker Application 00026-Alluvial Valley Floor Determination."
60. INFORM interview with Dennis Hemmer, Inspector, Montana DSL, April 23, 1980.
61. Written communication from Peter Kiewit Sons' Company to INFORM, December 26, 1979.
62. Montana DSL, Field Inspection Reports: Decker Coal Company, July 12, 1977, August 4, 1977, April 4, 1977, and October 13, 1977.

63. INFORM interviews with Montana DSL Inspectors Dennis Hemmer, Vegetation
    Specialist, Mike Bishop, Hydrogeologist, and Neil Harrington, Soil Scientist,
    April 2, 1979.
64. INFORM interview with Sam Scott, Reclamation Director, Sheridan Office,
    Peter Kiewit Sons' Company, June 21, 1979.
65. INFORM interviews with Montana DSL Inspectors Dennis Hemmer, Vegetation
    Specialist, Mike Bishop, Hydrogeologist, and Neil Harrington, Soil Scientist,
    April 2, 1979.
66. INFORM interview with Sam Scott, Reclamation Director, Sheridan Office,
    Peter Kiewit Sons' Company, June 21, 1979; U.S. Department of the Interior,
    USGS, Inspection Report: Decker Coal Company--Decker Mine, March 22,
    1977; Written communication from Peter Kiewit Sons' Company to INFORM,
    December 26, 1979.
67. INFORM interview with Sam Scott, Reclamation Director, Sheridan Office,
    Peter Kiewit Sons' Company, June 21, 1979.
68. *Idem.*
69. INFORM interviews with Dale Johnson, Reclamation Specialist, Peter Kiewit
    Sons' Company, and P. Russell Brown, NPRC, April 8, 1980.
70. Montana DSL, Field Inspection Report: Decker Coal Company, February 24,
    1977.
71. INFORM interview with Sam Scott, Reclamation Director, Sheridan Office,
    Peter Kiewit Sons' Company, June 21, 1979; Written communication from
    Peter Kiewit Sons' Company to INFORM, December 26, 1979.
72. INFORM interview with Sam Scott, Reclamation Director, Sheridan Office,
    Peter Kiewit Sons' Company, June 21, 1979; INFORM interviews with Dale
    Johnson, Reclamation Specialist, Peter Kiewit Sons' Company, and P. Rus-
    sell Brown, NPRC, April 8, 1980.
73. INFORM interview with Sam Scott, Reclamation Director, Sheridan Office,
    Peter Kiewit Sons' Company, June 21, 1979.
74. Decker Coal Company, Permit Application, Vol. I, April 25, 1977, p. 76;
    Montana DSL, Field Inspection Report: Decker Coal Company, November
    17, 1976.
75. Montana DSL, Field Inspection Report: Decker Coal Company, April 6, 1978.
76. INFORM interviews with Dale Johnson, Reclamation Specialist, Peter Kiewit
    Sons' Company, and P. Russell Brown, NPRC, April 8, 1980.
77. Montana DSL, Field Inspection Report: Decker Coal Company, November 28,
    1979.
78. U.S. Department of the Interior, OSM, Region V On-Site Inspection: Decker
    Coal Company--Decker Mine, December 17, 1979.
79. INFORM interview with Dennis Hemmer, Inspector, Montana DSL, April 23,
    1980.
80. U.S. Department of the Interior, BLM, *Decker North Extension Coal Lease
    Application--Final Technical Examination and Environmental Assessment,*
    pp. 2-3.
81. INFORM interview with Sam Scott, Reclamation Director, Sheridan Office,
    Peter Kiewit Sons' Company, June 21, 1979.
82. *Ibid.*
83. U.S. Department of the Interior, USGS, and Montana DSL, *Final Environ-
    mental Impact Statement, Proposed Plan of Mining and Reclamation, East
    Decker and North Extension Mines,* Appendix E.
84. U.S. Department of the Interior, BLM, *West Decker Coal Lease Application,
    Land Use Analysis--Draft Technical Examination and Environmental Assess-
    ment,* pp. II-30, II-31.
85. INFORM interviews with Dale Johnson, Reclamation Specialist, Peter Kiewit
    Sons' Company, and P. Russell Brown, NPRC, April 8, 1980.
86. *Idem.*
87. *Idem.*
88. *Idem.,* U.S. Department of the Interior, OSM, Region V On-Site Inspection:
    Decker Coal Company--Decker Mine, December 17, 1979; A. Kent Evans,
    E.W. Uhleman, and P.A. Eby, *Atlas of Western Surface-Mined Lands: Coal,
    Uranium and Phosphate* (Washington, D.C.: U.S. Department of the In-
    terior, Fish and Wildlife Service, Biological Services Program, January
    1978), p. 70.
89. Montana DSL, Field Inspection Reports: Decker Coal Company, May 13,

1976 and July 12, 1977; U.S. Department of the Interior, USGS, Inspection Report: Decker Coal Company--Decker Mine, June 6, 1978.

90. U.S. Department of the Interior, OSM, Region V On-Site Inspection: Decker Coal Company--Decker Mine, December 17, 1979.

91. INFORM interview with Sam Scott, Reclamation Director, Sheridan Office, Peter Kiewit Sons' Company, June 21, 1979.

92. *Idem.*

93. INFORM interviews with Dale Johnson, Reclamation Specialist, Peter Kiewit Sons' Company, and P. Russell Brown, NPRC, April 8, 1980.

94. Montana DSL, Field Inspection Reports: Decker Coal Company, June 8, 1977, March 22, 1978, and June 1, 1978; U.S. Department of the Interior, USGS, Inspection Report: Decker Coal Company--Decker Mine, June 6, 1978.

95. U.S. Department of the Interior, BLM, *Decker North Extension Coal Lease Application--Final Technical Examination and Environmental Assessment,* p. III-4.

96. INFORM interviews with Don Bailey, Nick Golden, and Wallace McRae, Rosebud Protective Association, April 4, 1980.

97. U.S. Department of the Interior, OSM, Region V On-Site Inspection: Decker Coal Company--Decker Mine, December 19, 1979.

98. INFORM interview with Sam Scott, Reclamation Director, Sheridan Office, Peter Kiewit Sons' Company, June 21, 1979.

99. INFORM interviews with Chris Cull, Senior Reclamation Engineer, and Earl Murray, Reclamation Engineer, Western Energy Company, June 23, 1979.

GASCOYNE

1. Knife River Coal Mining Company, "Mining and Reclamation Plan Pursuant to the Rules and Regulations of Title 30, Chapter II, Part 211, Coal Mining Regulations," p. 5.

2. U.S. Department of Agriculture, Forest Service, Paul E. Packer, *Rehabilitation Potentials and Limitations of Surface-Mined Land in the Northern Great Plains* (Ogden, Utah: U.S. Government Printing Office, July 1974).

3. Written communication from Curtis L. Blohm, Manager of Reclamation, Knife River Coal Mining Company, to INFORM, September 29, 1979.

4. U.S. Department of Commerce, NOAA, *Monthly Averages of Temperature and Precipitation for State Climatic Divisions 1941-70: North Dakota,* "Monthly and Annual Divisional Averages Precipitation (Inches)" (Asheville, N.C.: National Climtatic Center, July 1973), p. 6.

5. Knife River Coal Mining Company, "Mining and Reclamation Plan," p. 5.

6. U.S. Department of Commerce, NOAA, *Monthly Averages of Temperature and Precipitation: North Dakota,* p. 6.

7. National Academy of Sciences, *Rehabilitation Potential of Western Coal Lands,* Appendix A (Cambridge, Mass: Ballinger Publishing Co., 1974), p. 123.

8. Written communication from Knife River Coal Mining Company to INFORM, n.d.

9. Knife River Coal Mining Company, "Mining and Reclamation Plan," p. 10.

10. INFORM interview with David A. Jordan, Conservationist, Knife River Coal Mining Company, June 15, 1979.

11. INFORM interviews with Dean Peterson and James Deutsch, Environmental Scientists, North Dakota PSC, April 12, 1979.

12. INFORM interview with David A. Jordan, Conservationist, Knife River Coal Mining Company, June 15, 1979.

13. *Idem.*

14. INFORM interview with David A. Jordan, Conservationist, Knife River Coal Mining Company, February 12, 1980.

15. Written communication from Knife River Coal Mining Company to INFORM, n.d.

16. Knife River Coal Mining Company, "Mining and Reclamation Plan," p. 21.

17. U.S. Department of the Interior, OSM, Region V On-Site Inspection: Knife River Coal Company, Montana Dakota Utilities--Gascoyne Mine, August 24, 1978.

18. North Dakota PSC, Inspection Report: Knife River Coal Mining Company--
    Gascoyne Mine, August 25, 1978.
19. *Ibid.*
20. Written communication from Curtis L. Blohm, Manager of Reclamation, Knife
    River Coal Mining Company, to INFORM, September 29, 1979.
21. *Ibid.*
22. INFORM interview with David A. Jordan, Conservationist, Knife River Coal
    Mining Company, June 15, 1979.
23. INFORM interview with Dr. Edward Englerth, Director, Reclamation Division,
    North Dakota PSC, February 5, 1980.
24. Written communication from George F. Brown, Acting Conservation Manager,
    USGS, U.S. Department of the Interior, to Ron Lanoue, Research Associate,
    INFORM, March 27, 1979; U.S. Department of the Interior, OSM, Region V
    On-Site Inspection: Knife River Coal Mining Company, Montana Dakota Utili-
    ties--Gascoyne Mine, August 24, 1978.
25. Knife River Coal Mining Company, "Mining and Reclamation Plan," p. 22.
26. INFORM interview with David A. Jordan, Conservationist, Knife River Coal
    Mining Company, June 15, 1979.
27. North Dakota PSC, Inspection Report: Knife River Coal Mining Company--
    Gascoyne Mine, April 25, 1979.
28. Written communication from Curtis L. Blohm, Manager of Reclamation, Knife
    River Coal Mining Company, to INFORM, September 29, 1979.
29. U.S. Department of Agriculture, Forest Service, Paul E. Packer, *Rehabilita-
    tion Potentials and Limitations of Surface-Mined Land*, p. 26.
30. INFORM interview with David A. Jordan, Conservationist, Knife River Coal
    Mining Company, June 15, 1979.
31. *Idem.*
32. Knife River Coal Company, "National Pollutant Discharge Elimination System,
    Application for Permit to Discharge--Short Form C, Application Number ND-
    0023728," October 31, 1974.
33. INFORM interviews with Dean Peterson and James Deutsch, Environmental Sci-
    entists, North Dakota PSC, April 12, 1979.
34. Written communication from Curtis L. Blohm, Manager of Reclamation, Knife
    River Coal Mining Company, to INFORM, September 29, 1979.
35. INFORM interviews with Dean Peterson and James Deutsch, Environmental Sci-
    entists, North Dakota PSC, April 12, 1979.
36. U.S. Department of the Interior, OSM, Region V On-Site Inspection: Knife
    River Coal Mining Company, Montana Dakota Utilities--Gascoyne Mine, Au-
    gust 24, 1978.
37. North Dakota PSC, Inspection Report: Knife River Coal Mining Company--
    Gascoyne Mine, April 25, 1979.
38. U.S. Department of the Interior, USGS, Inspection Report: Knife River Coal
    Mining Company--Gascoyne Mine, November 28, 1977.
39. Knife River Coal Mining Company, NPDES Discharge Monitoring Reports--
    Gascoyne Mine, reporting periods from January 1, 1977 to January 1, 1978,
    and April 1, 1978 to July 1, 1978.
40. *Ibid.*; Knife River Coal Mining Company, NPDES Discharge Monitoring Re-
    ports--Gascoyne Mine, reporting period from July 1, 1978 to October 1,
    1978.
41. U.S. Department of the Interior, OSM, Region V On-Site Inspection: Knife
    River Coal Mining Company, Montana Dakota Utilities--Gascoyne Mine, Au-
    gust 24, 1978.
42. Knife River Coal Mining Company, "Mining and Reclamation Plan," p. 70.
43. Written communication from Curtis L. Blohm, Manager of Reclamation, Knife
    River Coal Mining Company, to INFORM, September 29, 1979.
44. Written communication from Knife River Coal Mining Company to INFORM,
    n.d.; Written communication from George F. Brown, Acting Conservation
    Manager, USGS, U.S. Department of the Interior, to Ron Lanoue, Research
    Associate, INFORM, March 27, 1979.
45. INFORM interview with David A. Jordan, Conservationist, Knife River Coal
    Mining Company, June 15, 1979.
46. INFORM interviews with Dean Peterson and James Deutsch, Environmental Sci-
    entists, North Dakota PSC, April 12, 1979.
47. INFORM interview with David A. Jordan, Conservationist, Knife River Coal

Mining Company, February 12, 1980.

48. Written communication from Knife River Coal Mining Company to INFORM, n.d.

49. INFORM interview with David A. Jordan, Conservationist, Knife River Coal Mining Company, June 15, 1979.

50. *Idem.*

51. *Idem.*

52. *Idem.*

53. *Idem.*

54. Written communication from Curtis L. Blohm, Manager of Reclamation, Knife River Coal Mining Company, to INFORM, September 29, 1979.

55. INFORM interview with David A. Jordan, Conservationist, Knife River Coal Mining Company, June 15, 1979.

56. Written communication from Knife River Coal Mining Company to INFORM, n.d.

57. INFORM interviews with Dean Peterson and James Deutsch, Environmental Scientists, North Dakota PSC, April 12, 1979.

58. INFORM interview with David A. Jordan, Conservationist, Knife River Coal Mining Company, June 15, 1979.

59. A. Kent Evans, E. W. Uhleman, and P. A. Eby, *Atlas of Western Surface-Mined Lands: Coal, Uranium, and Phosphate* (Washington, D.C.: U.S. Department of the Interior, Fish and Wildlife Service, Biological Services Program, January 1978), p. 259.

60. INFORM interview with David A. Jordan, Conservationist, Knife River Coal Mining Company, June 15, 1979.

61. U.S. Department of Agriculture, SCS, *Soil Survey of Mercer County, North Dakota* (Washington, D.C.: U.S. Government Printing Office, 1978), p. 3.

62. *Ibid.*

63. INFORM interview with David A. Jordan, Conservationist, Knife River Coal Mining Company, June 15, 1979.

64. Written communication from Curtis L. Blohm, Manager of Reclamation, Knife River Coal Mining Company, September 29, 1979.

65. INFORM interview with David A. Jordan, Conservationist, Knife River Coal Mining Company, February 12, 1980.

66. Letter from David A. Jordan, Conservationist, Knife River Coal Mining Company, to Susan Jakoplic, Research Assistant, INFORM, October 10, 1978.

67. INFORM interview with David A. Jordan, Conservationist, Knife River Coal Mining Company, June 15, 1979.

GLENHAROLD

1. U.S. Department of Commerce, NOAA, *Monthly Averages of Temperature and Precipitation for State Climatic Divisions 1941-70: North Dakota*, "Monthly and Annual Divisional Averages Precipitation (Inches)" (Asheville, N.C.: National Climatic Center, July 1973), p. 5.

2. National Academy of Sciences, *Rehabilitation Potential of Western Coal Lands* (Cambridge, Mass.: Ballinger Publishing Co., 1974), p. 123.

3. Written communication from Ken Wangerud, Manager of Reclamation, Glenharold Mine, to INFORM, June 8, 1979.

4. INFORM interviews with Ken Wangerud, Manager of Reclamation, and Rick Williamson, Range Scientist, Glenharold Mine, June 13, 1979.

5. Computer printout from H. Robert Moore, Assistant to the Director for Coal Management, BLM, U.S. Department of the Interior, November 28,1978.

6. INFORM interviews with Dean Peterson and James Deutsch, Environmental Scientists, North Dakota PSC, April 12, 1979.

7. INFORM interview with Ken Wangerud, Manager of Reclamation, Glenharold Mine, November 5, 1979.

8. U.S. Department of the Interior, OSM, Region V On-Site Inspection: Consolidated Coal Company--Glenharold Mine, August 23, 1978.

9. Letter from William J. Delmore, Director of Environmental Enforcement and Legal Services, North Dakota State Department of Health, to Daniel Wiener, Research Associate, INFORM, September 19, 1979.

10. INFORM interviews with Ken Wangerud, Manager of Reclamation, and Rick Williamson, Range Scientist, Glenharold Mine, June 13, 1979.
11. Written communication from Ken Wangerud, Manager of Reclamation, Glenharold Mine, to INFORM, June 8, 1979.
12. INFORM interviews with Dean Peterson and James Deutsch, Environmental Scientists, North Dakota PSC, April 12, 1979.
13. INFORM interviews with Ken Wangerud, Manager of Reclamation, and Rick Williamson, Range Scientist, Glenharold Mine, June 13, 1979.
14. *Idem.*; INFORM interview with Ken Wangerud, Manager of Reclamation, Glenharold Mine, November 14, 1979.
15. INFORM interviews with Ken Wangerud, Manager of Reclamation, and Rick Williamson, Range Scientist, Glenharold Mine, June 13, 1979.
16. U.S. Department of the Interior, OSM, Region V On-Site Inspection: Consolidated Coal Company--Glenharold Mine, August 23, 1978.
17. INFORM interviews with Ken Wangerud, Manager of Reclamation, and Rick Williamson, Range Scientist, Glenharold Mine, June 13, 1979.
18. *Idem.*
19. Memorandum from the Reclamation Division, North Dakota PSC, to Ray H. Walton, Esq., Commerce Counsel, North Dakota PSC, October 11, 1977.
20. INFORM interviews with Ken Wangerud, Manager of Reclamation, and Rick Williamson, Range Scientist, Glenharold Mine, June 13, 1979.
21. *Idem.*
22. *Idem.*
23. North Dakota State Public Service Commission regulations, Section 69-05.1-09-03(4) as per conversation with Dr. Edward Englerth, Director, PSC Reclamation Division, December 12, 1979.
24. INFORM interviews with Ken Wangerud, Manager of Reclamation, and Rick Williamson, Range Scientist, Glenharold Mine, June 13, 1979.
25. INFORM interviews with Dean Peterson and James Deutsch, Environmental Scientists, North Dakota PSC, April 12, 1979.
26. *Idem.*
27. INFORM interviews with Ken Wangerud, Manager of Reclamation, and Rick Williamson, Range Scientist, Glenharold Mine, June 13, 1979.
28. Written communication from Gary E. Slagel, Manager of Reclamation, CONSOL, to INFORM, October 10, 1979.
29. North Dakota State Department of Health, Division of Water Supply and Pollution Control, "Authorization to Discharge Under the National Pollutant Discharge Elimination System, Permit No. ND-0024546," February 25, 1977, p.3A.
30. INFORM interviews with Ken Wangerud, Manager of Reclamation, and Rick Williamson, Range Scientist, Glenharold Mine, June 13, 1979.
31. INFORM interviews with Dean Peterson and James Deutsch, Environmental Scientists, North Dakota PSC, April 12, 1979.
32. INFORM interviews with Ken Wangerud, Manager of Reclamation, and Rick Williamson, Range Scientist, Glenharold Mine, June 13, 1979.
33. U.S. Department of the Interior, OSM, Region V On-Site Inspection: Consolidated Coal Company--Glenharold Mine, August 23, 1978.
34. INFORM interviews with Ken Wangerud, Manager of Reclamation, and Rick Williamson, Range Scientist, Glenharold Mine, June 13, 1979.
35. INFORM interview with Ken Wangerud, Manager of Reclamation, Glenharold Mine, November 5, 1979.
36. Memorandum from Dennis L. Fewless, Environment Quality Specialist, North Dakota State Department of Health, to Norman L. Peterson, Director, Division of Water Supply and Pollution Control, North Dakota State Department of Health, October 6, 1977.
37. North Dakota State Department of Health, Division of Water Supply and Pollution Control, "Authorization to Discharge Under the National Pollutant Discharge Elimination System, Permit No. ND-0024546," February 25, 1977, p. 3A.
38. Consolidated Coal Company, NPDES Discharge Monitoring Report--Glenharold Mine, reporting period from September 1, 1977 to November 31, 1977.
39. Consolidated Coal Company, NPDES Discharge Monitoring Report--Glenharold Mine, reporting period from June 1, 1977 to August 31, 1977.
40. Consolidated Coal Company, NPDES Discharge Monitoring Report--Glenharold Mine, reporting period from June 1, 1978 to August 31, 1978.
41. North Dakota PSC, Frank Dennison, Terry J. Zich, and John Schultz, Envi-

ronmental Engineers, Inspection Report: Glenharold Mine, July 24, 1979.

42. Letter from Dennis L. Fewless, Environment Quality Specialist, North Dakota
     State Department of Health, to Ken Wangerud, Manager of Reclamation, Glen-
     harold Mine, November 2, 1978; Letter from William J. Delmore, Director of
     Environmental Enforcement and Legal Services, North Dakota State Depart-
     ment of Health, to Daniel Wiener, Research Associate, INFORM, July 13, 1979.

43. Letter from Dennis L. Fewless, Environment Quality Specialist, North Dakota
     State Department of Health, to Ken Wangerud, Manager of Reclamation, Glen-
     harold Mine, November 2, 1978.

44. Letter from Gerald J. Schissler, Regional Counsel, Consolidated Coal Compa-
     ny, to Ray H. Walton, Esq., Commerce Counsel, North Dakota PSC, June 28,
     1978.

45. U.S. Department of the Interior, OSM Reclamation and Enforcement, Notices
     of Violation No. 78-V-3-2, November 29, 1978.

46. Memorandum from Terry Clark, Staff Member, Dakota Resource Council, to
     Dakota Resource Council Board of Directors and Interested Parties, n.d.

47. Letter from William J. Delmore, Director of Environmental Enforcement and
     Legal Services, North Dakota State Department of Health, to INFORM, Sep-
     tember 19, 1979.

48. INFORM interviews with Steve Young, Vice President for Government Affairs,
     CONSOL, and Gary E. Slagel, Manager of Reclamation, CONSOL, November
     16, 1979.

49. INFORM interview with Frank Beaver, Inspector, North Dakota EPA, Novem-
     ber 20, 1979.

50. INFORM interview with Robert Walline, Chemical Engineer, Region 8,Environ-
     mental Protection Agency, October 3, 1979.

51. INFORM interview with Frank Beaver, Inspector, North Dakota EPA, Novem-
     ber 20, 1979.

52. INFORM interviews with Dean Peterson and James Deutsch, Environmental
     Scientists, North Dakota PSC, April 12, 1979.

53. Written communication from Ken Wangerud, Manager of Reclamation, Glen-
     harold Mine, to INFORM, June 8, 1979.

54. *Ibid.*

55. Written communication from Gary E. Slagel, Manager of Reclamation, CONSOL,
     to INFORM, October 10, 1979.

56. *Ibid.*

57. Written communication from Ken Wangerud, Manager of Reclamation, Glen-
     harold Mine, to INFORM, June 8, 1979.

58. INFORM interviews with Dean Peterson and James Deutsch, Environmental
     Scientists, North Dakota PSC, April 12, 1979.

59. Memorandum from the North Dakota PSC Reclamation Division to Ray H. Wal-
     ton, Esq., Commerce Counsel, North Dakota PSC, October 11, 1977.

60. Memorandum from Dennis L. Fewless, Environment Quality Specialist, North
     Dakota State Department of Health, to Norman L. Peterson, Director, Divi-
     sion of Water Supply and Pollution Control, North Dakota State Department of
     Health, October 6, 1977.

61. Memorandum from the North Dakota PSC Reclamation Division to Ray H. Wal-
     ton, Esq., Commerce Counsel, North Dakota PSC, October 11, 1977.

62. Letter from A. D. Klein, Department Head of Reclamation Division, North Da-
     kota PSC, to Mike Clarke, Mine Superintendent, Glenharold Mine, September
     27, 1977.

63. Written communication from Gary E. Slagel, Manager of Reclamation, CONSOL,
     to INFORM, October 10, 1979.

64. INFORM interview with Gordon Miller, Stanton rancher, January 31, 1979.

65. North Dakota PSC, Inspection Report: Glenharold Mine, September 13, 1978.

66. Written communication from Ken Wangerud, Manager of Reclamation, Glen-
     harold Mine, to INFORM, June 8, 1979.

67. North Dakota PSC, Inspection Report: Glenharold Mine, September 13, 1978.

68. U.S. Department of the Interior, OSM, Region V On-Site Inspection: Con-
     solidated Coal Company--Glenharold Mine, August 23, 1978.

69. INFORM interviews with Ken Wangerud, Manager of Reclamation, and Rick
     Williamson, Range Scientist, Glenharold Mine, June 13, 1979.

70. INFORM interviews with Dean Peterson and James Deutsch, Environmental
     Scientists, North Dakota PSC, April 12, 1979.

71. INFORM interview with Ken Wangerud, Manager of Reclamation, Glenharold Mine, to INFORM, June 8, 1979.
72. INFORM interviews with Ken Wangerud, Manager of Reclamation, and Rick Williamson, Range Scientist, Glenharold Mine, June 13, 1979.
73. Written communication from Ken Wangerud, Manager of Reclamation, Glenharold Mine, to INFORM, June 8, 1979.
74. INFORM interviews with Dean Peterson and James Deutsch, Environmental Scientists, North Dakota PSC, April 12, 1979.
75. INFORM interviews with Ken Wangerud, Manager of Reclamation, and Rick Williamson, Range Scientist, Glenharold Mine, June 13, 1979.
76. *Idem.*
77. *Idem.*
78. INFORM interviews with Dean Peterson and James Deutsch, Environmental Scientists, North Dakota PSC, April 12, 1979.
79. INFORM interviews with Ken Wangerud, Manager of Reclamation, and Rick Williamson, Range Scientist, Glenharold Mine, June 13, 1979.
80. U.S. Department of the Interior, OSM, Region V On-Site Inspection: Consolidated Coal Company--Glenharold Mine, August 23, 1978.
81. INFORM interviews with Ken Wangerud, Manager of Reclamation, and Rick Williamson, Range Scientist, Glenharold Mine, June 13, 1979.
82. *Idem.*
83. *Idem.*
84. *Idem.*
85. *Idem.*
86. *Idem.*
87. *Idem.*
88. Written communication from Ken Wangerud, Manager of Reclamation, Glenharold Mine, to INFORM, June 8, 1979.
89. U.S. Department of the Interior, OSM, Region V On-Site Inspection: Consolidated Coal Company--Glenharold Mine, August 23, 1978.
90. INFORM interview with Ken Wangerud, Manager of Reclamation, Glenharold Mine, December 18, 1979.
91. INFORM interview with Rick Williamson, Range Scientist, Glenharold Mine, November 5, 1979.
92. Written communication from Ken Wangerud, Manager of Reclamation, Glenharold Mine, to INFORM, June 8, 1979; INFORM interview with Rick Williamson, Range Scientist, Glenharold Mine, November 5, 1979.
93. INFORM interviews with Ken Wangerud, Manager of Reclamation, and Rick Williamson, Range Scientist, Glenharold Mine, June 13, 1979; INFORM interviews with Dean Peterson and James Deutsch, Environmental Scientists, North Dakota PSC, April 12, 1979; Written communication from Ken Wangerud, Manager of Reclamation, Glenharold Mine, to INFORM, June 8, 1979.
94. INFORM interviews with Ken Wangerud, Manager of Reclamation, and Rick Williamson, Range Scientist, Glenharold Mine, June 13, 1979.
95. *Idem.*
96. INFORM interviews with Dean Peterson and James Deutsch, Environmental Scientists, North Dakota PSC, April 12, 1979.
97. INFORM interviews with Ken Wangerud, Manager of Reclamation, and Rick Williamson, Range Scientist, Glenharold Mine, June 13, 1979.
98. U.S. Department of Agriculture, SCS, *Soil Survey of Mercer County, North Dakota* (Washington, D.C.: U.S. Government Printing Office, 1978), sheet number 60, pp. 128-132; U.S. Department of Agriculture, SCS, *Soil Survey of Oliver County, North Dakota* (Washington, D.C.: U.S. Government Printing Office, 1975), sheet number 8, pp. 74-75.
99. INFORM interview with Rick Williamson, Range Scientist, Glenharold Mine, November 5, 1979.
100. INFORM interviews with Ken Wangerud, Manager of Reclamation, and Rick Williamson, Range Scientist, Glenharold Mine, June 13, 1979.
101. U.S. Department of Agriculture, SCS, *Soil Survey of Mercer County, North Dakota*, sheet number 60, pp. 128-132; U.S. Department of Agriculture, SCS, *Soil Survey of Oliver County, North Dakota*, sheet number 8, pp. 74-75.
102. Written communication from Ken Wangerud, Manager of Reclamation, Glenharold Mine, to INFORM, June 8, 1979.

103. INFORM interviews with Ken Wangerud, Manager of Reclamation, and Rick
     Williamson, Range Scientist, Glenharold Mine, June 13, 1979.
104. Written communication from Ken Wangerud, Manager of Reclamation, Glen-
     harold Mine, to INFORM, June 8, 1979.
105. INFORM interviews with Ken Wangerud, Manager of Reclamation, and Rick
     Williamson, Range Scientist, Glenharold Mine, June 13, 1979.
106. Written communication from Ken Wangerud, Manager of Reclamation, Glen-
     harold Mine, to INFORM, June 8, 1979.

INDIAN HEAD

1.   U.S. Department of Commerce, NOAA, *Monthly Averages of Temperature and
     Precipitation for State Climatic Divisions 1941-70: North Dakota*, "Monthly
     and Annual Divisional Averages Precipitation (Inches)" (Asheville, N.C.:
     National Climatic Center, July 1973), p. 5.
2.   National Academy of Sciences, *Rehabilitation Potential of Western Coal Lands*
     (Cambridge, Mass.: Ballinger Publishing Co., 1974), Appendix A, p. 123.
3.   *Ibid.*
4.   Written communication from NACCO to INFORM, November 12, 1979.
5.   INFORM interview with James Brown, Manager of Reclamation, NACCO, No-
     vember 7, 1979.
6.   Computer printout from H. Robert Moore, Assistant to the Director for Coal
     Management, BLM, U.S. Department of the Interior, November 20, 1978;
     U.S. Department of Agriculture, SCS, *Soil Survey of Mercer County, North
     Dakota* (Washington, D.C.: U.S. Government Printing Office, 1978), p. 1;
     Written communication from George F. Brown, Acting Conservation Manager,
     USGS, U.S. Department of the Interior, to Ron Lanoue, Research Associate,
     INFORM, March 26, 1979; INFORM interviews with Dean Peterson and James
     Deutsch, Environmental Scientists, North Dakota PSC, April 12, 1979.
7.   INFORM interview with Gerold Groenewold, Hydrogeologist, North Dakota
     Geological Survey, December 5, 1977.
8.   INFORM interview with Peter Nielson, Director of Special Projects, NACCO,
     December 10, 1979.
9.   U.S. Department of the Interior, OSM, Region V On-Site Inspection: North
     American Coal Corporation--Indian Head Mine, August 24, 1978.
10. INFORM interview with Dr. Edward Englerth, Director, Reclamation Division,
     North Dakota PSC, February 5, 1980.
11. U.S. Department of the Interior, OSM, Region V On-Site Inspection: North
     American Coal Corporation--Indian Head Mine, August 24, 1978.
12. INFORM interviews with James Brown, Manager of Reclamation, and Dr. Kent
     Horne, former Executive Assistant, NACCO, June 14, 1979.
13. U.S. Department of the Interior, OSM, Region V On-Site Inspection: North
     American Coal Corporation--Indian Head Mine, August 24, 1978; North Dakota
     PSC, Inspection Report: Indian Head Mine, October 18, 1977.
14. North Dakota PSC, Inspection Report: Indian Head Mine, June 23, 1977;
     Written communication from NACCO to INFORM, November 12, 1979.
15. INFORM interviews with Dean Peterson and James Deutsch, Environmental
     Scientists, North Dakota PSC, April 12, 1979.
16. Written communication from NACCO to INFORM, November 12, 1979.
17. U.S. Department of the Interior, OSM, Region V On-Site Inspection: North
     American Coal Corporation--Indian Head Mine, November 29, 1978.
18. *Ibid.*
19. INFORM interviews with James Brown, Manager of Reclamation, and Dr. Kent
     Horne, former Executive Assistant, NACCO, June 14, 1979.
20. INFORM interview with Gerold Groenewold, Hydrogeologist, North Dakota
     Geological Survey, December 5, 1977.
21. Written communication from NACCO to INFORM, November 12, 1979.
22. INFORM interview with Gerold Groenewold, Hydrogeologist, North Dakota
     Geological Survey, December 5, 1979.
23. INFORM interview with James Brown, Manager of Reclamation, NACCO, No-
     vember 7, 1979.
24. INFORM interview with Peter Nielson, Director of Special Projects, NACCO,

December 10, 1979.

25. NACCO, NPDES Discharge Monitoring Report--Indian Head Mine, reporting period from October 1, 1977 to April 1, 1979.

26. *Ibid.*

27. Letter from Terry H. Brown, Senior Environmental Control Specialist, to Norman Peterson, Director, Division of Water Supply and Pollution Control, North Dakota State Department of Health, April 3, 1979.

28. *Ibid.*

29. INFORM interview with Gerold Groenewold, Hydrogeologist, North Dakota Geological Survey, December 5, 1979.

30. *Ibid.*

31. INFORM interview with Warner Benfit, Zap rancher, June 14, 1979; Written communication from NACCO to INFORM, November 12, 1979; INFORM interviews with Dean Peterson and James Deutsch, Environmental Scientists, North Dakota Public Service Commission, April 12, 1979.

32. Written communication from George F. Brown, Acting Conservation Manager, USGS, U.S. Department of the Interior, to Ron Lanoue, Research Associate, INFORM, March 26, 1979.

33. Written communication from NACCO to INFORM, November 12, 1979.

34. INFORM interview with Peter Neilson, Director of Special Projects, NACCO, December 10, 1979.

35. INFORM interviews with James Brown, Manager of Reclamation, and Dr. Kent Horne, former Executive Assistant, NACCO, June 14, 1979.

36. North Dakota PSC, Inspection Report: Indian Head Mine, September 14, 1977.

37. Written communication from NACCO to INFORM, November 12, 1979.

38. INFORM interview with Peter Neilson, Director of Special Projects, NACCO, December 10, 1979.

39. INFORM interviews with James Brown, Manager of Reclamation, and Dr. Kent Horne, former Executive Assistant, NACCO, June 14, 1979.

40. INFORM interview with Peter Neilson, Director of Special Projects, NACCO, December 10, 1979.

41. INFORM interviews with Dean Peterson and James Deutsch, Environmental Scientists, North Dakota PSC, April 12, 1979; North Dakota PSC, Inspection Report: Indian Head Mine, May 5, 1977.

42. INFORM interviews with James Brown, Manager of Reclamation, and Dr. Kent Horne, former Executive Assistant, NACCO, June 14, 1979.

43. *Idem.*; INFORM interviews with Dean Peterson and James Deutsch, Environmental Scientists, North Dakota PSC, April 12, 1979.

44. INFORM interview with James Brown, Manager of Reclamation, NACCO, November 7, 1979.

45. INFORM interview with Peter Neilson, Director of Special Projects, NACCO, December 5, 1979.

46. Written communication from NACCO to INFORM, November 12, 1979.

47. INFORM interviews with Peter Neilson, Director of Special Projects, NACCO, December 5 and 10, 1979.

48. Written communication from NACCO to INFORM, November 12, 1979.

49. *Ibid.*

50. U.S. Department of Agriculture, SCS, *Soil Survey of Mercer County, North Dakota*, p. 3.

51. INFORM interviews with Dean Peterson and James Deutsch, Environmental Scientists, North Dakota PSC, April 12, 1979.

52. Written communication from NACCO to INFORM, November 12, 1979.

53. INFORM interview with Peter Neilson, Director of Special Projects, NACCO, December 5, 1979.

54. Written communication from NACCO to INFORM, November 12, 1979.

55. INFORM interviews with James Brown, Manager of Reclamation, and Dr. Kent Horne, former Executive Assistant, NACCO, June 14, 1979.

COSTS

1. INFORM interview with Gary Grow, Vice President of Sales, Flagstaff Office, Peabody Coal Company, December 18, 1979.
2. "Will strip regulations cost Consol $500 million?" *Coal Outlook* (Washington, D.C.: Pasha Publications, 1979), January 15, 1979, pp. 4-5.
3. "Peabody assesses regulations' price impact," *Coal Outlook* (Washington, D.C.: Pasha Publications, 1978), September 18, 1978, p. 8.
4. INFORM interview with Russell Boulding, Reclamation Consultant, National Coal Policy Project, March 14, 1980.
5. N.B. Pundari, Manager of Engineering, and J.A. Coates, Mining Engineer, Gulf Energy & Minerals Co., "Estimate of reclamation costs resulting from federal law," *Coal Age* (New York: McGraw Hill, 1975), April 1975, pp. C135-136.
6. The President's Commission on Coal, *Staff Findings: The Acceptable Replacement of Imported Oil with Coal* (Washington, D.C.: U.S. Government Printing Office, March 1980), p. 33.
7. INFORM interview with Donald Koch, Analyst, Arizona Public Service Commission, March 21, 1980.
8. Pundari and Coates, "Estimate of reclamation costs," p. C136.
9. *Ibid.*
10. U.S. Department of the Interior, Bureau of Mines, Franklin H. Persse, David W. Lockard, and Alec E. Lindquist, *Coal Surface Mining Reclamation Costs in the Western United States,* Information Circular 8737 (Washington, D.C.: U.S. Government Printing Office, 1977), p. 7.
11. Kenneth L. Leathers, *Costs of Strip Mine Reclamation in the West,* Rural Development Research Report No. 19 (Washington, D.C.: U.S. Department of Agriculture, Economics, Statistics, and Cooperatives Service, February, 1980), p. 50.
12. U.S. Department of the Interior, Bureau of Mines, Franklin H. Persse, David W. Lockard, and Alec E. Lindquist, *Coal Surface Mining Reclamation Costs in the Western United States,* Information Circular 8737 (Washington, D.C.: U.S. Government Printing Office, 1977), pp. 18-27.
13. E.A. Nephew *et al.* "The Costs of Coal Surfacemining and Reclamation in Appalachia" (Paper delivered at the Third Symposium on Surface Mining and Reclamation, Louisville, Kentucky, October 21-23, 1975), Vol. II, pp. 119-130.
14. Pundari and Coates, "Estimate of Reclamation Costs," p. C134.
15. *Ibid.*
16. Leathers, *Costs of Strip Mine Reclamation in the West,* p. 48.
17. *Ibid.*
18. *Ibid.*
19. U.S. Department of the Interior, Bureau of Mines, Franklin H. Persse, David W. Lockard, and Alec E. Lindquist, *Coal Surface Mining Reclamation Costs in the Western United States,* Information Circular 8737 (Washington, D.C.: U.S. Government Printing Office, 1977), p. 8.
20. Leathers, *Costs of Strip Mine Reclamation in the West,* p. 111.

METHODOLOGY

1. National Coal Association, *Coal Facts 1978-1979* (Washington, D.C.: National Coal Association, 1979), p. 81; "Table 3 - New Coal Mines, Development and Expansion of Those Mines by States by Years," *Coal Age* (New York: McGraw Hill, 1979), February 1979, p. 92.

About the Author

Daniel Philip Wiener, author of *Reclaiming the West,* is advisor to INFORM's energy and land use program. He has conducted on-site reclamation analyses not only in the western United States (and Illinois, Kentucky and Indiana), but also in North Rhine Westphalia, West Germany. He assisted with research on cancer-causing trace elements for INFORM's occupational safety and health study, *At Work In Copper,* and helped produce two other INFORM publications, *Fluidized-Bed Energy Technology* and *Promised Lands* (Volume 3). A 1977 graduate of Tufts University, Wiener holds a bachelor of science degree in biology, with an emphasis on environmental sciences. He has been assistant to the director of educational programs at the New England Aquarium in Boston, and was a juror for energy films at the 1979 American Film Festival. A skiing, sailing and running enthusiast, he lives in New York City.

INFORM

PUBLICATIONS

NEWSLETTER:

INFORM News, a bimonthly newsletter featuring INFORM's research and publication activities. $25 per annum

AIR AND WATER POLLUTION:

A CLEAR VIEW: Guide to Industrial Pollution Control (June 1975)
A manual on procedures for monitoring and assessing industrial air and water pollution problems and controls. $6.95/$3.95

ENERGY:

ENERGY FUTURES: Industry and the New Technologies (July 1976)
An analysis of corporate research, commercial development and environmental impact of new energy sources and technologies. (Ballinger Press abridged version $10.95; hardcover edition out of print)

INDUSTRIAL ENERGY CONSERVATION: Where Do We Go From Here? (December 1977)
A report on the federal programs and industry progress toward achieving federal energy conservation targets. $5.00

FLUIDIZED-BED ENERGY TECHNOLOGY: Coming to a Boil (June 1978)
A report on the state of the art of fluidized-bed combustion technology (for cleaner and more efficient direct burning of coal). This report provides information on the companies now researching and developing fluidized-bed systems, the variety of possible large- and small-scale applications for the technology, its cost, and the barriers to widespread commercial use. $45.00 (nonprofit discount available)

LAND USE:

THE INSIDER'S GUIDE TO OWNING LAND IN SUBDIVISIONS: How To Buy, Appraise and Get Rid of Your Lot (January 1980)
A manual for lot buyers: what to know about the environmental practices of a developer and the investment and residential value of a lot--before buying. Also, the steps to take if you have already bought and find your lot not what you expected. $2.50

PROMISED LANDS 1: Subdivisions in Deserts and Mountains
(October 1976)
   A study of ten sites in the Southwest and West describing
   and evaluating the effects of U.S. land subdivision industry
   operations on consumers and the environment. $20.00
PROMISED LANDS 2: Subdivisions in Florida's Wetlands (March
1977)
   An analysis and evaluation of the environmental and consumer impact of nine Florida subdivisions. (Check availability
   before ordering)
PROMISED LANDS 3: Subdivisions and the Law (January 1978)
   An assessment of the effectiveness of land-sales and land-use
   laws in protecting the environment and consumers. $20.00

*BUSINESS AND PRESERVATION: A Survey of Business Conser-
vation of Buildings and Neighborhoods* (May 1978)
   An examination of corporation activities involving preservation and reuse of existing buildings and historic sites, and
   support of neighborhood redevelopment. $14.00 softcover

OCCUPATIONAL SAFETY AND HEALTH:

AT WORK IN COPPER, Volumes 1, 2, 3 (April 1979)
   A study of conditions at the 16 U.S. copper smelters, identifying worker safety and health risks, defining the best
   available worker protection techniques and evaluating company, government and union efforts to protect employees.
   Set $70.00

AT WORK IN COPPER, Volume 1 (April 1979)
   Provides the findings of overall industry performance with
   recommendations of feasible engineering controls. Explains
   criteria upon which smelter evaluations are based. $40.00
AT WORK IN COPPER, Volume 2 (April 1979)
   Profiles smelters owned by ARCO, ASARCO, Cities Service,
   Inspiration. $20.00
AT WORK IN COPPER, Volume 3 (April 1979)
   Profiles smelters owned by Kennecott, Louisiana Land, New-
   mont Mining, Phelps Dodge. $20.00

NEW ISSUE RESEARCH:

WHAT'S FOR DINNER TOMORROW? (1980)
   An analysis of where, when, and to what degree nutritional
   considerations are factored into corporate decisions to develop and market a new food product. $5.00

FUTURE PUBLICATIONS

ENERGY:

Energy Futures Profiles (1980-81)
A series of reports on cleaner, more efficient technology options for energy production and supply. Each report will define the status of government and corporate progress in research and demonstration of a particular new supply source, its environmental implications, actual and projected costs, and prospects for commercialization. Three studies are underway: coal-cleaning techniques, flue-gas desulfurization, and energy storage.

Conservation Packaging Options (1980)
An investigation of financing options to encourage conservation by small businesses and consumers. This report will also identify present institutional barriers to the successful marketing of conservation.

LAND USE:

Subdivision Planning Guide for County Planners (1980)
How to evaluate the planning for a new subdivision: land use, water, basic services.

OCCUPATIONAL SAFETY AND HEALTH:

Occupational Safety and Health Regulation (1980)
An analysis of the strengths and weaknesses of regulatory programs in occupational safety and health at the state and federal levels.